FUNDAMENTALS
OF
DIGITAL SIGNAL
PROCESSING

FUNDAMENTALS OF DIGITAL SIGNAL PROCESSING

Lonnie C. Ludeman
New Mexico State University

John Wiley & Sons
NEW YORK CHICHESTER BRISBANE TORONTO SINGAPORE

089016

FUNDAMENTALS OF DIGITAL SIGNAL PROCESSING

Copyright © 1987, by John Wiley & Sons, Inc.

Library of Congress Cataloging-in-Publication Data.

Ludeman, Lonnie C.
   Fundamentals of digital signal processing.

   Includes index.
   1. Signal processing—Digital techniques.   I. Title.
TK5102.5.L83   1986   621.38'043   85-21950

Printed in the Republic of Singapore

10 9 8 7 6 5 4 3 2 1

Dedicated to
My Mother and Father
Marybelle and Robert Ludeman

CONTENTS

Preface   xi

Introduction   1

0.1   Signal Processing Example   1
0.2   Structure of Special Digital Signal Processors   7
0.3   Other Realizations of Digital Filters   11
0.4   Implementation of Digital Filters   16
0.5   Advantages of Digital Filters and Processing   17
        Bibliography   17

CHAPTER 1
Fundamentals of Discrete-Time Systems   18

1.0   Introduction   18
1.1   Basic Definitions   19
1.2   Important Discrete-Time Signals   21
1.3   Discrete-Time Systems   26
1.4   Fourier Transform of Sequences   55
1.5   Sampling of Continuous-Time Signals   57
1.6   Digital Filter with A/D and D/A   61
1.7   Summary   67
        Bibliography   67
        Problems   68

CHAPTER 2
The $\mathcal{Z}$ Transform   76

2.1   Definition of the $\mathcal{Z}$ Transform   76
2.2   Inverse $\mathcal{Z}$ Transforms   87
2.3   Relationships Between System Representations   100
2.4   Computation of Frequency Response   104
2.5   Solution of Linear Constant Coefficient Difference Equations   105
2.6   Summary   110
        Bibliography   112
        Problems   112

CHAPTER 3
Analog Filter Design 120

3.0 Introduction 120
3.1 Butterworth Filters 122
3.2 Chebyshev Filters 134
3.3 Elliptic Filters 146
3.4 General Filter Forms 163
3.5 Summary 164
References 165
Problems 165

CHAPTER 4
Digital Filter Design 167

4.0 Discrete-Time Filters 167
4.1 Design by Using Numerical Solutions of Differential Equations 168
4.2 Analog Design Using Digital Filters 179
4.3 Design of Digital Filters Using Digital-to-Digital Transformations 181
4.4 Impulse Invariant Design 187
4.5 Minimization of Mean Squared Error IIR Filter Design 190
4.6 FIR Filter Design 191
4.7 Summary 206
References 206
Problems 207

CHAPTER 5
Realizations of Digital Filters 217

5.1 Direct Form Realizations of IIR Filters 217
5.2 Cascade Realizations of IIR Filters 224
5.3 Parallel Realizations of IIR Filters 228
5.4 State Variable Realizations 231
5.5 Realizations of FIR Filters 246
5.6 Implementation of Digital Filters 252
5.7 Summary 254
References 254
Problems 255

CHAPTER 6
The Discrete Fourier Transform 257

6.0 Introduction 257
6.1 Continuous-Time Fourier Series 258
6.2 Discrete-Time Fourier Series 259
6.3 The Discrete Fourier Transform 262

6.4   Computation of the Discrete Fourier Transform   270
6.5   Fast Fourier Transform   272
6.6   Interpretation of DFT Results   286
6.7   DFT–Fourier Transform Relationships   293
6.8   Discrete Fourier Transforms of Sinusoidal Sequences   295
6.9   Summary   301
        References   302
        Problems   302

        Appendix A: A Short Review of Linear System Theory   315
        Appendix B: Useful Formulas   316
        Appendix C: Table of $\mathcal{Z}$ Transform Pairs   321

Index   323

PREFACE

In the early and mid-1960s, computer simulations for problems in speech, seismology, medical research, oceanography, and radar were prominent. The area was scattered, however, and some of the techniques, although used, were not well understood. At that time, Bernard Gold and Charles M. Rader of the Massachusetts Institute of Technology realized the importance of providing a unified and comprehensive coverage of this important area. They presented in their book, *Digital Processing of Signals* (1969), foundational material that could be used by engineers in the design of special-purpose digital hardware and general-purpose digital structures to solve signal processing problems. In the control systems area, excellent pioneering presentations in sampled data systems were given by E. I. Jury and A. J. Monroe. At that time, real-time implementation using the techniques presented bordered on feasibility. However, within the last decade integrated circuit technology and hardware developments have gradually made those digital signal processing techniques not only feasible in real time but almost required in everyday engineering applications.

The books in 1975 by Oppenheim and Shafer and Gold and Rabiner entitled *Digital Signal Processing and Theory* and *Applications of Digital Signal Processing*, respectively, were monumental efforts in clarifying, expanding, and organizing the fundamentals and tools of digital signal processing. The texts provided a strong foundation for further efforts; however, they were written with the graduate student in mind and were sometimes considered "rough sledding" for undergraduates and practicing engineers.

During those same years of development it became apparent that discrete-time techniques implemented by digital hardware were fundamental in attacking many problems of engineering and basic science, especially those in electrical and computer engineering. The educational community gradually began making discrete-time systems a required part of their electrical engineering undergraduate programs, and the impetus is now on to include digital signal processing as a basic required undergraduate course as well.

The main objective of this book is to provide background and fundamental material in: discrete-time systems, basic digital processing techniques, design procedures for digital filters, and the discrete Fourier transform. This book is written for advanced junior- and senior-level students in electrical and electrical and computer engineering programs, as well as for practicing engineers.

STRUCTURE OF TEXT

The theory of continuous-time linear systems has played a very important role in engineering and the physical sciences. The reason for this is that many natural physical components and systems can be described mathematically by linear differential equations. Measurable physical parameters such as voltage, current, force, velocity, etc., can be thought of as excitations or inputs to the system, with the resulting reactions or outputs being described by the solution of the proper differential equations. We are now faced with systems that are not exactly "physical" in nature but are constructed from computers, digital hardware, and even computer programs. To make matters even worse, these systems will be "bred" with the so-called physical systems mentioned above.

It would be noble, and certainly false, to say that this text will offer a unified approach to "modeling," designing, and analyzing such systems in general. It is hoped, however, that the text can present a logical and precise development for analyzing and designing certain specific structures of digital systems, namely, those in the A/D–digital filter–D/A structure.

In Chapter 1, definitions are presented for special types of discrete-time signals and systems with special emphasis on frequency response characteristics and the aliasing property. Further background material including fundamentals of the $\mathcal{Z}$ transform and relationships among linear constant coefficient difference equations, system functions, and impulse response, are presented in Chapter 2.

Chapters 3 and 4 provide design procedures for both classical analog filters and digital filters, while Chapter 5 gives some insight into the hardware realization of such digital filters.

The discrete Fourier transform and its inverse are presented along with algorithms for fast calculation in Chapter 6. Special attention is given to interpreting the results obtained by using the discrete Fourier transform on sampled continuous-time signals.

SUGGESTED COVERAGE

The text allows several options with respect to the selection and order of topics covered depending on the emphasis desired and the interests of the students. First, for those who wish to emphasize digital filters the suggested coverage is Chapters 0–5 in normal order, while those primarily interested in pursuing the discrete Fourier transform and fast Fourier transform may wish to cover Chapters 0–2 in regular order, followed by Chapter 6 and a return to either Chapter 5 or Chapters 3 and 4. Another convenient order, if both the discrete Fourier transform and digital filtering areas are to be presented, is Chapters 0–6 in regular order, skipping Chapter 5.

In any of the above orders Chapter 3 may be covered quickly or omitted if the students have a strong background in classical analog filter design involving Chebyshev, Butterworth, and elliptic filters as well as filter transformation theory.

ACKNOWLEDGMENTS

This book was born out of a fundamental need to present digital signal processing methods to undergraduate students and practicing engineers. There are many people and organizations that have contributed in a myriad of ways to the writing of this text, the list of whom and which would be larger than the text itself.

In particular, a great debt is owed to the pioneering and masterful text *Digital Signal Processing,* written by Alan V. Openheim and Ronald W. Schafer. Their book provided the framework, main source of references, and basic need for this text.

The author greatly appreciates the atmosphere in the Electrical and Computer Engineering Department of New Mexico State University, and the understanding of its faculty who supported him in many ways throughout all stages of the project. A sincere special thanks goes to the many students who endured the suffering of the initial and final stages of this project. Their constant encouragement, suggestions for improvement, and enthusiasm were a driving force.

The opportunity to present various parts of the manuscript as course material to engineers at: the Naval Undersea Warfare Engineering Station, Keyport, Washington; RAYTHEON, White Sands Missile Range; and the Central Inertial Guidance and Test Facility, Holloman AFB, was very important and helpful during the beginning stages of organization and writing. Thanks also are extended to the Centre d'Etude des Phénomènes Aléatoires et Géophysiques, Institut National Polytechnique de Grenoble and its researchers for providing assistance during the final polishing of the manuscript.

Appreciation is also expressed to Lynn Kirlin, University of Wyoming; Ernest Baxa, Jr., Clemson University; and the following reviewers for their helpful comments: Jan Allebach, Purdue University; Gary Ford, University of California at Davis; Ken Jenkins, University of Illinois; Russell M. Mersereau, Georgia Institute of Technology; David C. Munson, Jr., University of Illinois; and James T. Kajiya, California Institute of Technology. The book could never have been written without the magic performed by Lois Griese in the typing of the manuscript.

I will be forever grateful for the influence and instruction of one of the best high school educators of mathematics in the world, Florence Krieger, Rapid City High School, South Dakota.

Last, and most important, is the love, understanding, and encouragement provided by family and friends, especially my parents, who gave me a sound and healthy outlook on life and the skills necessary to mature in a changing society, and my two daughters Laurie and Miranda, who kept me from growing up.

LONNIE C. LUDEMAN

FUNDAMENTALS OF DIGITAL SIGNAL PROCESSING

Introduction

What is digital signal processing? This question will be answered by carefully defining each of the words and then presenting a precise example.

Examples of signals are: a voltage as a function of time, a potential as a function of position in a three-dimensional space, a force as a function of time and position, and an intensity as a function of x and y coordinates and time. More precisely, a *signal* is a function of a set of independent variables, with time being perhaps the most prevalent single variable. The signal itself carries some kind of information available for observation.

By *processing* we mean operating in some fashion on a signal to extract some useful information. In many cases this processing will be a nondestructive "transformation" of the given data signal; however, some important processing methods turn out to be irreversible and thus destructive.

The word *digital* shall mean that the processing is done with a digital computer or special purpose digital hardware.

In the following very simple example, analog and digital processing are illustrated and compared.

0.1 SIGNAL PROCESSING EXAMPLE

To explore the parallels and divergences between digital and analog filtering design methodology, a very simple problem will be approached by both methods.

Assume that a low-frequency signal $s(t)$, band limited to f_s Hz, is observed in an additive noisy environment to give a received signal $x(t)$ given by

$$x(t) = s(t) + n(t) \qquad (0.1)$$

The noise signal $n(t)$ is assumed to be band-limited white noise, that is, its spectral content has equal per unit bandwidth power from dc to f_n Hz. The problem is to operate on $x(t)$ in some way to obtain an estimate $\hat{s}(t)$ of the signal $s(t)$. Two solutions are presented, the first being an analog signal processor, the second a digital signal processor. Both processors will be based on a direct approach of obtaining better estimates of $s(t)$ by using a low-pass filter to minimize the effect of the high-frequency components of the interfering noise. To be more precise, a set of frequency domain requirements is established, perhaps, in terms of specifying an acceptable passband attenuation, cutoff frequency, and minimum stopband attenuation. The analog and digital processors will be assumed to take the general forms shown in Figs. 0.1 and 0.2.

0.1.1 Analog Signal Processing

Using the given requirements above, classical techniques for filter design can be used to establish a transfer function $H(s)$—perhaps representing a Butterworth or Chebyshev low-pass filter. From the $H(s)$, a linear time invariant circuit could be synthesized; for example, the simple single pole filter consisting of an inductor and resistor as shown in Fig. 0.3 might be suggested. To actually build or construct the filter an approximation of the analytically determined L and R by actual available components would be required.

0.1.2 Digital Signal Processing

It can be established by methods described in later chapters that the prescribed filtering operation can be accomplished digitally using the structure shown in Fig. 0.2. The structure is composed of an analog prefilter, an analog-to-digital (A/D) converter, a digital filter represented by a transfer function, $H(z)$, a digital-to-analog (D/A) converter, and a reconstruction filter.

The analog prefilter specified by its transfer function $H_{pf}(s)$ in most cases is a low-pass or bandpass filter designed to reduce the effects of out of band interfering signals. Interfering signals could be extraneous noise or higher-fre-

$$x(t) \bullet\!\!-\!\!\boxed{H(s)}\!\!-\!\!\bullet\ \hat{s}(t)$$

Figure 0.1   Analog signal processing of analog signals.

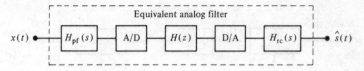

Figure 0.2 Digital signal processing of analog signals.

quency signals that when sampled produce lower-frequency signals (a phenomenon called "aliasing," which is fully described in the text). For this reason, the analog prefilter is sometimes called an "antialiasing" filter in that it combats the aliasing phenomenon.

The A/D converter is a device which will, upon command, give a binary code word corresponding to the quantized level of a continuous-time input signal at that time. Conceptually, it can be thought of as a combination of ideal sampler, quantizer, and encoder, as shown in Fig. 0.4. Normally the code words are strings of binary digits using ones-complement, twos-complement, offset binary, or sign and magnitude binary representations.

The digital filter represented symbolically by $H(z)$ is an algorithm that produces an output sequence $y(n)$ from the input sequence $x(n)$. A very widely used type of operation is one that produces the output $y(n)$ at time n as a weighted sum of the past input and output values.

The D/A converter is a device that operates on a sequence of input code words to produce a continuous-output signal usually of a staircase form. This staircase form is then smoothed by a reconstruction filter to produce the desired output signal $y(t)$.

An expanded version of the digital signal processor illustrating typical signals at various positions within the structure is shown in Figure 0.5. The input signal $x(t)$ shown on the left is filtered by the prefilter to provide the analog signal $x_a(t)$, which is sampled and coded to give the input sequence $x(n)$. This $x(n)$ is then operated on by the digital filter to produce an output sequence $y(n)$, which passes to the D/A converter producing a staircase function $y_a(t)$. Finally, $y_a(t)$ is smoothed by the reconstruction filter to produce the output $y(t)$, thus realizing an equivalent analog filter. For our example we desire $y(t)$ to be an estimate $\hat{s}(t)$ of the signal $s(t)$.

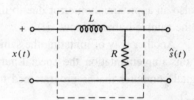

Figure 0.3 Simple analog processor.

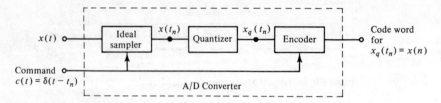

Figure 0.4 Block diagram of an analog-to-digital converter.

To complete the design of the processor for our example requires the specification of each block described above along with the sample rate for the A/D and D/A. Various techniques to be presented in this text can be used to determine the digital filter represented by $H(z)$. For the requirements of our example, it can be shown that the filter can be described by a difference equation as follows:

$$y(n) = b_0 x(n) + b_1 x(n-1) - a_1 y(n-1) \qquad (0.2)$$

Later in the text it will be shown that this corresponds to a system function $H(z)$ given by

$$H(z) = \frac{b_0 + b_1 z^{-1}}{1 + a_1 z^{-1}}$$

Implementation of the filter described by (0.2) requires digital circuitry to perform the multiplications and additions; memory for the storage of present and past values of the input $x(n)$; memory for the storage of present and past values of the output $y(n)$; and memory for storage of the coefficients a_1, b_0, and b_1. A block diagram illustrating the required multiplications, additions, and storage is shown in Fig. 0.6.

The approximations necessary for actual implementation of the design are of two basic types. First, the coefficients b_0, b_1, and a_1 must be quantized because of finite memory requirements and, second, the corresponding multiplications and additions must be performed using finite representations. There is also an approximation error due to the fact that $x(n)$ is a finite representation of the sampled analog input $x(t)$.

Another way of imitating the same analog processor is shown in Fig. 0.7. In this implementation the special-purpose digital hardware is replaced by a digital computer that is programmed to perform the calculations given in Eq. (0.2).

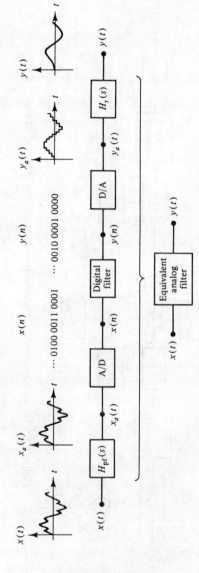

Figure 0.5 Expanded version of a digital signal processor.

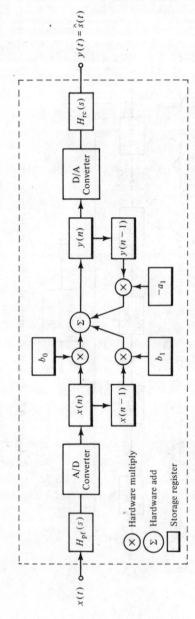

Figure 0.6 Digital signal processing equivalent to the analog processor of Figure 0.3.

6

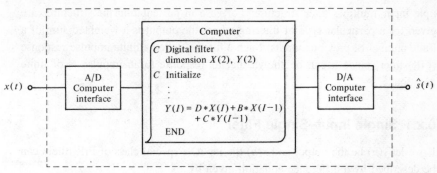

Figure 0.7   Implementation of digital filter with digital computer.

At this point you might, with good reason, ask "what have we gained?" We have managed to replace a resistor and inductor with an analog prefilter, A/D converter, a number of digital multipliers and adders, a D/A converter, and a reconstruction filter or, alternatively, a digital computer with A/D and D/A interfaces. For this simple example we have certainly lost in terms of overall complexity but, as we will see, have gained in terms of flexibility. A change of design simply requires a change in the constants b_0, b_1, and a_1 with no change in the actual components and structure. The advantage that analog signal processing has over digital signal processing concerning complexity fades when many complex filtering operations on many different inputs are required. Analog processing would require components for each filtering operation whereas the digital processor can perform all the operations within the same overall structure.

Since the same physical structure can be used to satisfy a wide variety of applications, the added complexity even for simple processing functions can be accepted because of the increased generality. Also, in many cases process operations difficult to obtain with conventional analog methods are easily performed using nonlinearities and time-varying parameters that allow adaption. Thus, digital signal processing can provide a simple fixed physical construction along with flexibility. The flexibility and structure are further explored in the following sections.

0.2 STRUCTURE OF SPECIAL DIGITAL SIGNAL PROCESSORS

The simplicity of the digital signal processing structure is now illustrated by beginning with a single input–single filtering operation and evolving to a multiple input–multiple filter structure. For ease in presentation the structures are given for a particular type of filtering where the output is a weighted sum of a finite number of previous inputs. Such a filter is called a finite impulse response (FIR) filter, since it can be shown that its response to an impulse is of finite duration.

0.2.1 Single Input–Single Filter

If we let $y(n)$ be the output and $x(n)$ the input, a special class of FIR filters can be described by a difference equation given by

$$y(n) = \sum_{k=0}^{L} b(k)x(n-k) \tag{0.3}$$

In this way the output at time n is determined by a weighted sum of past and present inputs as shown in Fig. 0.8. For its operation, the B register is first fixed with the coefficient values $b(k)$, $k = 0, 1, \ldots, L$ representing the filter coefficients.

The operation can be explained as follows: At time n, $x(n)$ is shifted into X, the input register, and $x(n - (L + 1))$ is pushed out of the X register. A term-by-term product of the contents of X and B is then completed and the sum of these products which gives the output $y(n)$ is placed in the output register Y. After $y(n)$ is calculated, another input value is shifted into the X register, with the last value on the right being shifted out and the multiplications and additions performed to obtain the next output value, etc. Thus, the determination of the

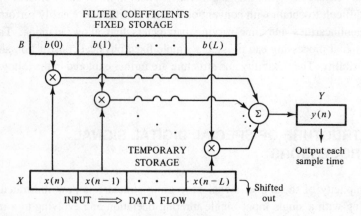

Figure 0.8   Single input–single FIR filter configuration.

output requires $L + 1$ multiplications and the addition of $L + 1$ terms. The $L + 1$ multiplications could be performed by a parallel bank of hardware multipliers or a sequential time-sharing operation on one multiplier. Hardware two-input adders could also be used in a sequential or parallel fashion. For the filter to operate in real time, all multiplications and additions must be performed before the arrival of the next input sample. The time to perform these operations then places a constraint on the minimum time between samples and the corresponding maximum frequency of sampling.

0.2.2 Single Input–Multiple Filters

In many cases we wish to process a single input signal to obtain different information about that signal. Such an operation may be realized as shown in Figure 0.9.

The desired unit sample response vector B_1, B_2, . . ., B_N for each of the N filtering operations on the input sequence are placed in the storage matrix as shown in Fig. 0.9. At time n, $x(n)$ is shifted into X, the input register, the contents of B_1 are read from the storage matrix into the register B, and the switch SW is placed in position 1. This switching operation can easily be done digitally with a multiplex chip but is drawn as a switch for ease of representation and understanding. The output $y_1(1)$ from the first filter is then calculated by a term-by-term multiplication of the contents of B and X, with the results summed and placed in the first location in Y. The contents of B_2 are then read from the storage matrix into B, SW put in position 2, and $y_2(n)$, the output of filter 2 at time n, is placed in the proper location of Y. This process is continued until the Nth output $y_N(n)$ is obtained. The previous operations are all repeated when another input value is shifted into X, and the outputs shifted from Y, to prepare for the new respective outputs to be calculated and placed in Y. For real-time operation, all N outputs must be calculated before a new input sample is obtained. Ignoring the reading and transfer times, the approximate minimum time T_{min} elapsed for these calculations with serial operations is

$$T_{min} = N((L + 1)T_m + LT_a) \qquad (0.4)$$

where T_m and T_a represent the corresponding two-input multiplication and addition times. For various degrees of parallelism the T_{min} would be less than the amount given in Eq. (0.4) and would depend on the specific structure. For a highly parallel structure using $N \log_2(L + 1)$ two-input adders and $N(L + 1)$ two-input multipliers, the T_{min} can be reduced to

$$T_{min} = T_m + T_a \log_2(L + 1) \qquad (0.5)$$

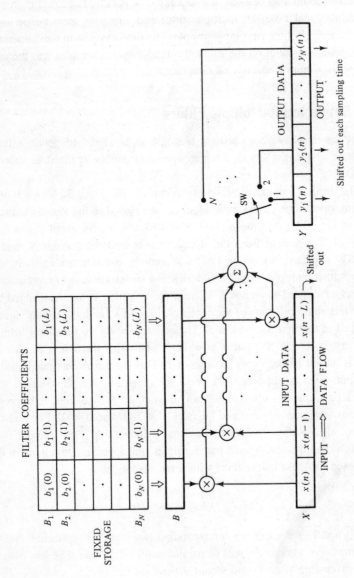

Figure 0.9 Single input–multiple FIR filter configuration.

0.2.3 Multiple Input–Single Filter Configuration

Frequently we desire to process many signals with the same filter. So-called time sharing of single analog filters is impossible, thus requiring duplication of filters, whereas time sharing digitally is easily accomplished with a structure similar to that of the single input–multiple filter configuration described previously. A digital structure to handle P input signals is shown in Fig. 0.10. The coefficients of the difference equation are stored in B. At time n, $x_1(n)$, $x_2(n)$, $\ldots$, and $x_N(n)$ are shifted into the first locations of the input registers X_1, X_2, $\ldots$, X_n and $x_1(n - (L + 1))$, $x_2(n - (L + 1))$, $\ldots$, $x_N(n - (L + 1))$ are pushed out the last locations, respectively. The X_1, X_2, $\ldots$, X_P form the input matrix X. The contents of X_1 are shifted into buffer register X, switch SW is placed in position 1, a term-by-term multiplication of X and B is obtained, followed by an addition to give $y_1(n)$ of the output register Y. Next, the contents of X_2 are shifted into register B, SW is placed in position 2, and term-by-term multiplication of X and B is obtained, added, and placed in $y_2(n)$. This process is continued until $y_P(n)$ has been calculated. At that time the filter is ready to accept another vector of input values. To operate in real time the T_{min} is seen to be identical to that of Eq. (0.4) and Eq. (0.5) with N replaced by P, i.e.,

$$T_{min} = P((L + 1)T_m + LT_a) \qquad \text{serial} \qquad (0.6)$$

$$T_{min} = T_m + T_a \log_2(L + 1) \qquad \text{parallel} \qquad (0.7)$$

0.2.4 Multiple Input–Multiple Filter Configuration

By combining Fig. 0.9 and Fig. 0.10 we are able to obtain a P input–N filter configuration shown in Fig. 0.11. With operations similar to the multiple input–single filter for each filter, a matrix of $N \cdot P$ outputs can be calculated. The minimum time between samples necessary to caculate all output values can be easily calculated depending upon how much serial and parallel multiplication and addition is performed.

0.3 OTHER REALIZATIONS OF DIGITAL FILTERS

Although the previous development has been for FIR digital filtering, the addition of another fixed vector or bank of coefficients (whichever is appropriate) for the delayed outputs and a vector or bank of memory for the delayed outputs, will realize infinite impulse response (IIR) filters represented by the following difference equation:

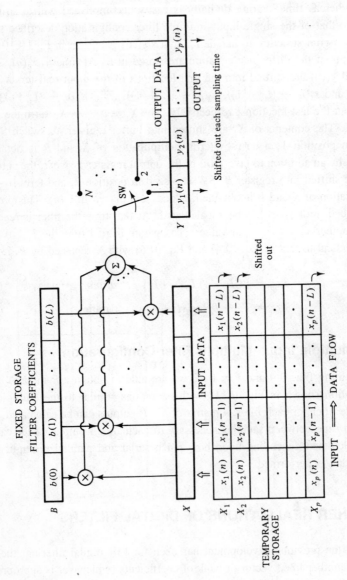

Figure 0.10 Multiple input–single FIR filter configuration.

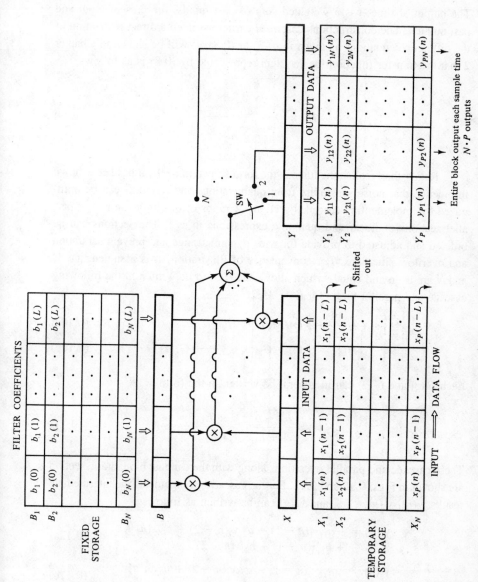

Figure 0.11 Multiple input–multiple FIR filter configuration.

13

$$y(n) = -\sum_{k=1}^{N} a_k y(n - k) + \sum_{k=0}^{M} b_k x(n - k) \tag{0.8}$$

The output at time n is a weighted sum of past inputs, the present input and past outputs. The constructions described earlier assumed a direct realization of the difference equations as given in (0.3) or (0.8). It will be shown in Chapter 2 that the transfer function for a system represented by (0.8) is as follows:

$$H(z) = \frac{\displaystyle\sum_{k=0}^{M} b_k z^{-k}}{1 + \displaystyle\sum_{k=1}^{N} a_k z^{-k}} \tag{0.9}$$

Rather than realize the filter in the form shown in (0.9), it has been shown in practice that noise sensitivity from quantizations and roundoff can be minimized by implementing cascaded sections of ratios of quadratics in z^{-1} or parallel sections of linear and quadratic expressions in z^{-1}. The sections can be ordered and adjusted to provide for wide dynamic range and prevent saturation and overflow situations. For convenience of illustration, it is assumed that M and N are even and equal, which allows the $H(z)$ to be written in the following cascade and parallel forms for $K = M/2$:

$$H(z) = H_1(z)H_2(z) \cdots H_K(z) \tag{0.10a}$$

$$H(z) = G_1(z) + G_2(z) + \cdots + G_K(z) \tag{0.10b}$$

Each $H_i(z)$ and $G_i(z)$ can in general be written in the form

$$H_i(z) = \frac{b_{0i} + b_{1i}z^{-1} + b_{2i}z^{-2}}{1 + a_{1i}z^{-1} + a_{2i}z^{-2}} \tag{0.11}$$

These cascade and parallel operations along with the canonic biquadratic section are shown in Figs. 0.12 and 0.13. For the cascade arrangement, the intermediate results $y_1(n)$, $y_2(n)$, . . ., and $y_K(n)$ can be written as follows:

$$\begin{aligned} y_1(n) = &-a_{11}y_1(n-1) - a_{21}y_1(n-2) + b_{01}x(n) \\ &+ b_{11}x(n-1) + b_{21}x(n-2) \end{aligned} \tag{0.12a}$$

$$\begin{aligned} y_2(n) = &-a_{12}y_2(n-1) - a_{22}y_2(n-2) + b_{02}y_1(n) \\ &+ b_{12}y_1(n-1) + b_{22}y_1(n-2) \end{aligned} \tag{0.12b}$$

.
.
.

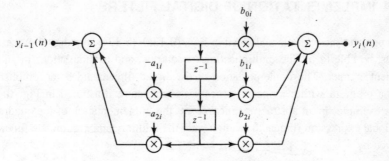

Figure 0.12 Canonic biquadratic section.

$$y_K(n) = -a_{1K}y_K(n-1) - a_{2K}y_K(n-2) + b_{0K}y_K(n)$$
$$+ b_{1K}y_{K-1}(n-1) + b_{2K}y_{K-1}(n-2) \qquad (0.12c)$$

$$y(n) = y_K(n)$$

Therefore, since successive evaluation of (0.12a)–(0.12c) is the same as the direct evaluation given in (0.8), these equations can be used as the basis for realization of the system instead of using (0.8). The successive evaluation could be implemented by a shared structure similar to those described earlier or replicated structures.

The parallel realization shown in Fig. 0.13(b) has the advantage in that the intermediate outputs can be performed simultaneously with replicated structrues or can be performed sequentially in a shared hardware structure.

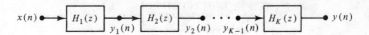

(a) Cascade

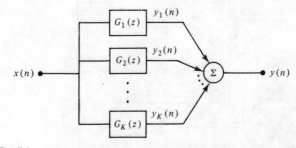

(b) Parallel

Figure 0.13 Cascade and parallel realizations for linear discrete time systems.

0.4 IMPLEMENTATION OF DIGITAL FILTERS

To perform the operations defined in Eqs. (0.12a)–(0.12c), the digital processor must be able to handle multiplications, additions, and accumulations in an efficient manner. The implementation of the digital filter could be accomplished by a program within a special processor architecture illustrated in Fig. 0.14. For example, in an abstract instruction set, the implementation of the quadratic rational expression (biquads) requires only the multiply and accumulate function given by

$$A = A + B \cdot K \qquad (0.13)$$

Each biquad section could then be implemented by repeated use of this instruction with different A, B, and K, taking five such instructions for each section. For example, the first biquad (0.12a) requires the values of $y_1(n)$, $y_1(n - 1)$, $y_1(n - 2)$, $x(n)$, $x(n - 1)$, and $x(n - 2)$ to be stored in a data RAM, whereas the coefficients a_{11}, a_{21}, b_{01}, b_{11}, and b_{21} could be stored in a ROM for non-time-varying filters or in a RAM for adaptive and time-varying applications. Similarly, space must be made available in memory for the other biquad coefficients and data given in (0.12a)–(0.12c).

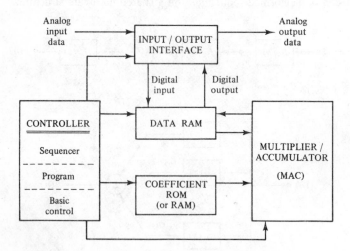

Figure 0.14   General structure of a digital signal processor.

0.5 ADVANTAGES OF DIGITAL FILTERS AND PROCESSING

In practice, digital processing has been shown to be stable, predictable, and repeatable. There is little degradation due to circuit interaction and noise is limited to digitization, arithmetic effects, and reconstruction. The basic overall structure remains fixed yet allows changes via programming. No costly components are needed for reasonable accuracy, and the overall structure offers reduced cost, size, weight, and maintenance.

The use of digital signal processors has, for the most part, traded the original hardware design problem for a programming problem within a fixed structure or the design of a special structure to reduce the programming load. The time is here when a program can be used to generate the coefficients for the digital filters from a set of specifications, a program written to write a program that can be used in the digital signal processor. Therefore, we can minimize the design time, minimize the size and cost of the physical components, and maximize reliability and flexibility. We are left, however, with a very important question: What types of processing are needed for our particular application?

BIBLIOGRAPHY

Gold, B., and C. Radar. *Digital Processing of Signals*, McGraw-Hill, New York, 1969.

Ludeman, L. C. "Digital Signal Processing Tutorial," Proceedings of the Jordan International Electrical and Electronic Engineering Conference, Amman, Jordan, April 25–28, 1983, pp. 267–273.

Monroe, A. J. *Digital Processes for Sampled-Data Systems*, Wiley, 1966.

Oppenheim, A. V., and R. W. Schafer. *Digital Signal Processing*, Prentice-Hall, Englewood Cliffs, NJ, 1975.

Peled, A., and B. Liu. *Digital Signal Processing*, Wiley, New York, 1976.

Rabiner, L. R., and B. Gold. *Theory and Application of Digital Signal Processing*, Prentice-Hall, Englewood Cliffs, NJ, 1975.

"Terminology in Digital Signal Processing," *IEEE Transactions on Audio and Electroacoustics*, Au-20 (December 1972): 322–337.

Chapter 1

Fundamentals of Discrete-Time Systems

1.0 INTRODUCTION

An important class of analog systems can be specified or modeled by differential equations. When the differential equations were linear with constant coefficients, the corresponding systems could also be represented by their transfer function, frequency response, or impulse response within a linear system framework.

The main objective of this chapter is to present definitions and theory for discrete-time systems and signals that are both self-consistent and independent from analog system theory. However, the similarities to and differences from continuous-time systems will be explored whenever possible. Particular emphasis will be placed on linear and time invariant discrete-time systems and their corresponding representations including linear constant coefficient difference equations, unit sample response, and frequency response. It will also become convenient to define the Fourier transform of a discrete-time signal to aid in the analysis of system dynamics and the description of signal characteristics.

The theory of discrete-time and continuous-time become coupled when discrete-time signals are obtained by sampling continuous signals or a discrete-time filter is imbedded in a special equivalent analog filter structure. When discrete-time signals are obtained by sampling continuous-time signals, special relationships exist between the time frequency formulations for each. These relationships, including the Nyquist sampling theorem, will be investigated closely.

If a discrete-time system is placed in an analog-to-digital (A/D), discrete-time system, digital-to-analog (D/A) structure, an equivalent analog system is specified. The coupling between the two theories of continuous-time and discrete-time systems is thus explored to present an equivalent frequency response for such systems.

1.1 BASIC DEFINITIONS

The concept of a continuous-time signal and system is made precise by the following definitions.

Definition.   A continuous-time signal is a function of time, that is, an assignment of a real value for every value of time.

Examples of continuous-time signals are shown in Fig. 1.1 and include periodic, positive time transient, sinusoidal, and random signals. Signal 1.1a is continuous-time signal even though it takes on only two values and has discontinuities, since it takes on a real value for every time.

Definition.   A continuous-time system is a mapping or an assignment of a continuous-time output signal for every continuous-time input signal.

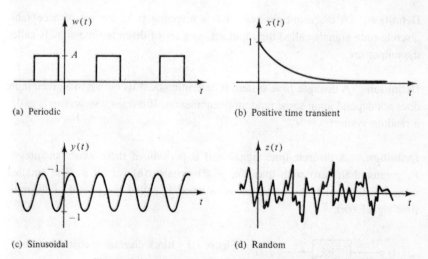

(a) Periodic

(b) Positive time transient

(c) Sinusoidal

(d) Random

Figure 1.1   Examples of continuous-time signals.

Classically, a system has been illustrated by the block diagram shown in Fig. 1.2, where $x(t)$ represents the input signal, $T[\cdot]$ represents the mapping, and $y(t)$ represents the output.

A short and certainly not complete review of linear continuous-time invariant systems is given in Appendix A to provide a basis for comparison with the discrete-time signals and system presentation.

The development of discrete-time systems and signal analysis presented in the text in some sense parallels the continuous-time development. When appropriate, analogies will be discussed, but in no way should we think that each discrete-time result has a corresponding continuous-time counterpart or vice versa. Although many discrete-time systems use sampled-in-time signals as inputs, we should not assume that a discrete-time system is an approximation of an analog system or that the input signals are necessarily sampled continuous-time signals, for the theory of discrete-time systems, including its peculiarities, stands on its own. This theory begins with the following definitions for discrete-time signals and systems.

Definition. A discrete-time signal is a sequence, that is, a function defined on the positive and negative integers.

A sequence $x(n)$ is complex if it can be written in the form $x_R(n) + jx_I(n)$, where $x_R(n)$ and $x_I(n)$ are the real and imaginary parts of the sequence, respectively, and $x_I(n)$ does not equal zero for all n. If $x_I(n)$ equals zero for all n, the sequence will be referred to as a real sequence.

Definition. A discrete-time system is a mapping from the set of acceptable discrete-time signals called the input set, to a set of discrete-time signals called the output set.

Definition. A discrete-time system is deterministic if its output to a given input does not depend upon some random phenomenon. If it does, the system is called a random system.

Definition. A discrete-time signal $x(n)$ is periodic if there exists an integer, P, greater than zero such that $x(n + P)$ equals $x(n)$ for all n. The smallest integer P for which the condition is satisfied is called the period of discrete-time signal $x(n)$.

$x(t)$ ○———————□ $T[\cdot]$ □———————○ $y(t)$

Figure 1.2 Block diagram of continuous-time system.

Examples of discrete-time signals are given in Fig. 1.3. It should be noted that signal $w(n)$ is periodic with period 4 and that $w(n)$ and $z(n)$ take on only a finite number of different values, while $x(n)$ and $y(n)$, with a little imagination, take on a countable infinite number of values. This difference motivates the following definitions for digital signals and digital systems.

Definition. A discrete-time signal whose values are from a finite set is called a digital signal.

Definition. A digital system is a mapping which assigns a digital output signal to every acceptable digital input signal.

The text is primarily involved with analyzing and designing deterministic discrete-time systems to operate on or process real discrete-time signals and special continuous-time systems that are formed by a combination of an analog-to-digital converter, a discrete-time system, and a digital-to-analog converter to process real continuous-time signals.

1.2 IMPORTANT DISCRETE-TIME SIGNALS

A real-valued sequence is a function defined on the integers, that is, $\{x(n): n = -\infty, \ldots, -1, 0, 1, \ldots, \infty\}$. If a continuous signal $x(t)$ is sampled every T seconds a sequence $\{x(nT)\}$ results. For convenience the $\{\ \}$ and sampling ref-

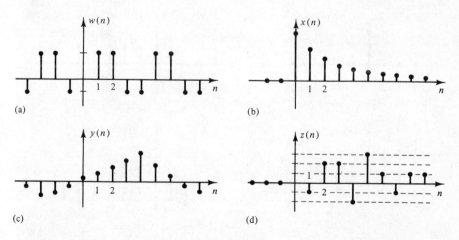

Figure 1.3 Examples of discrete-time signals: (1) periodic digital; (b) positive n; (c) discrete-time; (d) nonperiodic digital.

erence T will be dropped and $x(n)$ used to represent the sequence. A sequence will be displayed graphically by lines parallel to the ordinate axis at the integers on the abscissa with length proportional to the value of $x(n)$ as shown in Fig. 1.4. Sequences that play important roles in digital signal processing are the unit sample sequence, the unit step sequence, the real exponential sequence, and the sinusoidal sequence. These sequences are described analytically as follows and are graphically represented in Fig. 1.4.

Unit-Sample Sequence

$$\delta(n) = \begin{cases} 1, & n = 0 \\ 0, & n \neq 0 \end{cases}$$

Unit-Step Sequence

$$u(n) = \begin{cases} 1, & n \geq 0 \\ 0, & n < 0 \end{cases}$$

Real Exponential Sequence

$$x(n) = a^n \qquad \text{for all } n$$

Sinusoidal Sequence

$$x(n) = A \sin \omega_0 n \qquad \text{for all } n$$

The unit sample sequence is sometimes referred to as a discrete-time impulse and has a sifting property similar to that of the continuous-time impulse

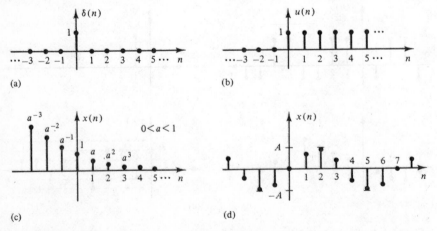

(a)

(b)

(c)

(d)

Figure 1.4 Graphical representation of: (a) unit sample sequence; (b) unit step sequence; (c) real exponential sequence; and (d) sinusoidal sequence.

function. It differs, however, by being well defined for all values of its independent variable. The sifting property for the unit sample sequence is given by

$$\sum_{n=-\infty}^{\infty} x(n)\delta(n-k) = x(k) \tag{1.1}$$

The unit step sequence and the real exponential sequence have properties similar to their continuous-time counterparts; however, the sinusoidal sequence has several major differences as follows.

(1) The sinusoidal sequence is periodic if $\omega_0/2\pi$ is rational. This can be seen by writing

$$\sin \omega_0 n = \sin\left[(\omega_0/2\pi)\cdot 2\pi n\right]$$

If $\omega_0/2\pi$ is rational, then there exist integers p and q such that $\omega_0/2\pi$ equals p/q and

$$\sin\left[(\omega_0/2\pi)\cdot 2\pi n\right] = \sin\left[(p/q)2\pi n\right]$$

The above is obviously periodic, for if $n = q$ we would start repeating the sequence. If p and q are in reduced fraction form then q is the period.

(2) If $\omega_0/2\pi$ is not rational, the sequence will not be periodic.

(3) The sinusoidal sequences $\sin \omega_0 n$ and $\sin\left[(\omega_0 + 2\pi k)n\right]$ for $0 \le \omega_0 < 2\pi$ are identical. This is easily shown using a trigonometric identity from Appendix B as follows:

$$\begin{aligned}
\sin\left[(\omega_0 + 2\pi k)n\right] &= \sin\left[\omega_0 n + 2\pi k n\right] \\
&= \sin \omega_0 n \cos 2\pi k n + \sin 2\pi k n \cos \omega_0 n
\end{aligned}$$

However, since $\cos 2\pi k n = 1$ and $\sin 2\pi k n = 0$, we have

$$\sin\left[(\omega_0 + 2\pi k)n\right] = \sin \omega_0 n$$

Similarly, $\cos \omega_0 n$ and $\cos\left[(\omega_0 + 2\pi k)n\right]$ can be shown to be the same sequence.

(4) Furthermore, if $\pi < \omega_0 < 2\pi$ then $\sin \omega_0 n$ is the negative of $\sin \omega_0' n$, where $\omega_0' = 2\pi - \omega_0$ and $0 < \omega_0' < \pi$. This result follows from the periodicity of the sine function as follows:

$$\begin{aligned}
\sin \omega_0 n &= \sin\left[\omega_0 n - 2\pi n\right] = \sin\left[(\omega_0 - 2\pi)n\right] \\
&= -\sin\left[-(\omega_0 - 2\pi)n\right] = -\sin \omega_0' n
\end{aligned}$$

where $\omega_0' = 2\pi - \omega_0$. Since ω_0 is between π and 2π; ω_0' must be between 0 and π, which is the desired result. In a similar fashion it can be shown that $\cos \omega_0 n$ and $\cos [(2\pi - \omega_0)n]$ are the same sequence. Therefore in considering sinusoidal sequences for analysis purposes, ω_0 can be restricted to the following range:

$$0 \leq \omega_0 \leq \pi$$

without any loss in generality.

Definition. The energy E of a given sequence $x(n)$ is given by

$$E = \sum_{n=-\infty}^{\infty} x(n)x^*(n) = \sum_{n=-\infty}^{\infty} |x(n)|^2 \tag{1.2}$$

where $x^*(n)$ is the complex conjugate of $x(n)$. If $x(n)$ is a real sequence, the energy E can be written as

$$E = \sum_{n=-\infty}^{\infty} x^2(n) \tag{1.3}$$

TABLE 1.1 IMPORTANT SEQUENCE OPERATIONS AND THEIR NOTATIONS

Operation	Definition	Notation
Sequence multiplication	$\{x(n)\} \cdot \{y(n)\} \overset{\triangle}{=} \{z(n) : z(n) = x(n)y(n)\}$	$x(n) \cdot y(n)$
Sequence addition	$\{x(n)\} + \{y(n)\} \overset{\triangle}{=} \{z(n) : z(n) = x(n) + y(n)\}$	$x(n) + y(n)$
Scalar multiplication	$a\{x(n)\} \overset{\triangle}{=} \{ax(n)\}$	$ax(n)$
Translation	$\{x(n - n_0)\} \overset{\triangle}{=} \{z(n) : z(n) = x(n - n_0)\}$	$x(n - n_0)$
Reflection (folding)	$\{x(-n)\} \overset{\triangle}{=} \{z(n) : z(n) = x(-n)\}$	$x(-n)$
Convolution	$\{x(n)\} * \{y(n)\} \overset{\triangle}{=} \{z(n) = \sum_{k=-\infty}^{\infty} x(k)y(n - k)\}$	$x(n) * y(n)$
Summation (indefinite)	$\sum_{k=-\infty}^{n} \{x(k)\} \overset{\triangle}{=} \{z(n) : z(n) = \sum_{k=-\infty}^{n} x(k)\}$	$\sum_{k=-\infty}^{n} x(k)$
Summation (definite)	$\sum_{k=k_0}^{k_1} \{x(k)\} \overset{\triangle}{=} \sum_{k=k_0}^{k_1} x(k)$	$\sum_{k=k_0}^{k_1} x(k)$
Backward difference	$\nabla[\{y(n)\}] \overset{\triangle}{=} \{z(n) : z(n) = y(n) - y(n - 1)\}$	$\nabla y(n)$
Forward difference	$\triangle[\{y(n)\}] \overset{\triangle}{=} \{z(n) : z(n) = y(n + 1) - y(n)\}$	$\triangle y(n)$

Operations on Sequences In analyzing discrete-time systems, operations on sequences occur frequently. The most important operations along with their notations are listed in Table 1.1 for arbitrary sequences $\{x(n)\}$ and $\{y(n)\}$.

In continuous-time analysis the derivative and integral play a very important role. For discrete-time systems the differences and summations are very important and play roles analogous to the derivative and integral, respectively. A table containing useful formulas for summations is presented in Appendix B for your convenience.

At times it is convenient to represent a discrete-time signal as a weighted sum of delayed unit sample sequences. For example, the $x(n)$ given in Fig. 1.5 can be written as

$$x(n) = x(-2)\delta(n + 2) + x(0)\delta(n) + x(1)\delta(n - 1) + x(4)\delta(n - 4)$$

In general, an arbitrary sequence $x(n)$ can be written as follows:

$$x(n) = \sum_{k=-\infty}^{\infty} x(k)\delta(n - k) \tag{1.4}$$

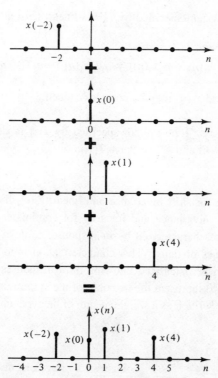

Figure 1.5 Discrete-time signal as a weighted sum of delayed unit sample sequences.

where $x(n)$ evaluated at k is the coefficient of a unit sample sequence $\delta(n - k)$. For example, the unit step sequence $u(n)$ can be written as

$$u(n) = \sum_{k=0}^{\infty} \delta(n - k) \tag{1.5}$$

1.3 DISCRETE-TIME SYSTEMS

A discrete-time system can be thought of as a transformation or operator that maps an input sequence $x(n)$ to an output sequence $y(n)$ as shown in Fig. 1.6. By placing various conditions on $T[\cdot]$ we can define different classes of systems, e.g., linear, nonlinear, shift invariant, etc.

Definition. A discrete-time system specified by $T[\cdot]$ is linear if the response to a weighted sum of inputs $x_1(n)$ and $x_2(n)$ is a weighted sum (with the same weights) of the responses of the two inputs separately for all weights and all acceptable signals $x_1(n)$ and $x_2(n)$.

Therefore, a system specified by $T[\cdot]$ is linear if for all a_1, a_2, $x_1(n)$, and $x_2(n)$, we have

$$T[a_1x_1(n) + a_2x_2(n)] = a_1T[x_1(n)] + a_2T[x_2(n)] \tag{1.6}$$

The a_1, a_2, $x_1(n)$, and $x_2(n)$ may be complex-valued.

Definition. A discrete-time system specified by $T[\cdot]$ is shift invariant (time invariant) if for all $x(n)$ and n_0, we have $T[x(n - n_0)] = y(n - n_0)$ where $y(n) = T[x(n)]$.

The definition for shift invariance and linearity for discrete-time systems corresponds to time invariance and linearity for continuous systems. A linear system can easily be characterized by its responses to delayed unit sample sequences for all values of delay. This characterization is to the extent that the system's output can be calculated knowing those responses and the input. To show this we let $h_k(n)$ represent the response of the system to an impulse at time k. Since $x(n)$ can be written as a weighted sum of delayed unit sample functions,

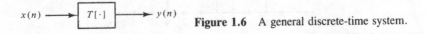

$x(n) \longrightarrow \boxed{T[\cdot]} \longrightarrow y(n)$ **Figure 1.6** A general discrete-time system.

we have the output $y(n)$ of the linear system given by

$$y(n) = T[x(n)] = T\left[\sum_{k=-\infty}^{\infty} x(k)\delta(n-k)\right]$$

Because $T[\cdot]$ is linear, Eq. (1.6) can be applied a countable number of times to give

$$y(n) = \sum_{k=-\infty}^{\infty} x(k)T[\delta(n-k)] = \sum_{k=-\infty}^{\infty} x(k)h_k(n) \qquad (1.7)$$

Thus, the output of a linear discrete-time system can be written as a weighted sum of $h_k(n)$, where $h_k(n)$ is the response of the system to a delayed unit sample sequence $\delta(n-k)$.

For a specified discrete-time system the definitions of linearity and shift invariance can be used directly to determine linearity and shift invariance; however, to prove that a system is not linear or not time varying, the construction of an acceptable counterexample is all that is necessary. The following example illustrates this concept.

EXAMPLE 1.1

A given system is represented by the following input–output formula:

$$y(n) = nx(n) = T[x(n)]$$

where $x(n)$ is the input and $y(n)$ the output. Show whether the system is (a) linear, (b) time varying.

Solution. (a) To show linearity requires that the output to a weighted sum of inputs is the weighted sum of the outputs from each input separately. Let $x(n) = a_1x_1(n) + a_2x_2(n)$.

Then the output $y(t)$ becomes

$$y(n) = n[a_1x_1(n) + a_2x_2(n)]$$

After multiplying and reorganizing, $y(n)$ can be written as

$$y(n) = a_1nx_1(n) + a_2nx_2(n)$$
$$= a_1T[x_1(n)] + a_2T[x_2(n)]$$

The results of the last equation establish the system as linear.

(b) To show that the system is time varying an input signal must be determined such that its translated version on the input does not produce a translated version of the output with the nontranslated input. If $x(n) = u(n)$,

then
$$y(n) = nu(n)$$

whereas if the input is $u(n - 2)$, the output $y(n)$ is

$$y(n) = nu(n - 2) = (n - 2)u(n - 2) + 2u(n - 2)$$
$$= T[nu(n)]|_{n \to n-2} + 2u(n - 2)$$

Therefore, the system is not time invariant, i.e., time varying because of the extra term $2u(n - 2)$.

Definition. A discrete-time system that satisfies both properties of linearity and shift invariance will be called a linear shift invariant system.

1.3.1 Convolution

The following theorem shows that a discrete-time linear shift invariant system can be characterized by its unit sample response $h(n)$. That is, $h(n)$ provides all the information needed to determine the response to any input.

Theorem 1. If $x(n)$ is the input to a linear shift invariant discrete-time system characterized by $T[\cdot]$, then $y(n)$, the output, is given by

$$y(n) = \sum_{k=-\infty}^{\infty} x(k)h(n - k) = \sum_{k=-\infty}^{\infty} x(n - k)h(k) \qquad (1.8)$$

where $h(n) = T[\delta(n)]$.

Proof. From the linearity property of $T[\cdot]$ and the decomposition of the input into a weighted sum of unit sample sequences, we have from Eq. (1.7) that

$$y(n) = \sum_{k=-\infty}^{\infty} x(k)T[\delta(n - k)] \qquad (1.9)$$

From the invariance property we have, if $T[\delta(n)] = h(n)$, that $T[\delta(n - k)] = h(n - k)$. Substituting this result into Eq. (1.9) gives the first summation of Eq. (1.8). The second summation is obtained by making a change of variables. If we let $m = n - k$ in the first summation we obtain

$$\sum_{k=-\infty}^{\infty} x(k)h(n-k) = \sum_{m=+\infty}^{-\infty} x(n-m)h(m) = \sum_{m=-\infty}^{\infty} x(n-m)h(m)$$

Since it makes no difference whether we sum the terms forward or backward we obtain the last part of the result above. Since m is a dummy variable of the summation, this result is equal to the last summation of Eq. (1.8) as desired.

Since the type of summation given in Eq. (1.8) occurs frequently in the analysis of linear shift invariant discrete-time systems, it is convenient to use the convolution operator $*$ defined by

$$x(n) * h(n) \triangleq \sum_{k=-\infty}^{\infty} x(k)h(n-k) = \sum_{k=-\infty}^{\infty} x(n-k)h(k)$$

with this notation we have shown in Theorem 1 that the response $y(n)$ of a linear shift invariant system, characterized by its impulse response $h(n)$ to an input $x(n)$, is

$$y(n) = x(n) * h(n) = h(n) * x(n) \qquad (1.10)$$

that is, the output is the convolution of the input with the system's impulse response. The type of summation given in Eq. (1.8) is sometimes referred to as the convolution sum.

If $x(n)$, $y(n)$, and $z(n)$ are sequences, then the following useful properties of the discrete-time convolution operator can be shown to be true:

1. Commutativity
   $x(n) * y(n) = y(n) * x(n)$
2. Associativity
   $x(n) * (y(n) * z(n)) = (x(n) * y(n)) * z(n) = x(n) * y(n) * z(n)$
3. Distributivity over sequence addition
   $x(n) * (y(n) + z(n)) = x(n) * y(n) + x(n) * z(n)$
4. The identity sequence for the convolution operator is $\delta(n)$
   $x(n) * \delta(n) = \delta(n) * x(n) = x(n)$
5. The convolution of a delayed unit sample sequence with $x(n)$
   $x(n) * \delta(n-k) = x(n-k)$

The following example illustrates the evaluation of the convolution sum by graphical methods. The operation consists of a folding, translation, multiplication, and summation for each value of n.

EXAMPLE 1.2

Let $x(n)$ be the input to a linear shift invariant system characterized by unit sample response $h(n)$, and call $y(n)$ the corresponding output. For the $x(n)$ and $h(n)$ given below, find the output $y(n)$.

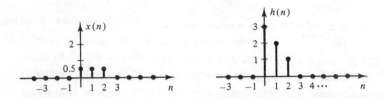

Solution. The output of the system can be found by convolving the input $x(n)$ with $h(n)$. It then becomes

$$y(n) = x(n) * h(n) = \sum_{k=-\infty}^{\infty} x(k)h(n-k)$$

For a given value of n, say n_0, the summand is a product of the two sequences $x(k)$ and $h(n_0 - k)$, where the $h(n_0 - k)$ can be viewed as a right translation of the folded sequence $h(-k)$ by n_0 indices as shown in Fig. 1.7.

With the input $x(n)$ and $h(n)$ as shown, it is easy to see that if n_0 is less than zero or greater than 4, then the product $x(k)h(n_0 - k)$ gives an all-zero sequence, and the sum of the components in that sequence is zero. Therefore, the value of the convolution sum $y(n)$, for $n < 0$ and $n > 4$, is zero. For other values of n_0 we carry out the multiplications and summation as shown in Fig. 1.8. As $h(n_0 - k)$ slides across $x(k)$, $y(n)$ is generated for each n and the results are shown at the bottom of Fig. 1.8.

In the previous example convolution was obtained graphically. In many cases convolution is best evaluated by using a combination of graphical and

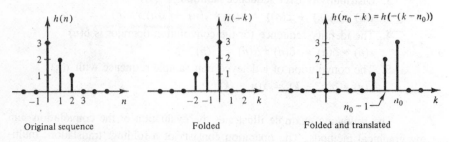

Original sequence   Folded   Folded and translated

Figure 1.7 Folding and translation operation for convolution.

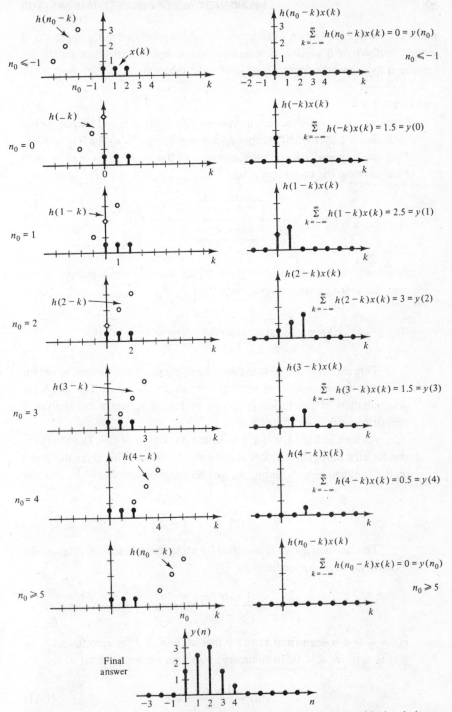

Figure 1.8 Example of the evaluation of the convolution sum by a graphical technique.

analytical methods as shown in Example 1.3. Also presented in Example 1.3 is another method of obtaining convolution which works best when either the impulse response or the input signal is of short duration.

EXAMPLE 1.3

Suppose a linear shift invariant system with input $x(n)$ and output $y(n)$ is characterized by its unit sample response $h(n) = a^n u(n)$ for $0 < a < 1$. Find the response $y(n)$ of such a system to the input signal $x(n) = u(n)$ by evaluating the convolution sum.

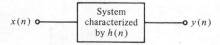

$$x(n) \circ\!\!-\!\!\!\begin{array}{|c|}\hline \text{System} \\ \text{characterized} \\ \text{by } h(n) \\ \hline \end{array}\!\!\!-\!\!\circ y(n)$$

Solution. The output $y(n)$ is given by the convolution of the input $x(n)$ with the unit sample response $h(n)$, i.e.,

$$y(n) = \sum_{k=-\infty}^{\infty} x(n-k)h(k) = \sum_{k=-\infty}^{\infty} u(n-k)a^k u(k)$$

This expression is best evaluated using a graphical method to obtain the limits and an analytical method to evaluate the sums. The graphical determination of the limits is shown in Fig. 1.9, while the analytical evaluation of the sum follows.

As seen in Fig. 1.9, for $n \geq 0$, the product of $u(n-k)$ and $u(k)$ is one for all k in region $0 \leq k \leq n$ and zero elsewhere. This gives the limits on the summation as 0 and n, so $y(n)$ becomes

$$y(n) = \sum_{k=0}^{n} a^k = \frac{1 - a^{n+1}}{1 - a}$$

The last step above is obtained by using the geometrical progression formula given in Appendix B by

$$\sum_{k=0}^{n} b^k = \frac{1 - b^{n+1}}{1 - b} \qquad \text{for } b \neq 1$$

For $n < 0$ it is seen from Fig. 1.9 that $u(k)u(n-k)$ is zero for all k, so $y(n)$ is zero for $n < 0$. In summary, $y(n)$ then becomes for all n

$$y(n) = \frac{1 - a^{n+1}}{1 - a} u(n) \qquad (1.11)$$

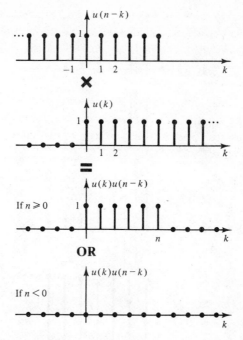

Figure 1.9   Graphical determination of limits for convolution.

Another way to approach the solution of this example is to rewrite the $u(n)$ in terms of a summation of unit sample sequences as follows:

$$u(n) = \sum_{k=0}^{\infty} \delta(n - k)$$

The convolution of $h(n)$ with $u(n)$ can then be written as

$$y(n) = h(n) * \sum_{k=0}^{\infty} \delta(n - k) = \sum_{k=0}^{\infty} h(n) * \delta(n - k) = \sum_{k=0}^{\infty} h(n - k)$$

Therefore, it is seen that the step response can be obtained by adding shifted versions of the impulse response as shown in Fig. 1.10. If the input was not a unit step, then the output would be a weighted sum of impulse responses. If the input or impulse response is of small duration, this method works quite easily.

Although for this example we were able to calculate a nice closed-form solution, Eq. (1.11), there are many times when the input to the system cannot be written in an analytical form, so a closed-form answer cannot be found. The convolution summation, however, in many cases can still be carried out to give the output at each value of n.

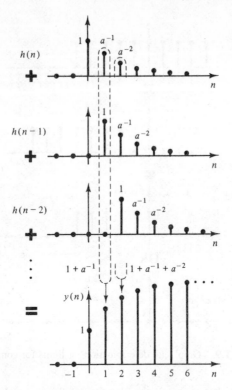

Figure 1.10 Another way of obtaining convolution sum for Example 1.3.

If sequences $x_1(n)$ and $x_2(n)$ are of finite duration N_1 and N_2, that is, they are nonzero only over respective intervals of lengths N_1 and N_2 indices, the convolution $x_1(n) * x_2(n)$ is easily seen to be also of finite duration. By considering the graphical calculation of the convolution of two such sequences beginning at zero it is seen that the length of the convolution is $N_1 + N_2 - 1$, and that the convolution is nonzero for the interval 0 to $N_1 + N_2 - 2$ as illustrated in Fig. 1.11. In Fig. 1.8 it is noticed that the duration of $x(n)$ and $h(n)$ are both 3, while the duration of the convolution is 5, which corresponds to $3 + 3 - 1$ determined from $N_1 + N_2 - 1$.

Using the associative and distributive properties, the equivalent unit sample responses for parallel and cascade linear shift invariant systems can be shown to be as in Fig. 1.12.

1.3.2 Bounded Input–Bounded Output Stability

For the most part, a desirable property of any system is that its output does not "blow up" when excited by reasonable input signals. This sense of stability is

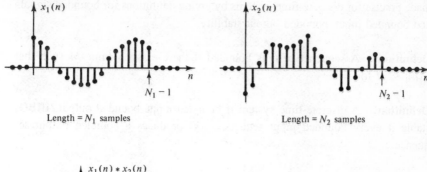

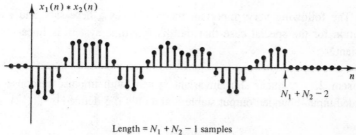

Figure 1.11 Length of convolution of two finite duration sequences.

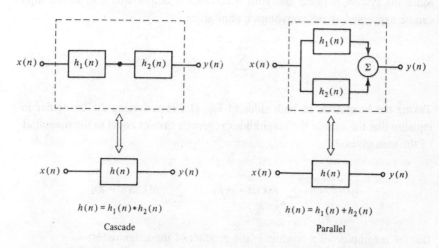

Figure 1.12 Equivalent systems for parallel and cascade combinations.

made precise for discrete-time systems by giving definitions for bounded signals and bounded input–bounded output stability.

Definition.   A sequence $x(n)$ is bounded if there exists a finite M such that $|x(n)| < M$ for all n.

Definition.   A discrete-time system is bounded input–bounded output (BIBO) stable if every bounded input sequence $x(n)$ produces a bounded output sequence.

The following very important theorem gives a necessary and sufficient condition for the special case that the discrete-time system is linear and shift invariant.

Theorem 2.   A linear shift invariant system with impulse response $h(n)$ is bounded input–bounded output stable if and only if S defined by (1.12) is finite:

$$S \triangleq \sum_{k=-\infty}^{\infty} |h(k)| \qquad (1.12)$$

Proof.   Suppose $x(n)$ is any bounded input, then there exists an M such that for all n

$$|x(n)| < M$$

Since the system is linear and shift invariant, its output with $x(n)$ as the input can be expressed as the convolution sum given by

$$y(n) = \sum_{k=-\infty}^{\infty} h(k)x(n-k) \qquad (1.13)$$

Taking the magnitude of both sides of Eq. (1.13) and using the triangular inequality that the sum of the magnitudes is greater than or equal to the magnitude of the sum gives

$$|y(n)| = \left| \sum_{k=-\infty}^{\infty} h(k)x(n-k) \right| \leq \sum_{k=-\infty}^{\infty} |h(k)x(n-k)|$$

But the magnitude of a product is the product of the magnitudes, so

$$|y(n)| \leq \sum_{k=-\infty}^{\infty} |h(k)| \, |x(n-k)|$$

Also, since $x(n)$ is bounded, the $|x(n-k)|$ can be replaced by M without changing the direction of the inequality, giving for all n

$$|y(n)| < \sum_{k=-\infty}^{\infty} M|h(k)| = M \cdot \sum_{k=\infty}^{\infty} |h(k)| = M \cdot S$$

Therefore, if S is finite, then $|y(n)|$ for all n is less than $M \cdot S$, which is finite, and we have shown that the system is BIBO stable since every bounded input produces a bounded output.

If $S = \infty$, then there exists a bounded input for which the output is not bounded, namely,

$$x(n) = \begin{cases} \dfrac{h^*(-n)}{|h(-n)|} & \text{for all } n \text{ such that } h(-n) \neq 0 \\[2mm] 0 & \text{for all in such that } h(-n) = 0 \end{cases} \tag{1.14}$$

To show this we need only find one value of n for which the output $y(n)$ is unbounded; at zero we have

$$y(0) = \sum_{k=-\infty}^{\infty} x(-k)h(k) = \sum_{k \in \mathscr{X}} \frac{h^*(k)}{|h(k)|} h(k)$$

where $\mathscr{X} = \{k : h(-k) \neq 0\}$. Recognizing $h^*(k)h(k)$ as $|h(k)|^2$, $y(0)$ can be written as

$$y(0) = \sum_{k \in \mathscr{X}} \frac{|h(k)|^2}{|h(k)|} = \sum_{k \in \mathscr{X}} |h(k)|$$

But the summation can be changed to include all k from $-\infty$ to $+\infty$, since the missing terms are zero, to give

$$y(0) = \sum_{k \in \mathscr{X}} |h(k)| = \sum_{k=-\infty}^{\infty} |h(k)| = S$$

Therefore, if S is infiinte, $y(n)$ will not be bounded since $y(0)$ is infinite.

At this time it should be mentioned that although BIBO stability has been defined with respect to input and output signals, it is a property of the system itself. Theorem 2 shows that stability of a linear shift invariant system can be determined from its unit sample response. In Chapter 2 it will be shown that BIBO stability for a linear shift invariant system can be determined from the poles of the transfer function, which is the $\mathcal{Z}$ transform of the unit sample response $h(n)$.

Although Theorem 2 provides a way of showing whether a system is stable, it may not always be easily applied; for example, when the summation cannot be analytically found. For some cases various comparison tests and ratio tests may be applied to show convergence or divergence of the summation. It is also worth noting that it is necessary to find only one bounded $x(n)$ that gives an unbounded output value to show a system is BIBO unstable. Therefore the definition is useful to determine instability but is not easily applied to show stability since you would have to show that *every* bounded input produced a bounded output.

Definition. A discrete-time system is *causal* if the output at $n = n_0$ depends only on the input for $n \leq n_0$.

Other words commonly used that are equivalent to *causal* are *realizable* and *nonanticipatory*. It is also common to define a sequence as being causal in the following way.

Definition. A discrete-time *sequence* is called *causal* if it has zero values for $n < 0$.

Causality of a linear shift invariant discrete-time system can also be described in terms of properties of its unit sample response as given in the following theorem.

Theorem 3. A linear shift invariant *system* with impulse response $h(n)$ is causal if and only if $h(n)$ is zero for $n < 0$.

Proof. The "if" part of the theorem follows directly from the convolution sum. It is known that the output of a linear shift invariant system is given by the convolution sum

$$y(n) = \sum_{k=-\infty}^{\infty} x(k)h(n-k)$$

If $h(n)$ is zero for $n < 0$, then $h(n - k)$ is zero for $n - k < 0$, which is equivalent to $k > n$. Using this fact, the terms in the infinite summation given above for $k > n$ are zero, and $y(n)$ can be rewritten as

$$y(n) = \sum_{k=-\infty}^{n} x(k)h(n - k)$$

From this expression it is seen that the output $y(n)$ at any time n is a weighted sum of the values of the input $x(k)$ for k less than or equal to n, that is, only the present and past inputs. Therefore the system is causal by use of the definition for causality. The "only if" part of the theorem is easily shown by contradiction.

The concepts of causality and stability are further explored in Examples 1.4a and 1.4b.

EXAMPLE 1.4a

A linear shift invariant system is characterized by its unit sample response $h(n)$ given by $h(n) = a^n u(n)$.

(a) Does this represent a causal system?
(b) Is the system BIBO stable?

Solution. (a) The system is *causal* by Theorem 3 since $h(n)$ is zero for $n < 0$.

(b) One way to show if the system is stable is to use Theorem 2, which requires the evaluation of S given by

$$S = \sum_{k=-\infty}^{\infty} |h(k)|$$

Replacing the n in the given $h(n)$ by k, substituting into the above equation for S, and taking the magnitude, gives

$$S = \sum_{k=-\infty}^{\infty} |a^k u(k)| = \sum_{k=0}^{\infty} |a|^k$$

The summation on the right converges to $1/(1 - |a|)$, which is finite provided $|a| < 1$, but which otherwise diverges. Therefore, the system is *BIBO stable* if $|a| < 1$ since S is finite, and the system is BIBO unstable for $|a| \geq 1$ since S diverges. The magnitude of a less than 1 implies that the a lies inside the unit circle of the complex plane.

EXAMPLE 1.4b

A given system is represented by the following input–output formula:

$$y(n) = nx(n) = T[x(n)]$$

where $x(n)$ is the input and $y(n)$ the output. Show whether the system is (a) causal (b) BIBO stable.

Solution. (a) To determine causality we can compute $y(n)$ at an arbitrary n, say n_0:

$$y(n_0) = n_0 x(n_0)$$

The output at time n_0 depends upon the input at time n_0 and no future input values; therefore, the system is causal.

(b) To show that the system is not BIBO stable requires the specification of a bounded input that produces an unbounded output. For example, $x(n) = u(n)$, the unit step function produces an output $y(n) = nu(n)$ that grows without bound; therefore, the system is not BIBO stable.

1.3.3 Linear Constant Coefficient Difference Equations

A subclass of linear shift invariant systems are those for which the input $x(n)$ and output $y(n)$ satisfy an Nth order linear constant coefficient difference equation, given by

$$\sum_{k=0}^{N} a_k y(n - k) = \sum_{r=0}^{M} b_r x(n - r), \qquad a_0 \neq 0 \qquad (1.15)$$

If the system is causal, then we can rearrange Eq. (1.15) to provide a computational realization of the system as follows:

$$y(n) = - \sum_{k=1}^{N} \frac{a_k}{a_0} y(n - k) + \sum_{r=0}^{M} \frac{b_r}{a_0} x(n - r) \qquad (1.16)$$

This realization allows the calculation of the present output at time n in terms of the present input at time n, the past inputs at times $n - 1, \ldots,$ $n - M$, and the past outputs at time $n - 1, \ldots, n - N$. If we think of the input as beginning at $n = 0$, then $y(n)$ would be specified for all $n \geq 0$ once $y(-1), y(-2), \ldots, y(-N)$ are specified. Example 1.5 shows how the solution can be generated recursively for an important difference equation, namely, a first-order linear constant coefficient difference equation.

EXAMPLE 1.5

Solve the following difference equation for $y(n)$ assuming $y(n) = 0$ for all $n < 0$ and $x(n) = \delta(n)$:

$$y(n) - ay(n-1) = x(n)$$

This corresponds to calculating the response of the system when excited by an impulse, assuming zero initial conditions.

Solution. The solution is obtained by simply rewriting the difference equation as in Eq. (1.16) and evaluating $y(n)$ at successive values of n starting at zero using the zero initial conditions.

Rewrite: $y(n) = ay(n-1) + x(n)$.

Evaluate: $n = 0$, $y(0) = ay(\overset{\nearrow 0}{-1}) + \overset{\nearrow 1}{x(0)}$
$= 1$

$n = 1$, $y(1) = ay(0) + \overset{\nearrow 0}{x(1)}$
$= a$

$n = 2$, $y(2) = ay(1) + \overset{\nearrow 0}{x(2)}$
$= a^2$

Continuing this process it is easy to see for all $n \geq 0$ that

$$y(n) = a^n$$

Since the response of the system for $n < 0$ is defined to be zero, the unit sample response becomes

$$h(n) = a^n u(n)$$

In general, the analytical expression is difficult to recognize, and we must be content with a numerical solution.

It is also possible to rearrange the difference equation above so we would have a realization of a noncausal or negative time system as shown in the following example.

EXAMPLE 1.6

Assume the difference equation of Example 1.5 and that $y(n) = 0$ for $n > 0$. Find the unit sample response of the system represented by the difference equation.

Solution. Rearrange the difference equation and evaluate backward using the fact that $x(n) = \delta(n)$. Solving for $y(n - 1)$ yields

$$y(n - 1) = (1/a)[y(n) - x(n)]$$

or

$$y(n) = (1/a)[y(n + 1) - x(n + 1)]$$

The term $y(n)$ can now be evaluated for zero and negative values of n as follows:

$$
\begin{aligned}
n = 0, \qquad y(0) &= (1/a)[y(1) - x(1)] \quad \overset{\to 0}{} \\
&= 0/a \\[6pt]
n = -1, \qquad y(-1) &= (1/a)[y(0) - x(0)] \quad \overset{\to -1}{} \\
&= -1/a \\[6pt]
n = -2, \qquad y(-2) &= (1/a)[y(-1) - x(-1)] \quad \overset{\to 0}{} \\
&= -1/a^2
\end{aligned}
$$

Continuing this process it is seen that for $n < 0$ we have $y(n) = -a^n$, and, because of the assumption that $y(n) = 0$ for $n > 0$, we can write the impulse response as

$$h(n) = -a^n u(-n - 1)$$

There are other techniques for solving difference equations. Among these is a method paralleling the procedure for solving linear constant coefficient differential equations which involves finding and combining particular and homogeneous solutions. Another method to be presented in Chapter 2 uses the $\mathcal{Z}$ transform and is analogous to solving linear constant coefficient differential equations using the Laplace transform. The state variable approach also provides another way of obtaining the solution.

1.3.4 FIR and IIR Filters

It was noted that the response of the system in Example 1.5 to a unit sample sequence was an exponential sequence that essentially lasts for all positive time. It is convenient to distinguish a system with this property from one whose impulse response lasts for only a finite number of time values.

Definition. If the unit sample response of a linear shift invariant system is of finite duration, the system is said to be a finite impulse response (FIR) system.

Definition. If the unit sample response of a linear shift invariant system is of infinite duration, the system is said to be an infinite impulse response (IIR) system.

A sufficient but not necessary condition for a linear shift invariant system to be an FIR filter is given in the following theorem.

Theorem 4. A causal linear shift invariant system characterized by the following difference equation

$$\sum_{k=0}^{N} a_k y(n-k) = \sum_{r=0}^{M} b_r x(n-r)$$

represents a finite impulse response (FIR) system if $a_0 \neq 0$ and $a_k = 0$ for $k = 1, 2, \ldots, N$; otherwise it could represent either an IIR or FIR system.

Proof. Setting $a_k = 0$ for $k = 1, 2, \ldots, N$ in the difference equation (1.17) yields

$$a_0 y(n-0) = \sum_{r=0}^{M} b_r x(n-r)$$

Dividing through by a_0 yields

$$y(n) = \sum_{r=0}^{M} (b_r/a_0) x(n-r)$$

Comparing this to terms of the convolution sum, the b_r/a_0 can be recognized as $h(r)$, the value of the unit sample response at time r. Therefore $h(n)$

is given by

$$h(n) = \begin{cases} b_n/a_0, & 0 \le n \le M \\ 0, & \text{otherwise} \end{cases}$$

which is obviously of finite duration.

It has been shown that the special class of linear shift invariant systems characterized by linear difference equations can also be characterized by their unit sample response.

Another way of characterizing a linear shift invariant system is by specifying how the system responds to a sinusoidal sequence. This topic and the relationship between the characterizations is explored in the following section.

1.3.5 Frequency Domain Representation

In continuous linear time invariant systems it was important to know the frequency response of the system. This information was available through the Fourier transform $H(j\Omega)$ of the impulse response $h(t)$ of the system. $H(j\Omega)$ could be used to determine the *steady state* response of the system to a sinusoid, that is, if the input $x(t)$ was cos $\Omega_0 t$, the steady state output was $|H(j\Omega_0)|$ cos $[\Omega_0 t + \text{arg } H(j\Omega_0)]$. The steady state response to a sum of sinusoids could then be obtained, by superposition, as the sum of the individual steady state responses.

In linear shift invariant systems the steady state response to sinusoidal sequences is just as important, and the relationship between the Fourier transform of the unit sample response and the frequency response information will now be explored.

Suppose we are given a discrete-time linear shift invariant system with unit sample response $h(n)$ and want to find the steady state response of this system to an input $x(n) = A \cos (\omega_0 n + \phi)$. It will be convenient to first find the response of a complex exponential sequence and then apply superposition, since the system is linear, to the sum of two complex exponential sequences.

Response to a Complex Exponential Sequence   Let the input to a linear shift invariant system be the complex exponential sequence $x(n)$ given by $x(n) = e^{j\omega n}$. The corresponding output of a linear shift invariant system with impulse response $h(n)$ is given by the convolution sum as follows:

$$y(n) = \sum_{k=-\infty}^{\infty} h(k)x(n-k) = \sum_{k=-\infty}^{\infty} h(k)e^{j\omega(n-k)} = e^{j\omega n} \sum_{k=-\infty}^{\infty} h(k)e^{-j\omega k} \quad (1.17a)$$

Let us define an $H(e^{j\omega})$ as ω varies from minus infinity to plus infinity to be

$$H(e^{j\omega}) \triangleq \sum_{k=-\infty}^{\infty} h(k)e^{-j\omega k} \qquad (1.18)$$

$H(e^{j\omega})$ is called the frequency response of the linear shift invariant system. Using Eq. (1.18) in (1.17a), the output $y(n)$ can be written in terms of $H(e^{j\omega})$ as follows:

$$y(n) = \overbrace{e^{j\omega n}}^{\text{Input}} \underbrace{H(e^{j\omega})}_{\substack{\text{Frequency} \\ \text{response}}} \qquad (1.18a)$$

That is, the output sequence $y(n)$ is a product of the input signal and the frequency response. In this way the definition $H(e^{j\omega})$ as a frequency response is justified since it is what the complex exponential input is multiplied by to give the output sequence $y(n)$. The magnitude of $H(e^{j\omega})$ multiplies the input magnitude while the argument or phase of $H(e^{j\omega})$ changes the phase of the complex exponential input.

In general, $H(e^{j\omega})$ is a complex variable for each ω and can be given in either rectangular or exponential form as

$$H(e^{j\omega}) = \underbrace{H_R(e^{j\omega})}_{\substack{\text{Real part}}} + \underbrace{jH_I(e^{j\omega})}_{\substack{\text{Imaginary} \\ \text{part}}} = \underbrace{\left|H(e^{j\omega})\right|}_{\substack{\text{Magnitude} \\ \text{response}}} \exp\left[\underbrace{j \arg H(e^{j\omega})}_{\substack{\text{Phase} \\ \text{response}}}\right] \qquad (1.19)$$

where the magnitude and phase of $H(e^{j\omega})$ are given by

$$\left|H(e^{j\omega})\right| = [H_R^2(e^{j\omega}) + H_I^2(e^{j\omega})]^{1/2} \qquad (1.20)$$

$$\arg H(e^{j\omega}) = \tan^{-1}[H_I(e^{j\omega})/H_R(e^{j\omega})] \qquad (1.21)$$

Response to a Sinusoidal Signal   Suppose that input $x(n)$ to a linear shift invariant system with real impulse response $h(n)$ is a sinusoidal signal given by

$$x(n) = A\cos(\omega_0 n + \phi) = Ae^{j\phi}e^{j\omega_0 n}/2 + Ae^{-j\phi}e^{-j\omega_0 n}/2$$

Because of linearity the response can be found by adding the responses of the complex exponential sequences of $Ae^{j(\omega_0 n + \phi)}/2$ and $Ae^{-j(\omega_0 n + \phi)}/2$.

Using Eq. (1.18a) for each of the positive and negative exponentials, the output $y(n)$ of the system with frequency response $H(e^{j\omega})$ becomes

$$y(n) = Ae^{j\phi}H(e^{j\omega_0}) e^{j\omega_0 n}/2 + Ae^{-j\phi}H(e^{-j\omega_0}) e^{-j\omega_0 n}/2$$

The second part of $y(n)$ is seen to be the complex conjugate of the first part; thus $y(n)$ becomes two times the real part of either, that is,

$$
\begin{aligned}
y(n) &= 2 \operatorname{Re} [(A/2)H(e^{j\omega_0}) e^{j\omega_0 n}e^{j\phi}] \\
&= A \operatorname{Re} [|H(e^{j\omega_0})| \exp [j(\phi + \omega_0 n + \arg H(e^{j\omega_0}))]] \\
&= A\underbrace{|H(e^{j\omega_0})|}_{\substack{\text{Change in} \\ \text{magnitude}}} \cos [\omega_0 n + \phi + \underbrace{\arg H(e^{j\omega_0})}_{\substack{\text{Change in} \\ \text{phase}}}]
\end{aligned}
\tag{1.22}
$$

Therefore, it has been shown that the output to a sinusoid is another sinusoid of the same frequency but with different phase and different magnitude. The magnitude of the input signal has been multiplied by $|H(e^{j\omega})|$ evaluated at ω_0, the digital frequency of the sinusoidal sequence, while the phase has been changed by an amount equal to the argument of $H(e^{j\omega})$ evaluated at the frequency ω_0.

Correspondingly, if a dc signal of magnitude A is applied to the system, the dc output is A times $H(e^{j0})$, where $H(e^{j0})$, called the dc gain, is the value of the frequency response at ω equals zero. The following examples find the frequency response and system response for the system with exponential impulse response.

EXAMPLE 1.7a

Find the frequency response of a linear shift invariant system characterized by unit sample response $h(n)$ given by $h(n) = a^n u(n)$ for $|a| < 1$. This filter is an IIR filter since $h(n)$ is nonzero for all $n \geq 0$.

Solution.   By definition the frequency response $H(e^{j\omega})$ is given by

$$H(e^{j\omega}) = \sum_{n=-\infty}^{\infty} h(n)e^{-j\omega n} = \sum_{n=0}^{\infty} a^n e^{-j\omega n} = \sum_{n=0}^{\infty} (ae^{-j\omega})^n = \frac{1}{1 - ae^{-j\omega}}$$

provided $|ae^{-j\omega}| < 1$ or equivalently $|a| < 1$.

Since the last step, that of summing the infinite series, converges if $|a| < 1$, the frequency response for $h(n)$ does not exist if $|a| \geq 1$.

The magnitude and phase of the frequency response are calculated as follows:

$$H(e^{j\omega}) = \frac{1}{1 - ae^{-j\omega}} = \frac{1}{(1 - a\cos\omega) + j(a\sin\omega)}$$

$$|H(e^{j\omega})| = \frac{1}{[(1 - a\cos\omega)^2 + (a\sin\omega)^2]^{1/2}} = \frac{1}{(1 + a^2 - 2a\cos\omega)^{1/2}}$$

$$\arg H(e^{j\omega}) = -\tan^{-1}\left[\frac{a\sin\omega}{(1 - a\cos\omega)}\right]$$

In obtaining these responses to a sinusoidal sequence, it was assumed that the sequence began at $-\infty$, that is, the sinusoid always excited the system. If the sinusoidal sequence begins at zero, the discrete-time system output goes through an initial transient period before arriving at a steady state. An illustration of these initial transient and steady state periods is given in the following example.

EXAMPLE 1.7b

Let $x(n)$ and $y(n)$ be the input and output, respectively, for a linear time invariant discrete-time filter that is specified by the following difference equation:

$$y(n) = 0.8y(n - 1) + x(n), \qquad n \geq 0$$

Calculate and plot the response for the case of zero initial conditions and an input $x(n)$ as follows:

$$x(n) = \cos(0.05\pi n)\, u(n)$$

Solution. The response, $y(n)$, shown in Fig. 1.13, is easily calculated from the difference equation with $\cos(0.05\pi n)u(n)$ as the input. The output $y(n)$ is seen to build up after several oscillations to be a cosine wave of the same frequency as the input having a peak value of approximately four and a phase relative to the input equivalent to between three and four samples. The region before the steady state is reached is called the transient period. From (1.22) we would expect a steady state cosine wave with a magnitude equal to $1|H(e^{j(0.05\pi)})|$ and a phase relative to the input of $\arg H(e^{j(0.05\pi)})$. These values are calculated from the frequency response determined in Example 1.7a to be

$$|H(e^{j(0.05\pi)})| = 4.09277$$
$$\arg H(e^{j(0.05\pi)}) = 0.537745\,\text{rad}$$

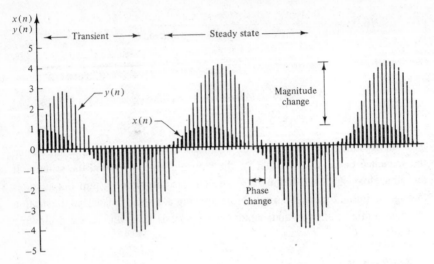

Figure 1.13 Steady state and transient response of a system specified by $y(n) = 0.8y(n-1) + x(n)$ to an input $x(n) = \cos(0.05\pi n)u(n)$.

The angle of 0.537745 rad is equivalent to $0.05\pi(3.4234)$ rad. These results of 3.4234 samples of phase change and magnitude of 4.09277 agree with our approximate answers determined visually from the plots of the output and input.

EXAMPLE 1.8

A discrete-time system has a unit sample response $h(n)$ given by

$$h(n) = \tfrac{1}{2}\delta(n) + \delta(n-1) + \tfrac{1}{2}\delta(n-2)$$

(a) Find the system frequency response $H(e^{j\omega})$; plot magnitude and phase.
(b) Find the steady state response of the system to $x(n) = 5\cos(\pi n/4)$.
(c) Find the steady state response of the system to $x(n) = 5\cos(3\pi n/4)$.
(d) Find the total response to $x(n) = u(n)$ assuming the system is initially at rest.

Solution.
(a)

$$H(e^{j\omega}) = \sum_{n=-\infty}^{\infty} h(n)e^{-j\omega n}$$

$$= \sum_{n=-\infty}^{\infty} [\tfrac{1}{2}\delta(n) + \delta(n-1) + \tfrac{1}{2}\delta(n-2)]e^{-j\omega n}$$

Using the sifting property, the above summation yields

$$H(e^{j\omega}) = \tfrac{1}{2}e^{-0} + 1e^{-j\omega} + \tfrac{1}{2}e^{-j\omega 2}$$
$$= e^{-j\omega}[\tfrac{1}{2}e^{j\omega} + 1 + \tfrac{1}{2}e^{-j\omega}]$$

By factoring, rearranging, and combining, $H(e^{j\omega})$ is finally given by

$$H(e^{j\omega}) = e^{-j\omega}(1 + \cos \omega)$$

Since $H(e^{j\omega})$ is complex and the magnitude of a product is the product of magnitudes, we have

$$|H(e^{j\omega})| = |e^{-j\omega}(1 + \cos \omega)| = |e^{-j\omega}||1 + \cos \omega| = 1 + \cos \omega$$

Since the argument of a product is the sum of the arguments, the arg $H(e^{j\omega})$ can be found to be

$$\arg H(e^{j\omega}) = \arg[e^{-j\omega}(1 + \cos \omega)] = \arg(e^{-j\omega}) + \arg(1 + \cos \omega)$$
$$= -\omega + \arg(1 + \cos \omega) = -\omega$$

These magnitude and phase plots are given in Fig. 1.14.

(b) The steady state response to the cosine wave found from Eq. (1.22) is

$$y_{ss}(n) = 5|H(e^{j\pi/4})|\cos[\pi n/4 + \arg H(e^{j\pi/4})]$$

From the plots in Fig. 1.13, the magnitude and phase at $\omega = \pi/4$ are

$$|H(e^{j\pi/4})| = 1.707$$
$$\arg H(e^{j\pi/4}) = -\pi/4$$

Therefore, $y_{ss}(n)$ becomes

$$y_{ss}(n) = 8.5355 \cos(\pi n/4 - \pi/4) = 8.5355 \cos[\pi(n - 1)/4]$$

(c) Similarly, for $\omega = 3\pi/4$, we have

$$|H(e^{j3\pi/4})| = 0.29289,$$

$$\arg H(e^{j3\pi/4}) = -3\pi/4$$

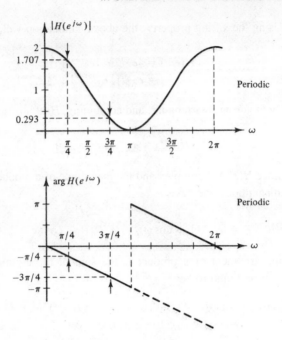

Figure 1.14 Magnitude and phase plots for a discrete-time system with unit sample response $h(n) = \frac{1}{2}\delta(n) + \delta(n-1) + \frac{1}{2}\delta(n-2)$.

and the steady state response becomes

$$y_{ss}(n) = 1.46447 \cos(3\pi n/4 - 3\pi/4)$$
$$= 1.46447 \cos[3\pi(n - 1)/4]$$

Notice that in part (b) the input signal with digital frequency $\omega = \pi/4$ rad is amplified while the input signal with $\omega = 3\pi/4$ rad is attenuated. Therefore the filter with the impulse response specified works like a low-pass amplifier. In (b) and (c) the steady state sinusoidal response is shifted by $\pi/4$ and $3\pi/4$ rad, respectively, resulting in a one-sample delay with respect to the corresponding inputs.

(d) The response of the system can be written in terms of the convolution as follows:

$$y(n) = x(n) * h(n)$$
$$= x(n) * [0.5\delta(n) + \delta(n - 1) + 0.5\delta(n - 2)]$$
$$= 0.5x(n) + x(n - 1) + 0.5x(n - 2)$$

n	$x(n)$	$y(n)$
-1	0	0
0	1	0.5
1	1	1.5
2	1	2
3	1	2
$\vdots$	$\vdots$	$\vdots$
n	1	2

Figure 1.15 Steady state dc response for Example 1.8.

This gives a difference equation realization of the discrete-time filter from which $y(n)$ can be calculated for any $x(n)$. Letting $x(n)$ equal $u(n)$, the output $y(n)$ shown in Fig. 1.15 is obtained. The response $y(n)$ reaches a steady state value of 2 (occurring for $n \geq 2$) which checks with $|H(e^{j0})|1$. In general $y(n)$ will only approach the steady state value as $n \to \infty$.

Properties of the Frequency Response   When $h(n)$ is a real sequence, the frequency response has the following easily proved general properties:

1. $H(e^{j\omega})$ takes on values for all ω, that is on a continuum of ω.
2. $H(e^{j\omega})$ is periodic in ω with period 2π.
3. $|H(e^{j\omega})|$ is an even function of ω and symmetrical about π.
4. Arg $H(e^{j\omega})$ is an odd function of ω and antisymmetrical about π.

Frequency Response of Rectangular Window   A very important discrete-time shift invariant filter is the so-called window filter characterized by the unit sample response $h(n)$ shown in Fig. 1.16 and given by

$$h(n) = \begin{cases} 1, & 0 \leq n \leq N - 1 \\ 0, & \text{elsewhere} \end{cases}$$

The frequency response of such a filter will be important in Chapter 4 in the design of FIR filters and is now established. Applying the definiton of the

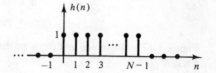

Figure 1.16 Unit sample response of a rectangular window filter.

frequency response given in Eq. (1.18) to the $h(n)$ shown in Fig. 1.15 yields

$$H(e^{j\omega}) = \sum_{k=-\infty}^{\infty} h(k)e^{-j\omega k} = \sum_{k=0}^{N-1} e^{-j\omega k} = (1 - e^{-j\omega N})/(1 - e^{-j\omega})$$

$$= \frac{e^{-j\omega N/2}(e^{j\omega N/2} - e^{-j\omega N/2})}{e^{-j\omega/2}(e^{j\omega/2} - e^{-j\omega/2})} = e^{-j\omega(N-1)/2} \frac{\sin(\omega N/2)}{\sin(\omega/2)} \qquad (1.23a)$$

The general expression for magnitude and phase of $H(e^{j\omega})$ are thus easily seen to be

$$|H(e^{j\omega})| = \left| \frac{\sin(\omega N/2)}{\sin(\omega/2)} \right| \qquad (1.23b)$$

$$\arg H(e^{j\omega}) = -(N-1)\omega/2 + \arg\left(\frac{\sin(\omega N/2)}{\sin(\omega/2)} \right) \qquad (1.23c)$$

The magnitude and phase plots for the special case of $N = 5$ are shown in Fig. 1.17.

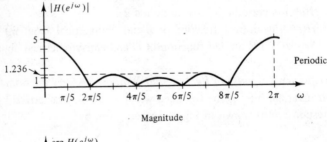

Magnitude

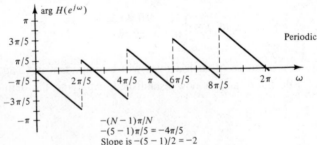

Phase

Figure 1.17 Magnitude and phase of the frequency response of a rectangular window filter, $N = 5$.

In general, for a rectangular window of size N, i.e., a string of N impulses beginning at zero and ending at $N - 1$, we see the following:

(a) The magnitude plot has zeros spaced $2\pi/N$ apart with the main lobe around 0 and side lobes reducing in amplitude as ω moves toward π.

(b) The slopes of all lines in the angle plot are $-(N - 1)/2$ with end of first line at $-(N - 1)\pi/N$ and end of the kth line at $-(N - k)\pi/N$. The angle plot is antisymmetric about π and piecewise linear.

A discrete-time shift invariant system with impulse response equaling the rectangular window acts like a low-pass filter passing signals with digital frequencies roughly between 0 and $2\pi/N$ and attenuating signals with frequencies between $2\pi/N$ and π. The important property that the phase is linear throughout the passband is useful in FIR digital filter design as will be shown in Chapter 4.

A very important theorem relating the frequency response and stability for linear shift invariant systems is the following:

Theorem 5. The frequency response $H(e^{j\omega})$ for a (BIBO) stable system will always converge.

In this way every BIBO stable system will have a frequency response and a describable steady state response to sinusoidal inputs. However, the converse of this statement is not true, that is, the fact that an $H(e^{j\omega})$ exists does not imply that the system is stable. A counterexample is given in Example 1.9.

Inverse Relationship The frequency response of a system was defined by

$$H(e^{j\omega}) = \sum_{k=-\infty}^{\infty} h(k)e^{-j\omega k}$$

We see that once $h(n)$ has been specified, $H(e^{j\omega})$ is obtained by evaluating the summation. A logical question is: How do we determine the unit sample sequence from the frequency response? It can be shown that $h(n)$ is obtained from $H(e^{j\omega})$ by the following:

$$h(n) = \frac{1}{2\pi} \int_{-\pi}^{\pi} H(e^{j\omega})e^{j\omega n}d\omega \tag{1.24}$$

Example 1.9 shows the process of obtaining the unit sample response from the frequency response for an ideal low-pass filter.

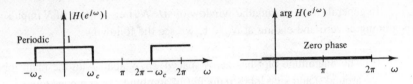

Figure 1.18 The frequency response of the ideal low-pass filter.

EXAMPLE 1.9

The frequency response $H(e^{j\omega})$ of the ideal low-pass filter is specified in Fig. 1.18. The magnitude of $H(e^{j\omega})$ is 1 for ω between $-\omega_c$ and ω_c and then periodically extended by necessity while the phase of $H(e^{j\omega})$ is zero for all ω. Find the unit sample response of such a filter and the plot of $h(n)$ for the special case where $\omega_c = \pi/2$.

Solution. From Eq. (1.24), $h(n)$ is given by the integral

$$h(n) = \frac{1}{2\pi}\int_{-\pi}^{\pi} H(e^{j\omega})e^{j\omega n}d\omega = \frac{1}{2\pi}\int_{-\omega_c}^{\omega_c} e^{j\omega n}d\omega = \frac{1}{2\pi}\frac{e^{j\omega_c n} - e^{-j\omega_c n}}{jn}$$

By recognizing the difference of the exponentials as a sine function, $h(n)$ can be reduced to

$$h(n) = \frac{\sin \omega_c n}{\pi n} \qquad \text{for all } n \qquad (1.25)$$

As both the numerator and denominator are zero at $n = 0$, $h(n)$ at $n = 0$ can be determined using l'Hôpital's rule to be

$$h(0) = \omega_c/\pi$$

The plot of $h(n)$ for the special case of $\omega_c = \pi/2$ is given in Fig. 1.19, showing $h(0)$ equal 0.5.

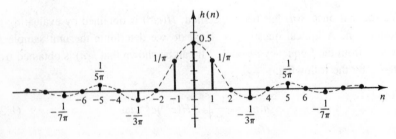

Figure 1.19 Plot of the impulse response $h(n)$ for an ideal low-pass filter with cutoff $\omega_c = \pi/2$.

From Eq. (1.25) the unit sample response $h(n)$ is not zero for $n <$ 0; therefore, the ideal low-pass filter is noncausal.

Although $h(n)$ tails off as n goes from 0 to ∞ and 0 to $-\infty$ it can be shown that $\sum_{n=-\infty}^{\infty} |h(n)|$ is not finite. This means that an ideal low-pass filter is not stable either.

1.4 FOURIER TRANSFORM OF SEQUENCES

The frequency response of a linear shift invariant system was defined by Eq. (1.18) and provides steady state information concerning response of the system to sinusoidal sequences. It would also be useful to know the frequency content of discrete-time signals so we could tell whether a given signal has most of its frequency content within the bandwidth of the system. This frequency content can be made available by defining the Fourier transform of sequences in general.

Definition. The Fourier transform $X(e^{j\omega})$ of a sequence $x(n)$ is given by

$$\mathcal{F}[x(n)] = X(e^{j\omega}) \triangleq \sum_{n=-\infty}^{\infty} x(n)e^{-j\omega n} \qquad (1.26a)$$

The sequence that represents a discrete-time signal can be determined from $X(e^{j\omega})$ by the inverse Fourier transform given by

$$\mathcal{F}^{-1}[X(e^{j\omega})] = x(n) = \frac{1}{2\pi} \int_{-\pi}^{\pi} X(e^{j\omega})e^{j\omega n} d\omega \qquad (1.26b)$$

Equations (1.26a) and (1.26b) are called the Fourier transform pair for sequences with $X(e^{j\omega})$ thought of as the frequency content of the sequence $x(n)$. The frequency response of a linear shift invariant system [Eq. (1.18)] is recognized as the Fourier transform of the system's impulse response $h(n)$.

A thorough treatment of the properties and theorems for the Fourier transform of sequences will not be presented, as they parallel, with a few exceptions, those given for the $\mathcal{Z}$ transform in Chapter 2. However, the following two theorems expressing the energy of a sequence in terms of its Fourier transform and the relationship between input and output transforms in terms of the frequency response of linear shift invariant systems are presented because of their importance.

Theorem 6. If $x(n)$ is any sequence which has a Fourier transform $X(e^{j\omega})$, then the energy of the signal can be written as follows:

$$E \triangleq \sum_{n=\infty}^{\infty} x(n)x^*(n) = \frac{1}{2\pi}\int_{-\pi}^{\pi} X(e^{j\omega})X^*(e^{j\omega})d\omega$$

The quantity $X(e^{j\omega})X^*(e^{j\omega})/2\pi$ can be viewed as energy density per unit of digital frequency for the sequence $x(n)$. We define the energy spectrum density $\phi_x(\omega)$ as follows:

$$\phi_x(\omega) \triangleq X(e^{j\omega})X^*(e^{j\omega})/2\pi$$

Theorem 7. If $x(n)$ and $y(n)$ are, respectively, the input and output of a linear shift invariant system with frequency response $H(e^{j\omega})$, then

$$Y(e^{j\omega}) = H(e^{j\omega})X(e^{j\omega}) \tag{1.27}$$

where $X(e^{j\omega})$ and $Y(e^{j\omega})$ are the Fourier transforms of $x(n)$ and $y(n)$, respectively.

Proof. The output of a linear system with impulse response $h(n)$ is given by the convolution sum

$$y(n) = \sum_{k=-\infty}^{\infty} h(n-k)x(k)$$

Taking the Fourier transform of $y(n)$ gives

$$Y(e^{j\omega}) = \sum_{n=-\infty}^{\infty}\left[\sum_{k=-\infty}^{\infty} h(n-k)x(k)\right]e^{-j\omega n}$$

Interchanging the order of the summations and multiplying by $e^{-j\omega k} \cdot e^{j\omega k}$ the above can be rearranged as

$$Y(e^{j\omega}) = \sum_{k=-\infty}^{\infty} x(k)e^{-j\omega k} \sum_{n=-\infty}^{\infty} h(n-k)e^{-j\omega(n-k)}$$

The second summation after translation is recognized as the Fourier transform of $h(n)$, while the first summation is the Fourier transform of $x(n)$. Therefore Eq. (1.27) is established.

By taking the inverse transform of both sides of Eq. (1.27), the output of the system can be seen to be

$$y(n) = \mathcal{F}^{-1}[H(e^{j\omega})X(e^{j\omega})]$$

thus providing an alternative method to the convolution sum for finding the output of a linear shift invariant system. For an arbitrary input signal $x(n)$, the convolution sum, however, provides a way of always calculating the output, whereas the above formula may not be convenient unless the $x(n)$ can be expressed in an analytical form.

In the next section the relationship between the Fourier transform of a sequence, formed by sampling a continuous-time signal, and the Fourier transform of the continuous-time signal is explored.

1.5 SAMPLING OF CONTINUOUS-TIME SIGNALS

The Fourier transform pair for continuous-time signals is defined by

$$X_a(j\Omega) \triangleq \int_{-\infty}^{\infty} x_a(t)e^{-j\Omega t}\, dt \tag{1.28a}$$

$$x_a(t) = \frac{1}{2\pi} \int_{-\infty}^{\infty} X_a(j\Omega)e^{j\Omega t}\, d\Omega \tag{1.28b}$$

If $x_a(t)$ is sampled uniformly at times T seconds apart from $-\infty$ to $+\infty$, a discrete-time signal $x(n)$ is obtained:

$$x(n) = x_a(t)\big|_{t=nT}$$

The Fourier transform of the resulting sequence $x(n)$ can be shown to be

$$X(e^{j\omega}) = (1/T) \sum_{r=-\infty}^{\infty} X_a\left[j\frac{1}{T}(\omega + 2\pi r)\right] \tag{1.29}$$

Thus $X(e^{j\omega})$ is the sum of an infinite number of amplitude-scaled, frequency-scaled, and translated versions of $X_a(j\Omega)$, the Fourier transform of the continuous-time signal $x_a(t)$, as illustrated in Fig. 1.20.

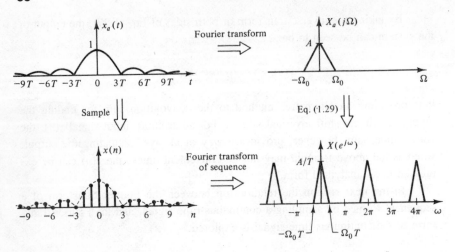

Figure 1.20 Fourier transform of a sequence resulting from sampling a continuous-time signal.

From Fig. 1.20 it is easy to see that the triangles of $X(e^{j\omega})$ will not overlap if $\Omega_0 T < \pi$. This inequality can be rearranged to give

$$1/T > \Omega_0/\pi$$

If we let Ω_0 equal $2\pi f_0$, where f_0 is in hertz, the above inequality becomes

$$1/T > 2f_0 \tag{1.30}$$

Therefore, if the sampling rate $1/T$ is greater than $2f_0$, no overlap occurs. If there is no overlap, the spectrum $X_a(j\Omega)$ can be found, and by the inverse transform the $x_a(t)$ can be reconstructed. If, however, $\Omega_0 T > \pi$ the triangles will overlap and the spectrum of the continuous signal cannot be reconstructed; therefore the original signal cannot be reconstructed.

If $X_a(j\Omega)$ remains nonzero over the infinite interval from zero to ∞, it is obvious that there will always be an overlap, and perfect reconstruction is impossible. A very important theorem relating band-limited signals and the sampling rate is now presented after formally defining band-limited signals.

Definition. A signal $x(t)$ is band-limited if there exists a finite radian frequency Ω_0 such that $X(j\Omega)$ is zero for all Ω such that $|\Omega| > \Omega_0$.

Theorem 8. A signal $x_a(t)$ can be reconstructed from its sample values $x_a(nT)$ if the sampling rate $1/T$ is greater than twice the highest frequency (f_0 in hertz) present in $x_a(t)$.

Definition. The sampling rate $2f_0$ for an analog band-limited signal is referred to as the *Nyquist rate*.

If $x_a(t)$ is not band limited, exact reconstruction from its sampled values is not possible. However, for the case that $x_a(t)$ is band-limited, $x_a(t)$ can be reconstructed from the samples $x(n)$ as shown in the following theorem.

Theorem 9. An analog signal $x_a(t)$, band-limited to $2\pi f_0$ rad/sec, can be reconstructed from its samples taken T seconds apart where $T > 1/(2f_0)$ as follows:

$$x_a(t) = \sum_{n=-\infty}^{\infty} x(n) \frac{\sin [\pi(t - nT)/T]}{\pi(t - nT)/T} \tag{1.30a}$$

Equation (1.30a) is referred to as an interpolation formula because it provides a way of calculating $x_a(t)$ between the sampled values of $x(n)$. This reconstruction or interpolation is shown in Fig. 1.21.

It is seen from Eq. (1.29) that $X(e^{j\omega})$ is obtained as an infinite sum of translated and scaled versions of $X_a(j\Omega)$. If T is such that these versions do not overlap, i.e., the signal is band-limited and $1/T$ is greater than the Nyquist rate, then one way to get the original spectrum $X_a(j\Omega)$ from $X(e^{j\omega})$ is to truncate $X(e^{j\omega})$ at π and invert the scaling procedure. This can be written as follows:

$$X_a(j\Omega) = \begin{cases} T \cdot X(e^{j\omega})|_{\omega = \Omega T}, & |\Omega| \le \pi/T \\ 0, & \text{otherwise} \end{cases}$$

The relationships existing between the Fourier transform pair for the continuous-time signals and the Fourier transform pair for sequences formed by sampling continuous-time signals are summarized in Fig. 1.22. The relationships shown in solid lines are true for any continuous signal $x(t)$ and sampled version $x(nT)$ that have Fourier transforms, while the relationships shown in dotted lines are true only for band-limited signals $x_a(t)$ where the sampling rate $1/T$ is equal to or greater than the Nyquist rate. The horizontal arrows describe the Fourier pair definitions for continuous- and discrete-time signals.

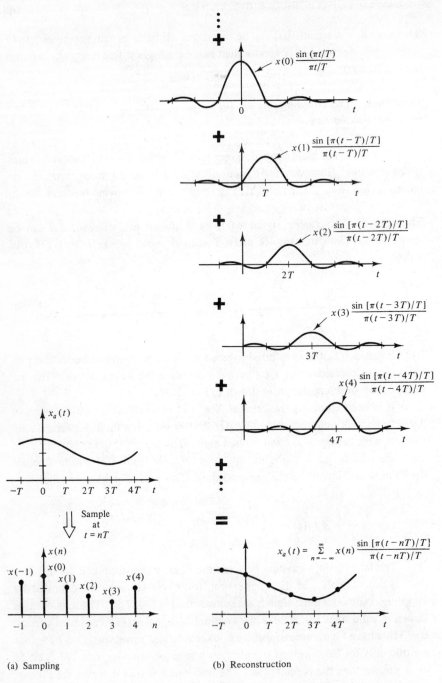

Figure 1.21 Reconstruction of a band-limited signal $x_a(t)$ from its samples $x(n)$ taken at $t = nT$; (a) sampling, (b) reconstruction.

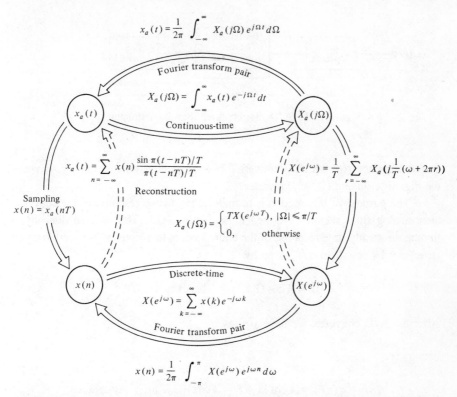

$$x_a(t) = \frac{1}{2\pi} \int_{-\infty}^{\infty} X_a(j\Omega) e^{j\Omega t} d\Omega$$

Fourier transform pair

$$X_a(j\Omega) = \int_{-\infty}^{\infty} x_a(t) e^{-j\Omega t} dt$$

Continuous-time

$x_a(t)$ $X_a(j\Omega)$

$$x_a(t) = \sum_{n=-\infty}^{\infty} x(n) \frac{\sin \pi(t-nT)/T}{\pi(t-nT)/T}$$

$$X(e^{j\omega}) = \frac{1}{T} \sum_{r=-\infty}^{\infty} X_a(j\frac{1}{T}(\omega + 2\pi r))$$

Reconstruction

Sampling
$x(n) = x_a(nT)$

$$X_a(j\Omega) = \begin{cases} TX(e^{j\omega T}), & |\Omega| \leq \pi/T \\ 0, & \text{otherwise} \end{cases}$$

Discrete-time

$x(n)$ $X(e^{j\omega})$

$$X(e^{j\omega}) = \sum_{k=-\infty}^{\infty} x(k) e^{-j\omega k}$$

Fourier transform pair

$$x(n) = \frac{1}{2\pi} \int_{-\pi}^{\pi} X(e^{j\omega}) e^{j\omega n} d\omega$$

Figure 1.22 Relationships between Fourier transforms of a continuous signal $x_a(t)$ and a discrete-time signal $x(n)$ that is obtained by sampling the continuous signal every T sec. The up paths shown in dotted lines do not hold if $x_a(t)$ is not band limited or if $x_a(t)$ is band limited and the sampling rate $1/T$ is less than the Nyquist rate.

1.6 DIGITAL FILTER WITH A/D AND D/A

Consider a linear shift invariant system placed in an A/D–filter–D/A structure shown in Fig. 1.23. The A/D converter normally samples the incoming signal at a uniform rate with time between samples of T seconds to give a sequence $x(n)$:

$$x(n) = x_a(t)|_{t=nT} = x_a(nT) \tag{1.31}$$

The discrete-time filter then operates on the incoming signal $x(n)$ to give an

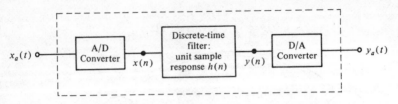

Figure 1.23 A/D–discrete-time filter–D/A structure.

output sequence $y(n)$ which is converted to a continuous-output signal $y_a(t)$ via the digital-to-analog (D/A) converter.

The purpose of this section is to find the frequency response of the equivalent analog filter shown by dotted lines in Fig. 1.23. This will be done by finding the steady state response of the entire system to a sinusoid of an arbitrary frequency Ω_0 less than π/T given by

$$x_a(t) = A \cos \Omega_0 t, \qquad \text{where } \Omega_0 < \pi/T$$

After the A/D converter we have $x(n)$ as

$$x(n) = x_a(nT) = A \cos \Omega_0 nT = A \cos \left[\overset{\overset{\omega_0}{\uparrow}}{(\Omega_0 T)} n \right] = A \cos \omega_0 n$$

The ω_0, equal to $\Omega_0 T$, is conventionally called a digital frequency even though its units are in radians/sample.‡ The steady state output $y(n)$ of the discrete-time filter can then be written using Eq. (1.22) as

$$y_{ss}(n) = A|H(e^{j\omega})|\Big|_{\omega = \omega_0 = \Omega_0 T} \cos \left[(\Omega_0 T)n + \arg H(e^{j\omega})\Big|_{\omega = \omega_0 = \Omega_0 T} \right] \quad (1.32)$$

Assuming a perfect digital-to-analog converter that is a perfect reconstruction from the samples, the steady state output $y_a(t)$ can be written as follows:

$$y_a(t) = A|H(e^{j\Omega_0 T})| \cos [\Omega_0 t + \arg H(e^{j\Omega_0 T})] \quad (1.33)$$

Therefore the equivalent analog steady state response to a cosine wave is another cosine wave of the same frequency but whose amplitude has been scaled by $|H(e^{j\Omega_0 T})|$ and whose phase has been changed by $\arg H(e^{j\Omega_0 T})$. Thus the mag-

‡Throughout the text the "/sample" will be dropped for simplicity and just radians will be used.

nitude and phase are determined by evaluating the digital frequency response $H(e^{j\omega})$ at $\omega = \Omega_0 T$. In Fig. 1.24 the magnitude of the frequency response of a discrete-time filter by itself is shown along with the corresponding magnitude of the equivalent analog frequency response of the A/D–$H(z)$–D/A structure. Similarly, corresponding phases of the frequency responses can be drawn for the discrete-time filter alone and the equivalent analog filter.

It is worth emphasizing at this point that the equivalent analog frequency response is not periodic in the ordinary sense. Analog input sinusoids of radian frequency Ω_1 and $\Omega_1 + k2\pi/T$ give the same output signal, which is not too surprising when we consider that the sampled versions of those analog signals give the same discrete-time sequences. Therefore, for an arbitrary analog radian frequency Ω we have the steady state output for the equivalent analog filter as

$$y_a(t) = A|H(e^{j\omega'})| \cos\left[\frac{\omega' t}{T} + \arg H(e^{j\omega'})\right] \tag{1.34}$$

where $\omega' = (\Omega T) \bmod 2\pi$ if $(\Omega T) \bmod 2\pi \leq \pi$ and $2\pi - (\Omega T) \bmod 2\pi$ if $(\Omega T) \bmod 2\pi > \pi$.

The equivalent analog filter is not a linear time invariant filter for all input signals. It will act as a linear time invariant filter if all the analog signals considered as inputs have frequencies below π/T. If input analog signals contain analog frequencies greater than π/T, then these frequencies will be reduced to

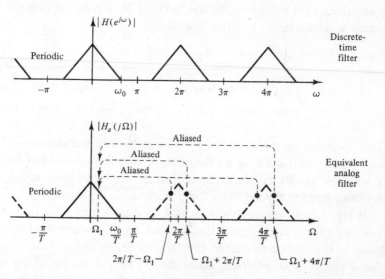

Figure 1.24 Equivalent analog filter frequency response for D/A–digital filter–A/D structure.

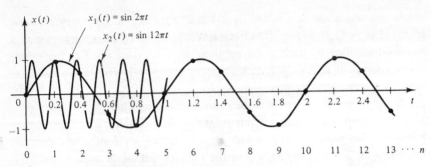

Figure 1.25 Two continuous sinusoidal signals whose sampled values give the same sequence as shown by sample points every 0.2 sec.

$((\Omega T) \bmod 2\pi)/T$, and the A/D–$H(z)$–D/A structures will no longer behave as a time invariant filter.

In order to clarify this concept consider the two analog signals $x_1(t)$ and $x_2(t)$ shown in Fig. 1.25 and given by

$$x_1(t) = \sin 2\pi t$$
$$x_2(t) = \sin 12\pi t$$

Suppose that each of these signals is sampled every 0.2 sec ($T = 0.2$) to give sequences $x_1(n)$ and $x_2(n)$. The dots in Fig. 1.24 indicate the values of the sequences tabulated below as n runs from 0 to 5.

n	0	1	2	3	4	5	...
$x_1(t)\|_{t=0.2n} \overset{\triangle}{=} x_1(n)$	0.0000	0.9510	0.5878	-0.5878	-0.9510	0.000	periodic
$x_2(t)\|_{t=0.2n} \overset{\triangle}{=} x_2(n)$	0.0000	0.9510	0.5878	-0.5878	-0.9510	0.00	periodic

It is easy to see that $x_1(n)$ and $x_2(n)$ are really the same sequence. If the discrete signals $x_1(n)$ and $x_2(n)$ are the inputs to a linear shift invariant system, they will give exactly the same output, that is, the higher-frequency analog signal when sampled appears like a lower frequency (i.e., aliased).

In Fig. 1.26 the output signals for an A/D–discrete-time filter–D/A structure are shown for input signals $\sin 2\pi t$, and $\sin 8\pi t$, where the frequency response of the discrete-time filter, $H(e^{j\omega})$, is as shown in Fig. 1.27. Using a sampling rate of 5 samples/sec, notice that all the outputs are of the same frequency. The $\sin 12\pi t$ signal has been aliased with preservation of the phase while the $\sin 8\pi t$ signal has been aliased but with a phase reversal. The phase

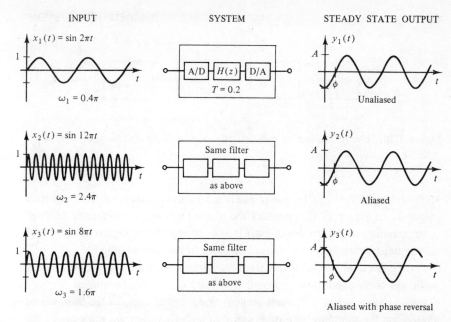

Figure 1.26 Illustration of aliasing for A/D–$H(z)$–D/A filter with $H(z)$ having frequency response given in Fig. 1.27. Any input analog frequency greater than π/T, in this case 5π, will be aliased. Therefore 8π and 12π rad/sec analog frequencies will be aliased as shown.

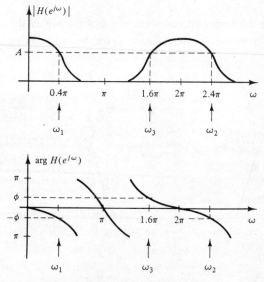

Figure 1.27 Frequency response of digital filter $H(z)$ used in Fig. 1.26.

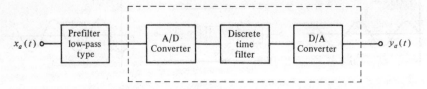

Figure 1.28 Use of prefilter to reduce aliasing in an A/D–discrete-time filter–D/A structure.

reversal for sin $8\pi t$ can be easily predicted by examination of the frequency response. In Fig. 1.27 the value of the phase for a digital frequency of 1.6π (corresponding to the sin $8\pi t$ signal) is $-\phi$ or opposite in sign from the ϕ for the digitial frequency of 0.4π (corresponding to the sin $2\pi t$ signal).

In general, any input analog frequency greater than π/T, called the aliasing point and corresponding to a digital frequency of π, will be aliased. For the example shown in Fig. 1.25, any analog signal greater than 5π rad/sec will be aliased for the sampling rate of 5 samples/sec. Therefore we must select the sampling rate high enough so that all input signals when sampled do not act like lower-frequency inputs. In many cases the input may have extraneous signals with frequency content above that specified by the sampler making it necessary to use a prefilter for their elimination. This filtering must be done before the signal is sampled, since nothing can be done after aliasing has occurred. Therefore, most digital filter implementations use an analog prefilter as shown in Fig. 1.28.

In selecting the sampling rate we notice that the higher the sampling rate the further out in frequency the aliasing point becomes; however, a higher

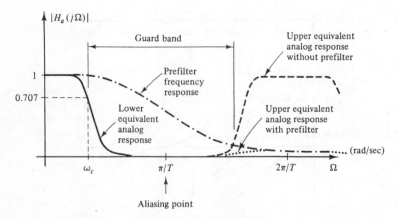

Figure 1.29 Guard band and prefilter frequency response.

sampling rate increases the number of calculations needed for the implementation. Therefore a compromise is usually made by selecting a sampling rate that results in a respectable guard band as shown in Fig. 1.29. This allows a suitable simple prefilter selection to minimize the effects of the aliased frequency components.

1.7 SUMMARY

In this chapter discrete-time signals and systems have been carefully defined and discussed. The similarities to continuous-time signals and systems were explored as well as the differences. By restricting the discrete-time systems to be linear and shift or time invariant, the important concepts of frequency response of systems and frequency content of signals were presented and illustrated. The realization and modeling of linear shift invariant discrete-time systems with linear constant-coefficient difference equations was discussed along with the impulse response and frequency response characterizations. The Nyquist theorem explored the relationships between the Fourier transform of continuous-time signals and the Fourier transform of sequences formed by uniformly sampling those continuous signals. Finally, the important concept of aliasing was described and the steady state analysis of the ever-prevalent equivalent analog filter of the A/D–discrete-time system–D/A structure was presented.

It is important to emphasize that the frequency domain interpretation for discrete-time systems is limited to the class of linear time invariant systems. The concepts can be extended to include the time-varying case; however, in general, nonlinear systems cannot be defined using these fundamental concepts. Some of the important nonlinear operations that prove useful in signal processing, although not describable in a frequency domain framework, are various forms of instantaneous nonlinearities including limiters and power law devices; minimum, maximum, and peak detectors; and homomorphic filtering. These systems are best investigated one by one since a unifying nonlinear processing structure is not available.

BIBLIOGRAPHY

Jong, M. T. *Methods of Discrete Signal and System Analysis*, McGraw-Hill, New York, 1982.

Oppenheim, A. V., and R. W. Schafer. *Digital Signal Processing*. Prentice-Hall, Englewood Cliffs, NJ, 1975.

Stanley, W. D. *Digital Signal Processing*, Reston Publishing, Reston, VA, 1975.

Stearns, S., *Digital Signal Analysis*. Hayden, Rochelle Park, NJ, 1975.

Tretter, S., *Introduction to Discrete-Time Signal Processing*. Wiley, New York, 1976.

PROBLEMS

1.1 Graph the following discrete-time signals:

(a) $x_1(n) = 2\delta(n - 3) - 3\delta(n + 2)$ (f) $x_6(n) = \delta(n) + u(n - 1)$

(b) $x_2(n) = 3 \sin (0.2\pi n)\, u(n)$ (g) $x_7(n) = u(n - 2)$

(c) $x_3(n) = 3 \sin (2.2\pi n)\, u(n)$ (h) $x_8(n) = u(n + 2)$

(d) $x_4(n) = 3 \sin (1.8\pi n)\, u(n)$ (i) $x_9(n) = u(-n + 2)$

(e) $x_5(n) = (1/2)^n u(n)$ (j) $x_{10}(n) = u(-n - 2)$

1.2 For the $x(n)$ shown, sketch the following:

(a) $x(n - 2)$ (d) $x(-n + 1)$

(b) $x(n + 2)$ (e) $x(-n - 1)$

(c) $x(-n)$ (f) $x(2n)$

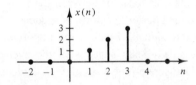

1.3 Calculate the period for the following sinusoidal sequences:

(a) $x_1(n) = 3 \sin (0.05\pi n)$

(b) $x_2(n) = -2 \sin (0.055\pi n)$

(c) $x_3(n) = 2 \sin (0.05\pi n) + 3 \sin (0.12\pi n)$

(d) $x_4(n) = 5 \cos (0.6n)$

1.4 Using the sequences $x(n)$ and $y(n)$ shown below, represent each of the following discrete-time signals by a graph, analytically as a sum of discrete-time impulses, and as a sequence of numbers:

(a) $x_1(n) = x(n) \cdot y(n)$

(b) $x_2(n) = x(n) + y(n)$

(c) $x_3(n) = 3x(n)$

(d) Calculate the energy E of $x(n)$

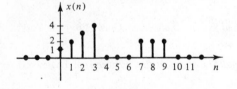   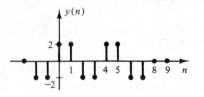

1.5 **(a)** Prove that the unit sample response of two linear shift invariant systems in cascade is the convolution of the two unit sample responses.

(b) Prove that two linear shift invariant systems in parallel are equivalent to a single system whose unit sample response is the sum of the individual unit sample responses.

(c) Are the above two statements true if the systems are linear but no longer shift invariant? Explain your answers.

1.6 Let $x(n)$ and $y(n)$ represent the input and output, respectively, for a given system. Determine for each system defined below whether it is (i) causal, (ii) linear, (iii) time invariant, and (iv) BIBO stable.

(a) $y(n) = ax^2(n)$

(b) $y(n) = ax(n) + b$

(c) $y(n) = e^{-x(n)}$

(d) $y(n) = ax(n + 1) + bx(n - 1)$

(e) $y(n) = ax(n) \cdot x(n - 1)$

(f) $y(n) = \sum_{k=n-3}^{n} e^{x(k)}$

(g) $y(n) = $ median of $[x(n), x(n - 1), x(n - 2)]$

(h) $y(n) = $ maximum of $[x(n), x(n - 1), x(n - 2)]$

(i) $y(n) = $ average of $[x(n + 1), x(n), x(n - 1)]$

(j) $y(n) = $ minimum $[x(n), x(n - 3)]$

(k) $y(n) = ax(n) + bx^2(n - 1)$

(l) $y(n) = \cosh[nx(n) + x(n - 1)]$

1.7 If a system is represented by the following difference equation:

$$y(n) = 3y^2(n - 1) - nx(n) + 4x(n - 1) - 2x(n + 1), \qquad n \geq 0$$

(a) Is the system linear? Explain.

(b) Is the system shift invariant? Explain.

(c) Is the system causal? Why or why not?

1.8 Compute the convolution of each pair of discrete-time signals given below using combination of graphical and analytical techniques.

(a) $x_1(n) = u(n) - u(n - N)$ $\qquad$ $x_2(n) = nu(n)$

(b) $x_3(n) = u(n - 1) - u(n - 3)$ $\qquad$ $x_4(n) = u(n + 3) - u(n + 1)$

(c) $x_4(n) = (\frac{1}{2})^n u(n)$ $\qquad$ $x_5(n) = (\frac{1}{4})^n u(n)$

(d) $x_6(n) = (\frac{1}{2})^n u(n)$ $\qquad$ $x_7(n) = u(n) - u(n - 10)$

(e) $x_8(n) = u(-n - 1)$ $\qquad$ $x_9(n) = (\frac{1}{2})^n u(n)$

(f) $x_{10}(n) = 2^n u(-n - 1)$ $\qquad$ $x_{11}(n) = 4^n u(-n - 1)$

1.9 Given

$$x_1(n) = u(n) - u(n - 5)$$
$$x_2(n) = 2[u(n) - u(n - 3)]$$

(a) Plot $x_1(n)$ and $x_2(n)$.

(b) Calculate and plot $y(n)$, the convolution of $x_1(n)$ and $x_2(n)$ for all n.

(c) What are the nonzero lengths of $x_1(n)$ and $x_2(n)$?

(d) What is the nonzero length of the convolution $x_1(n) * x_2(n)$?

1.10 Given

$$x_1(n) = \delta(n) + 2\delta(n - 1) + 3\delta(n - 2) + 4\delta(n - 3)$$
$$x_2(n) = \delta(n + 1) - 2\delta(n) + \delta(n - 1)$$

find $x_1(n) * x_2(n)$ at $n = -1, 0$, and $+1$.

1.11 Given $x_1(n)$ and $x_2(n)$ as shown, calculate the values of the convolution of $x_1(n)$ and $x_2(n)$ at
(a) $n = -1$
(b) $n = 2$
(c) $n = 6$

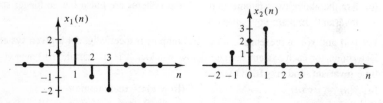

1.12 Given the following linear constant coefficient difference equation (LCCDE),

$$y(n) + 2y(n - 1) + y(n - 2) = x(n)$$

with $y(n) = 0$ for all $n < 0$
(a) Calculate and plot $y(n)$ for $x(n) = \delta(n)$ at $n = 0, 1, 2, 3, 4,$ and 5.
(b) Calculate and plot $y(n)$ for $x(n) = u(n)$.
(c) Calculate and plot $y(n)$ for $x(n) = u(n) - u(n - 2)$.
(d) What is the unit sample response $h(n)$ for this system?
(e) Is the system specified by $h(n)$ stable? Why or why not?
(f) Solve the LCCDE above if $x(n) = (\frac{1}{2})^n u(n)$ by finding the particular and homogeneous solutions using the initial conditions specified.
(g) Repeat (f) if $x(n) = nu(n)$.

1.13 Consider a system with unit sample response $h(n)$ given by

$$h(n) = 2^{-n} u(n)$$

If the input to the system is $x(n) = 2\delta(n) + 4\delta(n - 1) + 4\delta(n - 2)$,
(a) Calculate and plot $y(n)$ for $n = 0, 1, 2, 3, 4, 5,$ and 6.
(b) Is this system BIBO stable? Is it causal?

1.14 A linear shift invariant system is characterized by its impulse response $h(n)$ given by

$$h(n) = 3(-1/4)^n u(n - 1)$$

(a) Is the system causal? Why or why not?
(b) Is the system BIBO stable? Why or why not?
(c) Is the system an FIR or an IIR filter? Why?

Repeat (a), (b), (c) for $h(n) = u(n) - u(n - 5)$.

1.15 Give three methods for determining BIBO stability for a linear shift invariant system. Which of these can be used for general linear systems?

1.16 Let $x(n)$ represent an input to a linear shift invariant system with impulse response $h(n)$. For $x(n) = \delta(n) + 2\delta(n - 1)$ and $h(n) = 2\delta(n) - \delta(n - 1)$
(a) Calculate $X(e^{j\omega})$ and $H(e^{j\omega})$, the Fourier transforms of $x(n)$ and $h(n)$.

(b) Find $y(n)$, the output, by taking the inverse transform of the product of these transforms.

(c) Check your answer by using the convolution sum formula.

1.17 Determine in terms of $x(n)$ the sequence corresponding to

(a) $X(e^{j(\omega - \omega_0)})$

(b) $X^*(e^{j\omega})$, where * denotes complex conjugate.

(c) $\text{Re}[X(e^{j\omega})]$, where Re means "real part of."

1.18 Let $x(n)$ and $X(e^{j\omega})$ represent a sequence and its transform. Do not assume that $x(n)$ is real or that $x(n)$ is causal. Determine in terms of $X(e^{j\omega})$ the transform of each of the following sequences:

(a) $cx(n)$ for $c = $ any constant   **(d)** $x(-n)$

(b) $x(n - n_0)$ for n_0 a real integer   **(e)** $x^2(n)$

(c) $g(n) = \begin{cases} x(n/2), & n \text{ even} \\ 0, & n \text{ odd} \end{cases}$ **(f)** $x(2n)$

1.19 Verify Parseval's theorem for sequences, i.e.,

$$\sum_{n=-\infty}^{\infty} x(n)x^*(n) = \frac{1}{2\pi}\int_{-\pi}^{\pi} X(e^{j\omega})X^*(e^{j\omega})d\omega$$

for $x(n) = (\frac{1}{2})^n u(n)$.

1.20 A system has the unit sample response $h(n)$ given by

$$h(n) = -\tfrac{1}{4}\delta(n + 1) + \tfrac{1}{2}\delta(n) - \tfrac{1}{4}\delta(n - 1)$$

(a) Is the system BIBO stable? Why or why not?

(b) Is the filter causal? Why or why not?

(c) Find the frequency response $H(e^{j\omega})$.

(d) Plot $|H(e^{j\omega})|$ and arg $H(e^{j\omega})$.

(e) What type of filter is $h(n)$? Choices are low-pass, band-pass, highpass, band-stop.

1.21 Given a linear shift invariant system specified by the following difference equation:

$$y(n) + \tfrac{1}{2}y(n - 1) = x(n) - \tfrac{1}{2}x(n - 1), \qquad n \geq 0$$

(a) Find the frequency response $H(e^{j\omega})$ for the system.

(b) Find the $|H(e^{j\omega})|$.

(c) Find arg $H(e^{j\omega})$.

1.22 A system with input $x(n)$ and output $y(n)$ is characterized by the following equation:

$$y(n) = x(n + 1) + x(n - 1)$$

(a) Find the impulse response of this system.

(b) Is the system causal? Why or why not?

(c) Is the system linear? Why or why not?

(d) Is this an FIR or IIR filter?

(e) Find the frequency response of the system.

(f) Is this system BIBO stable? Why or why not?

1.23 A digital filter is realized by the following difference equation:

$$y(n) = x(n) - [2\cos\omega_0]x(n-1) + x(n-2)$$

  (a) Is the filter causal?
  (b) Is it an IIR or an FIR filter?
  (c) What is the frequency response of this filter.
  (d) Find $20|\log H(e^{j\omega})|$ at $\omega = \omega_0$.
  (e) Give a formula for arg $H(e^{j\omega})$.

1.24 Given $H(e^{j\omega})$, the frequency response for a digital filter as shown,
  (a) Determine $h(n)$, the impulse response?
  (b) Does $h(n)$ represent an FIR and IIR filter? Explain why.

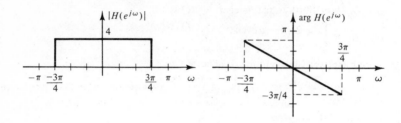

1.25 Prove that the discrete-time ideal low-pass filter is BIBO unstable using the impulse response given in Eq. (1.25).

1.26 The ideal discrete-time high-pass filter is defined by the frequency response $H(e^{j\omega})$ given below.
  (a) Find the unit sample response of such a system in general and plot for $\omega_c = \pi/2$.
  (b) Is this filter causal?
  (c) Is the filter BIBO stable?

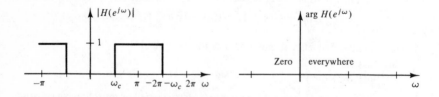

1.27 A signal $x_a(t)$ and its transform are shown below.

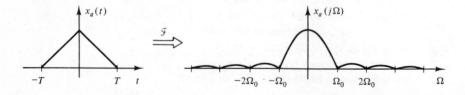

Suppose $x_a(t)$ is sampled every $0.1T$ sec to give a signal $x(n)$.

(a) Draw a picture of $X(e^{j\omega})$ where $X(e^{j\omega}) = \mathscr{F}[x_a(nT)]$ and give a formula for its determination.

(b) Are we able to reconstruct $x_a(t)$ from $x(n)$? Why or why not?

(c) What is the Nyquist sampling rate for $x_a(t)$? Explain your answer.

1.28 A certain analog signal $x_a(t)$ has the following real Fourier transform:

$$X_a(j\Omega) = \begin{cases} 1, & |\Omega| \leq 8\pi \\ 0, & \text{elsewhere} \end{cases}$$

(a) Illustrate $X_a(j\Omega)$.

(b) Carefully illustrate the discrete-time Fourier transform of the discrete-time signal $x(n)$ defined by

$$x(n) = x_a(nT)$$

for the cases $T = 1/16$ sec and $T = 1/4$ sec.

(c) Is $x_a(t)$ band limited? If so, find the Nyquist rate. If not, tell why not.

1.29 Given the frequency response for a hypothetical linear shift invariant discrete-time system as shown, find the steady state output for the following input:

$$x(n) = 2\cos(0.02\pi n + 0.5\pi)\,u(n)$$

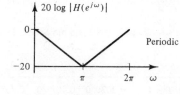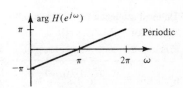

1.30 Given the frequency response $H(e^{j\omega})$ as indicated, find the steady state output for each of the following inputs

(a) $x(n) = 5\cos(\pi n/4)$

(b) $x(n) = 5\cos(9\pi n/4)$

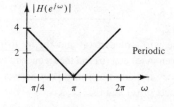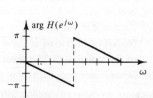

1.31 A given noncausal system has a unit sample response given by

$$h(n) = [\sin(n\pi/4)]/\pi n$$

(a) Illustrate the discrete-time Fourier transform of this sequence.

(b) Find the steady state output of such a system to the following inputs: (i) $x(n) = 2 \sin (\pi n/2)$; (ii) $x(n) = \sin (\pi n/8)$; (iii) $x(n) = \cos (33\pi n/16)$; (iv) $x(n) = 3 \sin (31\pi n/16)$; and (v) $x(n) = \sin (5\pi n/2)$.

1.32 Suppose $x(n)$, representing a sinusoid as shown below, is the input to a digital filter with output $y(n)$ described by the difference equation

$$y(n) = x(n) + \tfrac{1}{2}y(n - 1)$$

(a) Is the filter IIR or FIR? Why?

(b) What is the digital frequency of $x(n)$?

(c) What is the magnitude of the steady state response to the $x(n)$ above using the digital filter?

(d) If $x(n)$ represents the sampled version of an analog signal $x_a(t)$, using 5000 sample/sec, what is the corresponding analog frequency?

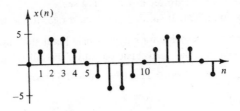

1.33 Define a sequence $x(n)$ to be zero for all $n < -2$ and $n > 10$ and have the following values for $-2 \geq n \geq 10$:

n	-2	-1	0	1	2	3	4	5	6	7	8	9	10
$x(n)$	0	0.707	1	0.707	0	-0.707	-1	-0.707	0	0.707	1	0.707	0

If $y(n)$ is the output of a linear discrete-time invariant system with impulse response $h(n)$ to the input $x(n)$ specified above,

(a) Calculate and plot $y(n)$ for
$h(n) = \tfrac{1}{3} \delta(n) + \tfrac{1}{2} \delta(n - 1) + \tfrac{1}{3} \delta(n - 2)$
at $n = -2, -1, \ldots, 9, 10, 11, 12$.

(b) Calculate and plot $y(n)$ for
$h(n) = -\tfrac{1}{3} \delta(n) + \tfrac{1}{2} \delta(n - 1) - \tfrac{1}{3} \delta(n - 2)$
at $n = -2, -1, \ldots, 9, 10, 11, 12$.

(c) Explain the differences between the $y(n)$'s of part (a) and (b) by using the frequency responses of the $h(n)$'s given in (a) and (b) and the frequency content of the signal $x(n)$. Plot these frequency responses to verify your answer.

(d) Discuss any peculiarities in the outputs of (a) and (b).

1.34 Suppose we have the analog filter shown which is composed of an A/D, discrete-time filter, and D/A.

(a) What is the digital frequency of $x(n)$ if $x(t) = 3 \cos 20\pi t$?

(b) Find the steady state output, $y_{ss}(t)$, if $x(t)$ is the same as in (a).

(c) Find the steady state output $y_{ss}(t)$ if $x(t) = 3u(t)$.

(d) Find two other $x(t)$'s, different analog frequency, that will give the same steady state output as $x(t) = 3 \cos 20\pi t$.

(e) To prevent aliasing effects a prefilter would be used on $x(t)$ before it passes to the A/D. What type (low-pass, high-pass, band-pass, band-reject) of filter would be used and what is the largest cutoff frequency that would work for the above specified configuration?

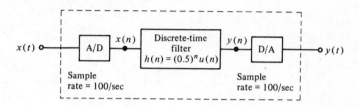

Chapter 2

The $\mathcal{Z}$ Transform

In solving linear constant coefficient differential equations, used for modeling linear electrical networks, the Laplace transform played a very important role. It transformed the differential equations into the complex s plane where algebraic operations and an inverse transform could be performed to yield the solution.

The $\mathcal{Z}$ transform occupies the same position in the theory of the solution of linear constant coefficient difference equations that the Laplace transform does in the theory of linear constant coefficient differential equations. In this chapter, several results analogous to the continuous case and, as before, some particular results that are characteristic of discrete-time systems only are presented.

2.1 DEFINITION OF THE $\mathcal{Z}$ TRANSFORM

The $\mathcal{Z}$ transform of a discrete-time signal or sequence $x(n)$ is defined by

$$\mathcal{Z}[x(n)] \triangleq \sum_{n=-\infty}^{\infty} x(n)z^{-n} = X(z) \qquad (2.1)$$

The z in the above definition is a complex variable and the set of z values for which the summation converges is called the *region of convergence* (ROC) for the transform. In general this region of convergence will be an annular region of the entire complex z plane given by

$$R_{x-} < |z| < R_{x+}$$

The lower limit R_{x-} may be zero and R_{x+} could possibly be ∞. Evaluating $X(z)$ at the complex number $z = re^{j\omega}$ gives

$$X(z)|_{z=re^{j\omega}} = \sum_{n=-\infty}^{\infty} x(n)(re^{j\omega})^{-n} = \sum_{n=-\infty}^{\infty} [r^{-n}x(n)]e^{-j\omega n}$$

Therefore, if $r = 1$, the $\mathcal{Z}$ transform evaluated on the unit circle gives the Fourier transform of the sequence $x(n)$, Eq. (1.26a). Also, note that for positive time signals the r^{-n} provides a convergence factor and thus it is expected that there are sequences for which the $\mathcal{Z}$ transform exists and the Fourier transform in the ordinary sense does not. One such sequence is the unit step function $u(n)$. If singularity functions are admitted, however, a Fourier transform can be found for such sequences.

2.1.1 $\mathcal{Z}$ Transforms of Exponential Sequences

One of the basic signals in continuous time is the exponential function $e^{-\alpha t}$ where α can be real, imaginary, or a complex number. If this signal is sampled the resulting sequence $x(n)$ is given by

$$x(n) = e^{-\alpha t}|_{t=nT} = e^{-\alpha nT} = (e^{-\alpha T})^n$$

It will not be a surprise then that one of the basic sequences used for discrete-time systems is the form a^n. If $x(t)$ had been a complex exponential sequence then $x(n)$ would be a complex sequence expressible as a power of a complex number a. Of major importance then are the $\mathcal{Z}$ transforms of both the positive and negative time exponential sequences. These transforms, including regions of convergence, are presented in Examples 2.1–2.3.

EXAMPLE 2.1

Find the $\mathcal{Z}$ transform including the region of convergence of

$$x(n) = \begin{cases} a^n, & n \geq 0 \\ 0, & n < 0 \end{cases}$$

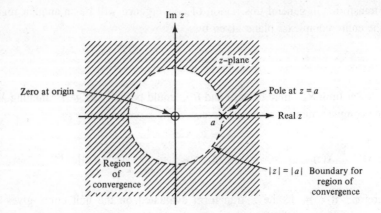

Figure 2.1 Region of convergence for the Z transform of $x(n) = a^n u(n)$.

Solution. By definition, the Z transform of $x(n)$ becomes

$$X(z) = Z[a^n u(n)] = \sum_{n=-\infty}^{\infty} a^n u(n) z^{-n} = \sum_{n=0}^{\infty} (az^{-1})^n$$

By the geometric progression formula (Appendix B), the last sum gives us

$$X(z) = \frac{1}{1 - az^{-1}} = \frac{z}{z - a}$$

The above result converges if $|az^{-1}| < 1$, which is equivalent to $|z| > |a|$. This region of convergence is shown in Fig. 2.1. Values of z for which $X(z) = 0$ are called zeros of $X(z)$, while values for z for which $X(z) \to \infty$ are called poles of $X(z)$. Poles are usually indicated by an $\times$ as seen in Fig. 2.1 at $z = a$ and zeros by an $\bigcirc$ as at $z = 0$.

EXAMPLE 2.2

Find the Z transform including region of convergence of

$$x(n) = -b^n u(-n - 1)$$

Solution. The transform of $x(n)$ is found from the basic definition as follows:

$$X(z) = Z[-b^n u(-n - 1)] = \sum_{n=-\infty}^{\infty} -b^n u(-n - 1) z^{-n} = -\sum_{n=-\infty}^{-1} (b/z)^n$$

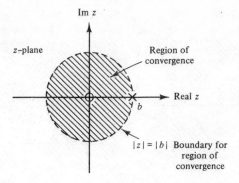

Figure 2.2 Region of convergence for the $\mathcal{Z}$ transform of $x(n) = -b^n u(-n-1)$.

By letting $n = -m$ in the above summation, that is, a change of variables, and since interchanging the order of summation does not change the sign, $X(z)$ becomes

$$X(z) = -\sum_{m=1}^{\infty} (z/b)^m = 1 - \sum_{m=0}^{\infty} (z/b)^m$$

The last step was obtained by adding the $n = 0$ term to the sum and subtracting it from the total. By using the geometric progression formula to evaluate the infinite sum, $X(z)$ becomes

$$X(z) = 1 - \frac{1}{1 - z/b} \quad \text{with } |z/b| < 1$$

Simplifying $X(z)$ above gives the transform and region of convergence as

$$X(z) = \frac{z}{z - b}, \quad \text{ROC } |z| < |b|$$

The region of convergence and pole zero plot for this negative time sequence are shown in Fig. 2.2. Note that the positive time exponential sequence has a transform with region of convergence outside a circle in the z plane, while the negative time exponential sequence has a transform with region of convergence inside a circle.

EXAMPLE 2.3

Find the $\mathcal{Z}$ Transform and region of convergence of $y(n)$, the sum of the positive and negative time sequences given in Examples 2.1 and 2.2:

$$y(n) = a^n u(n) - b^n u(-n - 1)$$

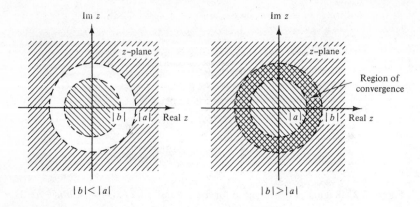

Figure 2.3   Region of convergence for the $\mathcal{Z}$ transform of $x(n) = a^n u(n) - b^n u(-n-1)$.

Solution.   The derivation proceeds from the fundamental definition as follows:

$$Y(z) = \mathcal{Z}[y(n)] = \mathcal{Z}[a^n u(n) - b^n u(-n-1)]$$

$$= \sum_{n=-\infty}^{\infty} [a^n u(n) - b^n u(-n-1)]z^{-n}$$

$$= \sum_{n=0}^{\infty} a^n z^{-n} - \sum_{n=-\infty}^{-1} b^n z^{-n}$$

The first summation obtained in Example 2.1 is $z/(z - a)$ with region of convergence $|z| > |a|$, while the second sum determined in Example 2.2 is $z/(z - b)$ with region of convergence $|z| < |b|$. Thus, $Y(z)$, the desired transform, is their sum with region of convergence equal to the intersection of the regions of convergence:

$$Y(z) = \frac{z}{z - a} + \frac{z}{z - b} \qquad \text{with ROC } \{z > |a|\} \cap \{z < |b|\}$$

If $|b| < |a|$ the above intersection is the empty set, i.e., the transform does not converge, while if $|b| > |a|$ the transform converges in the annular region shown in Fig. 2.3.

2.1.2 Important Properties of $\mathcal{Z}$ Transforms

In the preceding examples the $\mathcal{Z}$ transform of positive and negative time exponential sequences were obtained by using the fundamental definition of the $\mathcal{Z}$ transform. The regions of convergence were obtained naturally in the evaluation of the summations. In many cases we will be able to obtain the $\mathscr{Z}$ transform of sequences, including regions of convergence, without using the basic definition. The following properties give us additional leverage in obtaining $\mathscr{Z}$ transforms of various operations on sequences whose transforms are known. Fundamental operations included are the addition of sequences, translation of sequences, various multiplication of sequences, and convolution of sequences. The proofs are not presented but are easily obtained by using the basic $\mathscr{Z}$ transform definition and transformations in the summations.

1. **Linearity**

   If $\quad \mathscr{Z}[f_1(n)] = F_1(z) \quad$ with ROC $R_{1-} < |z| < R_{1+}$

   and $\quad \mathscr{Z}[f_2(n)] = F_2(z) \quad$ with ROC $R_{2-} < |z| < R_{2+}$

   then $\quad \mathscr{Z}[a_1 f_1(n) + a_2 f_2(n)] = a_1 F_1(z) + a_2 F_2(z)$ $\qquad$ (2.2)

   with ROC the intersection of the above ROCs

2. **Translation**

   If $\quad \mathscr{Z}[x(n)] = X(z) \quad$ with ROC $R_{x-} < |z| < R_{x+}$

   then $\quad \mathscr{Z}[x(n - n_0)] = z^{-n_0} X(z) \quad$ with ROC $R_{x-} < |z| < R_{x+}$ $\quad$ (2.3)

3. **Multiplication by an Exponential**

   If $\quad \mathscr{Z}[x(n)] = X(z) \quad$ with ROC $R_{x-} < |z| < R_{x+}$

   then $\quad \mathscr{Z}[a^n x(n)] = X(z)|_{z \to z/a} \quad$ with ROC $|a| R_{x-} < |z| < |a| R_{x+}$ $\quad$ (2.4)

4. **Multiplication by a Ramp**

   If $\quad \mathscr{Z}[x(n)] = X(z) \quad$ with ROC $R_{x-} < |z| < R_{x+}$

   then $\quad \mathscr{Z}[nx(n)] = -z \dfrac{dX(z)}{dz} \quad$ with ROC $R_{x-} < |z| < R_{x+}$ $\quad$ (2.5)

5. **Convolution (Time Domain)**

   Given $x(n) * y(n) = \displaystyle\sum_{k=-\infty}^{\infty} x(k)y(n - k)$

   and $\quad \mathscr{Z}[x(n)] = X(z) \quad$ with ROC $z \in \mathscr{R}_x$

   $\qquad\quad \mathscr{Z}[y(n)] = Y(z) \quad$ with ROC $z \in \mathscr{R}_y$

   then $\quad \mathscr{Z}[x(n) * y(n)] = X(z)Y(z) \quad$ with ROC $z \in \mathscr{R}_x \cap \mathscr{R}_y$ $\quad$ (2.6)

6. **Convolution (z Domain)**

   If $\quad \mathscr{Z}[x(n)] = X(z) \quad$ with ROC $R_{x-} < |z| < R_{x+}$

   and $\quad \mathscr{Z}[y(n)] = Y(z) \quad$ with ROC $R_{y-} < |z| < R_{y+}$

   then $\quad \mathscr{Z}[x(n) \cdot y(n)] = \dfrac{1}{2\pi j} \oint_{C_2} X(v)Y(z/v)v^{-1}\, dv$ $\qquad$ (2.7) where $\oint_{C_2}$ is a complex contour integral and C_2 is a closed contour in the intersection of the ROCs of $X(v)$ and $Y(z/v)$. The region of convergence for the resulting transform is $R_{x-}R_{y-} < |z| < R_{x+}R_{y+}$

7. **Initial Value Theorem**

   If $x(n)$ is a causal sequence with $\mathcal{Z}$ transform $X(z)$, then

   $$x(0) = \lim_{z\to\infty} X(z) \tag{2.8}$$

8. **Final Value Theorem**

   If the $\mathcal{Z}[x(n)] = X(z)$ and the poles of $X(z)$ are all inside the unit circle, then the value of $x(n)$ as $n \to \infty$ is given by

   $$\lim_{n\to\infty} x(n) = \lim_{z\to 1}\left[\left(\frac{z-1}{z}\right)X(z)\right] \tag{2.9}$$

The properties just presented allow the calculation of $\mathcal{Z}$ transforms of sequences formed by various operations on sequences whose transforms are known. For these properties to be useful, a catalog of transforms of fundamental or building sequences must be established. In the following discussion a mini-table will be determined for special positive and negative time sequences, while a more complete table for positive time sequences is given in Appendix C.

2.1.3 Transforms of Some Useful Sequences

In Section 2.1 the $\mathcal{Z}$ transform of a positive time exponential sequence was found. Using this result and the properties given in Section 2.1.2, the $\mathcal{Z}$ transforms of certain useful sequences, shown in Table 2.1, are easily established.

The first entry in the table is found by applying the definition [Eq. (2.1)] directly to the unit sample sequence. As the only nonzero value is at $n = 0$, the summation contains the single term 1, which converges everywhere. A unit sample sequence translated by an amount k to the right results in a negative power of z and thus we cannot have convergence at $z = 0$.

Entry 3 of Table 2.1 was obtained in Example 2.1. Also, since the unit step is just an exponential $a^n u(n)$ with $a = 1$, the $\mathcal{Z}$ transform of $u(n)$ can be written as $z/(z - 1)$ with region of convergence $|z| > 1$, thus establishing entry 2 of Table 2.1. To prove the results of entry 4 requires breaking up the sinusoidal sequences into sums of complex exponential sequences, applying the linearity property 1 given in Section 2.1.2 to the known transforms of the exponential sequences, and simplifying the complex expressions.

TABLE 2.1 TABLE OF THE z TRANSFORM OF A FEW USEFUL SEQUENCES

	Sequence	z Transform	Region of covergence
1. Unit sample	$\delta(n)$	1	All z
	$\delta(n - k) \quad k > 0$	z^{-k}	$\|z\| > 0$
	$\quad\quad\quad\quad k < 0$	z^{-k}	$\|z\| < \infty$
2. Unit step	$u(n)$	$z/(z - 1)$	$\|z\| > 1$
	$-u(-n - 1)$	$z/(z - 1)$	$\|z\| < 1$
3. Exponential	$a^n u(n)$	$z/(z - a)$	$\|z\| > \|a\|$
	$-b^n u(-n - 1)$	$z/(z - b)$	$\|z\| < \|b\|$
4. Sinusoidal	$\sin \omega_0 n\, u(n)$	$\dfrac{z \sin \omega_0}{z^2 - 2z \cos \omega_0 + 1}$	$\|z\| > 1$
	$\cos \omega_0 n\, u(n)$	$\dfrac{z^2 - z \cos \omega_0}{z^2 - 2z \cos \omega_0 + 1}$	$\|z\| > 1$
5. Unit ramp	$nu(n)$	$\dfrac{z}{(z - 1)^2}$	$\|z\| > 1$

The z transform of the unit ramp can be obtained directly from the definition and series evaluation using summation formulae from Appendix B or by using the z transform of the unit step sequence and property 4. Since $nu(n)$ is n times $u(n)$, property 4 gives the transform of $nu(n)$ as

$$Z[nu(n)] = -z\frac{d}{dz}Z[u(n)] = -z\frac{d}{dz}\left(\frac{z}{z - 1}\right) = \frac{z}{(z - 1)^2}$$

with the same region of convergence as that for the transform of the unit step sequence, i.e., $|z| > 1$.

The transforms of more complex analytical representations that can be decomposed into cascade operations can be easily found by repeated application of the properties in Section 2.1. The transforms shown in Table 2.1 are used as the basis as shown in the following example.

EXAMPLE 2.4

Find the z transform of the following $x(n)$ by using the properties of the z transforms and Table 2.1:

$$x(n) = (n - 2)a^{(n - 2)} \cos [\omega_0(n - 2)] u(n - 2)$$

Solution. Applying the translation property gives

$$\mathbb{Z}[x(n)] = z^{-2}\mathbb{Z}[na^n \cos \omega_0 n \, u(n)] \qquad \text{no change in ROC}$$

Applying the multiplication by a ramp property yields

$$\mathbb{Z}[x(n)] = z^{-2}\left\{ -z\frac{d}{dz}\mathbb{Z}[a^n \cos \omega_0 n \, u(n)] \right\} \qquad \text{no change in ROC}$$

Using the multiplication by an exponential property results in

$$\mathbb{Z}[x(n)] = -z^{-1}\frac{d}{dz}\{\mathbb{Z}[\cos \omega_0 n \, u(n)]|_{z \to z/a}\}$$

where the ROC is

$$|a|R_{x-} < |z| < |a|R_{x+}$$

The R_{x-} and R_{x+} above signify the region of convergence for the transform of $\cos \omega_0 n \, u(n)$, which is $|z| > 1$. Using Table 2.1 to obtain the z transform of $\cos \omega_0 n \, u(n)$ yields

$$\mathbb{Z}[x(n)] = -z^{-1}\frac{d}{dz}\left(\left.\frac{z^2 - z\cos \omega_0}{z^2 - 2z\cos \omega_0 + 1}\right|_{z \to z/a} \right) \qquad \text{for } |z| > |a|$$

To arrive at the final answer requires replacing z by z/a, taking the derivative, and simplifying.

2.1.4 Regions of Convergence for Special Sequences

The region of convergence for the z transform is very important. We have already seen that the z transforms of $a^n u(n)$ and $-a^n u(-n-1)$ both give $z/(z-a)$, the only distinguishing characteristic being the region of convergence. The purpose of this section is to establish some general results for convergence of transforms of sums of right- and left-hand sequences.

Definition. A right-hand sequence $x(n)$ is one for which $x(n) = 0$ for all $n < n_0$, where n_0 is positive or negative but finite. If n_0 is greater than or equal to zero, the resulting sequence is causal or a positive time sequence.

Definition. A left-hand sequence $x(n)$ is one for which $x(n) = 0$ for all $n \geq n_0$, where n_0 is positive or negative but finite. If $n_0 \leq 0$, the resulting sequence is negative time or anticausal sequence.

Suppose $x(n)$ is a causal sequence that can be written as a sum of complex exponentials. This takes in a wide class of functions including sinusoids, exponentials, and products thereof. Let

$$x(n) = \sum_{i=1}^{N} (a_i)^n u(n)$$

Taking the Z transform of $x(n)$ gives

$$Z[x(n)] = X(z) = \sum_{i=1}^{N} \frac{z}{z - a_i}$$

The region of convergence $\mathcal{R}$ is the intersection of the regions of convergence for each exponential as follows:

$$\mathcal{R} = \bigcap_{i=1}^{N} \mathcal{R}_i, \qquad \text{where } \mathcal{R}_i = \{z : |z| > |a_i|\}$$

Therefore, $\mathcal{R} = \{z : |z| > \text{largest of } |a_i|\}$ as shown in Fig. 2.4.

Since the region of convergence for a translated exponential remains the same as that for the original exponential Eq. (2.3), all right-hand sequences that are sums of translated exponentials have regions of convergence similar to that expressed above.

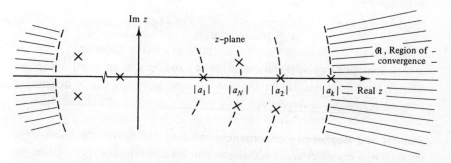

Figure 2.4 Region of convergence for the Z transform of the sum of causal complex exponential sequences.

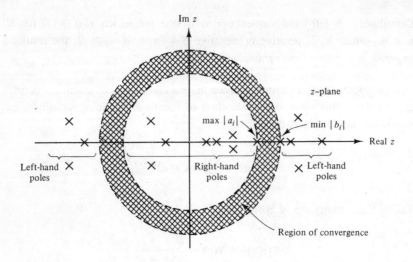

Figure 2.5 Region of convergence for a sequence containing both positive and negative time parts.

By a similar argument we can show that all left-hand sequences expressible as a sum of translated complex exponentials have a region of convergence $\mathcal{L}$ given by

$$\mathcal{L} = \{z : |z| < \text{smallest of } |b_i|\}$$

If we have a combination of right- and left-hand sequences, the corresponding region of convergence is the intersection of $\mathcal{R}$ and $\mathcal{L}$. Therefore the total region of convergence becomes an annular region as shown in Fig. 2.5, and given by

$$\mathcal{R}_{\text{Total}} = \mathcal{R} \cap \mathcal{L} = \{z : \text{largest of } |a_i| < |z| < \text{smallest of } |b_i|\}$$

It is useful to know that the stability of a system with an impulse response which is the sum of translated right- and left-hand sequences can be determined from the region of convergence.

Stability and Region of Convergence   Assume that $h(n)$ is the unit sample response of a causal or noncausal linear shift invariant system. Let $H(z)$ be the $\mathcal{Z}$ transform of $h(n)$, the so-called system function. Stability of the system and the region of convergence for $H(z)$ are related in the following theorem.

Theorem 1. A linear shift invariant system with system function $H(z)$ is BIBO stable iff the region of convergence for $H(z)$ contains the unit circle.

This theorem can be used to determine stability for a given $H(z)$ without obtaining the impulse response or checking outputs for all bounded input signals.

2.2 INVERSE $\mathcal{Z}$ TRANSFORMS

If we have a $\mathcal{Z}$ transform with its region of convergence and would like to have the sequence associated with the transform, there are several approaches that can be used in obtaining that sequence. These include (a) the complex inversion integral, (b) partial fraction expansion, and (c) division for obtaining the inverse $\mathcal{Z}$ transform.

2.2.1 The Complex Inversion Integral.

We know that the forward transform of a sequence $x(n)$ is given by

$$X(z) = \sum_{n=-\infty}^{\infty} x(n)z^{-n} \quad \text{with ROC } \mathcal{R}$$

Multiplying $X(z)$ by $z^k/(2\pi jz)\, dz$ and integrating around a closed contour C which lies entirely within the region of convergence $\mathcal{R}$ gives

$$\frac{1}{2\pi j}\oint_C X(z)\frac{z^k}{z}\,dz = \frac{1}{2\pi j}\oint_C \sum_{n=-\infty}^{\infty} x(n)z^{-n+k-1}\,dz$$

$$= \frac{1}{2\pi j}\sum_{n=-\infty}^{\infty} x(n)\oint_C z^{-n+k-1}\,dz \qquad (2.10)$$

Each contour integral can then be evaluated by using the Cauchy integral theorem which states that if C encircles the origin in a counterclockwise direction, that

$$\frac{1}{2\pi j}\oint_C z^{k-1}\,dz = \begin{cases} 1, & k=0 \\ 0, & k\neq 0 \end{cases} \quad \text{or} \quad \frac{1}{2\pi j}\oint_C z^n\,dz = \begin{cases} 1, & n=-1 \\ 0, & n\neq -1 \end{cases} \qquad (2.11)$$

Therefore, the only term in the summation of the right side of (2.10) that is not zero is the one where the exponent of z is -1, that is,

$$-n + k - 1 = -1 \quad \text{or} \quad n = k$$

Thus Eq. (2.10) can be rearranged and written as follows:

$$\frac{1}{2\pi j} \oint_C X(z) \frac{z^n}{z} dz = x(n) \tag{2.12}$$

Equation (2.12) is known as the complex inversion integral for the evaluation of the inverse z transform. When evaluating the complex inversion integral it becomes convenient to use Cauchy's formula.

Cauchy's Formula Let C be a simple closed path and $f'(z)$ exist on and inside C; then

$$\frac{1}{2\pi j} \oint_C \frac{f(z)}{(z - z_0)} dz = \begin{cases} f(z_0), & \text{for } z_0 \text{ inside } C \\ 0, & \text{for } z_0 \text{ outside } C \end{cases} \tag{2.13}$$

For poles of multiplicity k that are enclosed by the contour C, $f(z)$'s with $k + 1$ order derivatives, and $f(z)$ with no poles inside C, we have the following:

$$\frac{1}{2\pi j} \oint_C \frac{f(z)}{(z - z_0)^k} dz = \begin{cases} \dfrac{1}{(k - 1)!} \dfrac{d^{k-1}}{dz^{k-1}} [f(z)] \Big|_{z=z_0} & \text{for } z_0 \text{ inside } C \\ 0 & \text{for } z_0 \text{ outside } C \end{cases} \tag{2.14}$$

The right-hand side of the equation is sometimes called the residue of the pole at z_0. If there is more than one pole inside the contour, the integral can be evaluated by adding up the results from each pole; for example, if there are n unique poles $z_1, z_2, \ldots, z_n$ and $f(z)$ contains no poles inside C, it can be shown that

$$\frac{1}{2\pi j} \oint_C \frac{f(z)\, dz}{(z - z_1)(z - z_2) \cdots (z - z_n)} = \sum_{i=1}^{n} \lim_{z \to z_i} [(z - z_i)\psi(z)] \tag{2.15}$$

where
$$\psi(z) = \frac{f(z)}{(z - z_1)(z - z_2) \cdots (z - z_n)}$$

If any of the poles z_i are of multiplicity greater than one, the formula given in (2.14) is used to determine their contribution to the integral while (2.15) is used for the simple poles. The value of the integral is then obtained by simply adding the results from all poles.

EXAMPLE 2.5

Evaluate the following contour integral for C defined by the circle $|z| = 3$:

$$I = \frac{1}{2\pi j} \oint_C \frac{z + 2}{(z - 1)^2 (z - 2)(z - 5)} dz$$

Solution. The poles at 1 and 2 are inside the contour C and will each contribute to the value of the integral. The pole at 5 will not have a contribution since it is outside the contour. For the pole of multiplicity 2 at $z = 1$ we have from (2.13)

$$I_1 = \frac{1}{(1-1)!} \left\{ \frac{d^{2-1}}{dz^{2-1}} \left[\frac{z+2}{(z-2)(z-5)} \right] \right\} \Bigg|_{z=1} = 19/16$$

For the unique pole at $z = 2$, we have from (2.15)

$$I_2 = \left\{ (z-2) \cdot \frac{(z+2)}{(z-1)^2(z-2)(z-5)} \right\} \Bigg|_{z=2} = -4/3$$

Therefore the value of the integral I is

$$I = I_1 + I_2 = 19/16 - 4/3 = -7/48$$

The inverse Z transform which is a contour integral was shown to be

$$x(n) = \frac{1}{2\pi j} \oint_C X(z) \frac{z^n}{z} dz \qquad (2.16)$$

where C is a closed simple contour within the region of convergence. Using Eq. (2.14), $x(n)$ can be written as

$$x(n) = \sum_{\substack{\text{all poles} \\ z_i \text{ inside } C}} \text{residue of } X(z) \frac{z^n}{z} \text{ at } z = z_i \qquad (2.17)$$

If for a given n there are no poles of $X(z)x^n/z$ inside C, then $x(n)$ is zero for those values of n.

EXAMPLE 2.6

Using the complex inversion integral, find the inverse z transform of $X(z)$ given by

$$X(z) = \frac{z}{z - a} \qquad \text{with region of convergence } |z| > |a|$$

Solution. From the complex inversion integral Eq. (2.12), $x(n)$ can be written as

$$x(n) = \frac{1}{2\pi j} \oint_C \frac{z}{z - a} \frac{z^n}{z} \, dz = \frac{1}{2\pi j} \oint_C \frac{z^n}{z - a} \, dz \qquad (2.18)$$

The evaluation for $n \geq 0$ is easily accomplished by using Eq. (2.13) and recognizing that since C is in the region of convergence, the pole at a is in the interior of C. This yields

$$x(n) = a^n \qquad \text{for } n \geq 0$$

For $n < 0$ the z^n term gives a pole of order $(-n)$ at the origin, which is also inside the contour so there will be two parts to the answer: one for the pole at a and one for the pole at zero. For example, if $n = -1$ then $x(-1)$ becomes

$$x(-1) = \frac{1}{2\pi j} \oint_C \frac{z^{-1}}{z - a} \, dz = \frac{1}{2\pi j} \oint_C \frac{1}{z(z - a)} \, dz$$

$$= \frac{1}{z - a} \bigg|_{z=0} + \frac{1}{z} \bigg|_{z=a} = -\frac{1}{a} + \frac{1}{a} = 0$$

If $n = -2$ then $x(-2)$ becomes

$$x(-2) = \frac{1}{2\pi j} \oint_C \frac{z^{-2}}{z - a} \, dz = \frac{1}{2\pi j} \oint_C \frac{1}{z^2(z - a)} \, dz$$

$$= \frac{d}{dz} \left(\frac{1}{z - a} \right) \bigg|_{z=0} + \frac{1}{z^2} \bigg|_{z=a} = -\frac{1}{a^2} + \frac{1}{a^2} = 0$$

In each of the above evaluations the first term is from the pole or multiple pole at $z = 0$, while the second is from the pole at $z = a$.

If this process is continued for all negative n we arrive at the result that $x(n)$ is zero for all negative n and that $x(n)$ is given by

$$x(n) = a^n u(n)$$

Another way of showing the result for negative n is to perform a transformation on the complex integral. The transformation that will allow an easy evaluation is to define a new complex variable p by

$$p = 1/z \qquad (2.19)$$

Substituting (2.19) into Eq. (2.18) gives

$$x(n) = \frac{1}{2\pi j} \oint_{C'} \frac{(1/p)^n}{1/p - a} \left(-\frac{1}{p^2} \right) dp$$

By changing the contour to a counterclockwise direction, a minus sign is picked up, and upon simplifying we get

$$x(n) = \frac{1}{2\pi j} \oint_{C'} - \frac{(1/p)^{n+1}}{a(p - 1/a)} dp$$

For negative n the $(1/p)$ term does not give rise to any poles and the only pole of the integrand is at $1/a$. Since C contained the pole at a, the new contour C' defined from $p = 1/z$ is easily seen not to contain the pole at $1/a$. Therefore, since there are no poles inside the contour, the value of the integral is zero.

This particular approach is convenient in evaluating negative time answers for right-sided sequences and evaluating positive time results for left-sided sequences.

2.2.2 Partial Fraction Expansions

When $X(z)$ is expressed as a rational function of z, that is, a ratio of two polynomials in z, the inverse Z transform can be obtained conveniently by using the partial fraction expansion approach. Upon examining Table 2.1, it is noticed that z appears in the numerator of all the functions, so if z is factored out first, the same procedure used with Laplace transforms to get partial fraction expansions can be used. The expansion obtained is then multiplied by z to obtain the expansion of the original $X(z)$.

EXAMPLE 2.7

Find the inverse transform of $X(z) = z/(3z^2 - 4z + 1)$ for the regions of convergence shown in Fig. 2.6 as (a) $|z| > 1$, (b) $|z| < \frac{1}{3}$, and (c) $\frac{1}{3} < |z| < 1$.

Solution.   (a) Define $F(z)$ by $X(z)/z$; therefore

$$F(z) = \frac{X(z)}{z} = \frac{1}{3z^2 - 4z + 1} = \frac{1}{3(z^2 - \frac{4}{3}z + \frac{1}{3})}$$

$$= \frac{1}{3(z-1)(z-\frac{1}{3})} = \frac{A}{z-1} + \frac{B}{z-\frac{1}{3}}$$

where
$$A = F(z)(z-1)\bigg|_{z=1} = \frac{1}{3(z-\frac{1}{3})}\bigg|_{z=1} = \frac{1}{2}$$

$$B = F(z)(z-\tfrac{1}{3})\bigg|_{z=1/3} = \frac{1}{3(z-1)}\bigg|_{z=1/3} = -\frac{1}{2}$$

Therefore,
$$\frac{X(z)}{z} = \frac{\frac{1}{2}}{z-1} + \frac{-\frac{1}{2}}{z-\frac{1}{3}}$$

and
$$X(z) = \frac{\frac{1}{2}z}{z-1} + \frac{(-\frac{1}{2})z}{z-\frac{1}{3}}$$

From the pole plot of $X(z)$ we see that the region of convergence is outside the largest pole signifying right-hand sequences for each pole. Therefore, using entries 2 and 3 of Table 2.1, $x(n)$ becomes

$$x(n) = \mathcal{Z}^{-1}[X(z)] = \mathcal{Z}^{-1}\left[\frac{\frac{1}{2}z}{z-1}\right] + \mathcal{Z}^{-1}\left[-\frac{\frac{1}{2}z}{z-\frac{1}{3}}\right]$$

$$= \tfrac{1}{2}u(n) - \tfrac{1}{2}(\tfrac{1}{3})^n u(n)$$

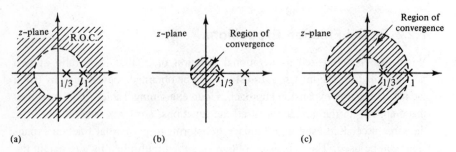

(a)   (b)   (c)

Figure 2.6   Region of convergence for (a) Example 2.6, (b) Example 2.7, and (c) Example 2.8.

(b) The partial fraction expansion obtained in (a) does not change. Therefore, $x(n)$ can be written

$$x(n) = Z^{-1}[X(z)] = Z^{-1}\left[\frac{\frac{1}{2}z}{z-1}\right] + Z^{-1}\left[-\frac{\frac{1}{2}z}{z-\frac{1}{3}}\right]$$

Because the poles at 1 and 1/3 are outside the ROC the inverses give negative time sequences. Therefore

$$x(n = -\tfrac{1}{2}(1)^n u(-n-1) + (-\tfrac{1}{2})(-1)(\tfrac{1}{3})^n u(-n-1)$$
$$= -\tfrac{1}{2} u(-n-1) + \tfrac{1}{2}(\tfrac{1}{3})^n u(-n-1)$$

(c) The partial fraction expansion remains the same; however, because of the region of convergence, the pole at 1/3 corresponds to positive time and the pole at 1 corresponds to negative time. Therefore, $x(n)$ becomes

$$x(n) = Z^{-1}[X(z)] = Z^{-1}\left[\frac{\frac{1}{2}z}{z-1}\right] + Z^{-1}\left[-\frac{\frac{1}{2}z}{z-\frac{1}{3}}\right]$$
$$= -\tfrac{1}{2} u(-n-1) - \tfrac{1}{2}(\tfrac{1}{3})^n u(n)$$

The preceding examples clearly illustrate the point that the inverse transform is unique only when the region of convergence is specified.

In taking the inverse Z transform of a given $X(z)$ by the partial fraction expansion method, $X(z)$ is first divided by z. This process may create a pole for $X(z)/z$ at the origin, which leads to an impulse at the origin upon inversion as shown in the following example.

EXAMPLE 2.8

Find the inverse Z transform of the following $X(z)$ by the partial fraction expansion method for the region of convergence $|z| > 1$.

$$X(z) = \frac{z+1}{3z^2 - 4z + 1}$$

Solution. We desire the partial fraction expansion of $X(z)/z$, which is

$$\frac{X(z)}{z} = \frac{z+1}{z(3z^2 - 4z + 1)} = \frac{z+1}{3z(z-1)(z-\frac{1}{3})}$$
$$= \frac{A}{z} + \frac{B}{z-1} + \frac{C}{z-\frac{1}{3}}$$

where A, B, and C are easily determined as follows:

$$A = \frac{z + 1}{3(z - 1)(z - \frac{1}{3})}\bigg|_{z=0} = 1$$

$$B = \frac{z + 1}{3z(z - \frac{1}{3})}\bigg|_{z=1} = 1$$

$$C = \frac{z + 1}{3z(z - 1)}\bigg|_{z=1/3} = -2$$

Therefore, $X(z)$ obtained by multiplying $X(z)/z$ by z is

$$X(z) = 1 + \frac{z}{z - 1} - \frac{2z}{z - \frac{1}{3}}$$

Taking the inverse z transform of $X(z)$, $x(n)$ becomes

$$x(n) = \delta(n) + u(n) - 2(\tfrac{1}{3})^n u(n)$$

In Examples 2.7 and 2.8, $X(z)$ did not have any repeated roots. In general, the partial fraction expansion approach with repeated poles requires taking inverses of terms $X_k(z)$ given by

$$X_k(z) = \frac{z}{(z - a)^k} \qquad \text{with ROC } |z| > |a|$$

By using the complex inversion integral, the inverse of $X_k(z)$ is easily found to be

$$x_k(n) = \frac{1}{2\pi j}\oint_C \frac{z^n \, dz}{(z - a)^k} \qquad \text{where } C \text{ is inside } |z| > |a| \qquad (2.20)$$

For $n \geq 0$ the integration has only one pole inside the contour C, namely, the one at a. Since this pole is of order k, we can use Cauchy's formula to give the following for $k \geq 2$:

$$x_k(n) = \frac{1}{(k - 1)!}\frac{d^{k-1}}{dz^{k-1}}(z^n)\bigg|_{z=a} = \frac{n(n - 1)\cdots(n - (k - 2))a^{n-k+1}}{(k - 1)!}$$

Furthermore, it can be shown by letting $z = 1/p$ in Eq. (2.20) and evaluating for negative n, that $x(n)$ is zero for all n less than zero. This result is shown in Table 2.2 for different values of k.

TABLE 2.2 USEFUL INVERSE TRANSFORMS FOR PARTIAL FRACTION EXPANSION METHOD

k	$X_k(z)$	$x_k(n) = z^{-1}(X_k(z))$ for ROC $\|z\| > \|a\|$
1	$z/(z - a)$	$a^n\, u(n)$
2	$z/(z - a)^2$	$na^{n-1}\, u(n)$
3	$z/(z - a)^3$	$\dfrac{n(n - 1)a^{n-2}}{2!}\, u(n)$
4	$z/(z - a)^4$	$\dfrac{n(n - 1)(n - 2)a^{n-3}u(n)}{3!}$
.	.	
.	.	
.		
k	$z/(z - a)^k$	$\dfrac{n(n - 1)\cdots(n - (k - 2))a^{n-k+1}u(n)}{(k - 1)!}$

The general procedure for the partial fraction expansion of any rational $X(z)$ for the purpose of obtaining the inverse transform is now presented. Since $X(z)/z$ must be rational, it takes the following form:

$$\frac{X(z)}{z} = \frac{a_0 + a_1 z + \cdots + a_N z^N}{b_0 + b_1 z + \cdots + b_M z^M} \tag{2.21}$$

If $N > M$, no adjustment need be made to $X(z)/z$; however, if $N > M$, we must divide until the remainder polynomial in z has a degree of $M - 1$ or less as shown below:

$$\frac{X(z)}{z} = c_{N-M}z^{N-M} + c_{N-M-1}z^{N-M-1} + \cdots + c_1 z + c_0 \tag{2.22}$$
$$+ \frac{d_0 + d_1 z + \cdots + d_{M-1}z^{M-1}}{b_0 + b_1 z + \cdots + b_M z^M}$$

The last part of the above expression is then factored into its product of poles form and expanded using ordinary partial fraction expansion methods. Assume for purpose of illustration that we have one repeated pole of order k, call it z_r, and that all the rest are unique. Call them $z_{k+1}, z_{k+2}, \ldots, z_M$. Then

$$\psi(z) = \frac{d_0 + d_1 z + \cdots + d_{M-1}z^{M-1}}{b_0 + b_1 z + \cdots + b_M z^M}$$
$$= \frac{A_{1k}}{(z - z_r)^k} + \frac{A_{1k-1}}{(z - z_r)^{k-1}} + \cdots + \frac{A_{11}}{z - z_r} + \sum_{j=k+1}^{M} \frac{A_j}{z - z_j} \tag{2.23}$$

The coefficients A_{1j} and A_j are found as follows:

$$A_{1j} = \frac{1}{(k-j)!} \left\{ \frac{d^{k-j}}{dz^{k-j}} [(z-z_j)^k \psi(z)] \right\} \bigg|_{z=z_j}, \qquad j = 1, 2, \ldots, k$$

$$A_j = \{(z-z_j)\psi(z)\}|_{z=z_j}, \qquad\qquad\qquad j = k+1, k+2, \ldots, M$$

Substituting Eq. (2.23) into Eq. (2.22), multiplying by z, and taking the inverse transform gives us

$$Z^{-1}[X(z)] = x(n)$$
$$= Z^{-1}(c_{N-M}z^{N-M+1} + c_{N-M+1}z^{N-M} + \cdots + c_1 z^2 + c_0 z) \qquad (2.24)$$
$$+ Z^{-1}\left[\sum_{j=1}^{k} A_{1j}z/(z-z_r)^j \right] + Z^{-1}\left[\sum_{j=k+1}^{M} A_j z/(z-z_j) \right]$$

Taking the inverse of the first term gives a series of unit sample sequences, the second, a number of exponentials multiplied by n, $n-1$, etc., and the third, a number of complex exponentials. This is shown in Eq. (2.25).

$$x(n) = \sum_{m=M}^{N} C_{N-m}\delta(n + (N-M+1))$$
$$+ \left[A_{11}z_r{}^n + A_{12}n(z_r)^{n-1} + \cdots + \frac{A_{1k}n(n-1)\cdots(n-(k-2))(z_r)^{n-k+1}}{(k-1)!} + \sum_{j=k+1}^{M} A_j(z_j)^n \right] u(n)$$
$$(2.25)$$

The use of Eq. (2.25) is shown in Example 2.9 for unique poles and $N > M$, and in Example 2.10 for multiple poles. It is the spirit of this equation that is important, and the use is much easier than the complex form indicates.

EXAMPLE 2.9
Find the inverse Z transform of

$$X(z) = \frac{z^4 + z^2}{(z - \frac{1}{2})(z - \frac{1}{4})}$$

for region of convergence $\frac{1}{2} < |z| < \infty$.

Solution.

$$\frac{X(z)}{z} = \frac{z^3 + z}{z^2 - \frac{3}{4}z + \frac{1}{8}} = z + \frac{3}{4} + \frac{\frac{23}{16}z - \frac{3}{32}}{z^2 - \frac{3}{4}z + \frac{1}{8}}$$

The rational expression in z is then expanded by partial fraction expansion

$$\frac{\frac{23}{16}z - \frac{3}{32}}{z^2 - \frac{3}{4}z + \frac{1}{8}} = \frac{A_1}{z - \frac{1}{2}} + \frac{A_2}{z - \frac{1}{4}}$$

where

$$A_1 = \frac{\frac{23}{16}z - \frac{3}{32}}{z - \frac{1}{4}}\bigg|_{z=1/2} = \frac{\frac{5}{8}}{\frac{1}{4}} = \frac{5}{2}$$

$$A_2 = \frac{\frac{23}{16}z - \frac{3}{32}}{z - \frac{1}{2}}\bigg|_{z=1/4} = \frac{\frac{17}{64}}{-\frac{1}{4}} = -\frac{17}{16}$$

Therefore

$$\frac{X(z)}{z} = z + \frac{3}{4} + \frac{\frac{5}{2}}{z - \frac{1}{2}} - \frac{\frac{17}{16}}{z - \frac{1}{4}}$$

Multiplying through by z and taking the inverse transform with ROC $\frac{1}{2} < |z| < \infty$ gives

$$x(n) = \mathcal{Z}^{-1}\left[z^2 + \frac{3}{4}z\right] + \mathcal{Z}^{-1}\left[\frac{\frac{5}{2}z}{z - \frac{1}{2}} - \frac{\frac{17}{16}z}{z - \frac{1}{4}}\right]$$

$$= \delta(n + 2) + \frac{3}{4}\delta(n + 1) + [\frac{5}{2}(\frac{1}{2})^n - \frac{17}{16}(\frac{1}{4})^n]\, u(n)$$

Note that although the annular ROC is infinite in extent, the answer has values for $n = -1$ and $n = -2$ because of the pole at ∞, and thus $x(n)$ is not a causal sequence.

EXAMPLE 2.10

Find the inverse $\mathcal{Z}$ transform of

$$X(z) = \frac{z^2 + z}{(z - \frac{1}{2})^3(z - \frac{1}{4})}$$

for region of convergence $|z| > \frac{1}{2}$.

Solution. By using the partial fraction expansion method $X(z)/z$ can be written in terms of factors as follows:

$$\frac{X(z)}{z} = \frac{z + 1}{(z - \frac{1}{2})^3(z - \frac{1}{4})} = \frac{A_{11}}{z - \frac{1}{2}} + \frac{A_{12}}{(z - \frac{1}{2})^2} + \frac{A_{13}}{(z - \frac{1}{2})^3} + \frac{A_4}{z - \frac{1}{4}}$$

where A_{11}, A_{12}, A_{13}, and A_4 are calculated as

$$A_{11} = \frac{1}{2!} \frac{d^2}{dz^2} \left(\frac{z+1}{z-\frac{1}{4}} \right) \bigg|_{z=1/2} = 80$$

$$A_{12} = \frac{d}{dz} \left(\frac{z+1}{z-\frac{1}{4}} \right) \bigg|_{z=1/2} = -20$$

$$A_{13} = \frac{z+1}{z-\frac{1}{4}} \bigg|_{z=1/2} = 6$$

$$A_4 = \frac{z+1}{(z-\frac{1}{2})^3} \bigg|_{z=1/4} = -80$$

Therefore, upon multiplying by z we arrive at

$$X(z) = \frac{80z}{z-\frac{1}{2}} - \frac{20z}{(z-\frac{1}{2})^2} + \frac{6z}{(z-\frac{1}{2})^3} - \frac{80z}{z-\frac{1}{4}}$$

Using the results from Table 2.2, $x(n)$ can be written as

$$x(n) = \{80(\tfrac{1}{2})^n - 20n(\tfrac{1}{2})^{n-1} + 6[n(n-1)/2](\tfrac{1}{2})^{n-2} - 80(\tfrac{1}{4})^n\} u(n)$$

2.2.3 Inverse z Transforms by Division

The z transform of a given sequence can be written as follows from the basic definition [Eq. (2.1)]:

$$X(z) = \cdots + x(-2)z^2 + x(-1)z^1 + x(0)z^0 + x(1)z^{-1} + x(2)z^{-2} + \cdots$$

$$(2.26)$$

Therefore, if the $X(z)$ can be expanded in a power series, the coefficients represent the inverse sequence values. Thus, the coefficient of z^{-k} is the kth term in the sequence. This approach will not provide an analytic solution, and the region of convergence will determine whether the series has positive or negative exponents. For right-hand sequences the $X(z)$ will be obtained with primarily negative exponents, while left-hand sequences will have primarily positive exponents. For annular regions of convergence, a Laurent expansion would give both the positive and negative time terms. Example 2.11 illustrates the inverse z transform by the division method for different regions of convergence. The method is useful in obtaining a quick look at the first couple of values of the inverse of a given $X(z)$.

EXAMPLE 2.11

Find the inverse transform by division of

$$X(z) = \frac{z}{3z^2 - 4z + 1}$$

where the ROC is (a) $|z| > 1$, and (b) $|z| < \frac{1}{3}$.

Solution. (a) for $|z| > 1$ we must divide to obtain negative powers of z since $|z| > 1$ indicates a right-hand sequence. We have

$$
\begin{array}{r}
\frac{1}{3}z^{-1} + (\frac{1}{3}) \cdot (\frac{4}{3})z^{-2} + \cdots \\
3z^2 - 4z + 1 \overline{\big)\, z } \\
\underline{z - \frac{4}{3} + \frac{1}{3}z^{-1}} \\
\frac{4}{3} - \frac{1}{3}z^{-1} \\
\underline{\frac{4}{3} - \frac{16}{9}z^{-1} + \frac{4}{9}z^{-2}} \\
\frac{13}{9}z^{-1} - \frac{4}{9}z^{-2} \quad \text{etc.}
\end{array}
$$

Therefore,

$$X(z) = \overset{x(1)}{\overline{\tfrac{1}{3}z^{-1}}} + \overset{x(2)}{\overline{\tfrac{4}{9}z}} + \cdots$$

and we can recognize that

$$
\begin{aligned}
x(n) &= 0 \qquad \text{for } n \le 0 \\
x(1) &= \tfrac{1}{3} \\
x(2) &= \tfrac{4}{9}
\end{aligned}
$$

.

.

.

(b) for $|z| > \frac{1}{3}$ we must divide to get positive powers, i.e., for negative n

$$
\begin{array}{r}
\overset{x(-1)}{\overline{1z}} + \overset{x(-2)}{\overline{4z^2}} + \cdots \\
1 - 4z + 3z^2 \overline{\big)\, z } \\
\underline{z - 4z^2 + 3z^3} \\
4z^2 - 3z^3 \\
\underline{4z^2 - 16z^3 + 12z^4} \\
13z^3 - 12z^4 \quad \text{etc.}
\end{array}
$$

Therefore, ·

$$x(n) = 0 \qquad \text{for } n \geq 0$$
$$x(-1) = 1$$
$$x(-2) = 4$$
$$\cdot$$
$$\cdot$$
$$\cdot$$

2.3 RELATIONSHIPS BETWEEN SYSTEM REPRESENTATIONS

We have seen that a linear shift invariant system can be characterized by its unit sample response, a difference equation, a system function, or a frequency response. In this section, procedures and formulas are established for converting one representation to another. As a starting place assume that a system is described by a linear constant coefficient difference equation given by

$$\sum_{k=0}^{N} a_k y(n-k) = \sum_{r=0}^{M} b_r x(n-r) \qquad (2.27)$$

The corresponding system function, impulse response, and frequency response are found in the following way.

To obtain the system function we first take the z transform of both sides of (2.27) using the linearity property to obtain

$$\sum_{k=0}^{N} a_k z[y(n-k)] = \sum_{r=0}^{M} b_r z[x(n-r)] \qquad (2.28)$$

The z transforms of the delayed versions of $x(n)$ and $y(n)$ can be written from the translation property [Eq. (2.3)] as

$$z[x(n-r)] = z^{-r}X(z) \qquad (2.29a)$$
$$z[y(n-k)] = z^{-k}Y(z) \qquad (2.29b)$$

Substituting (2.29) into (2.28) gives

$$\sum_{k=0}^{N} a_k z^{-k}Y(z) = \sum_{r=0}^{M} b_r z^{-r}X(z) \qquad (2.30)$$

Solving for $Y(z)/X(z)$, the Z transform of the output over the Z transform of the input, gives the desired system function $H(z)$ as follows:

$$\frac{Y(z)}{X(z)} = H(z) = \frac{\sum_{r=0}^{M} b_r z^{-r}}{\sum_{k=0}^{N} a_k z^{-k}} \tag{2.31}$$

The unit sample response, $h(n)$, easily obtained from the system function by taking the inverse Z transform, is

$$h(n) = Z^{-1}[H(z)] \tag{2.32}$$

That this represents the unit sample response can be easily seen by first writing the output $Y(z)$ as

$$Y(z) = H(z) \cdot X(z) \tag{2.33}$$

Then if $x(n)$ is a unit sample response, $X(z)$ is just 1, $Y(z)$ equals $H(z)$, and $y(n)$ equals the inverse transform of $H(z)$. Since $y(n)$ is the response to a unit sample excitation, (2.32) is established. With the additional specification of causality, a unique impulse response is established.

If the system function $H(z)$ is known, a difference equation characterization can be found by reversing the operations shown in (2.27)–(2.31). $H(z)$ is first written in terms of negative powers of z and set equal to $Y(z)/X(z)$. Then cross multiplying gives (2.30) and the inverse transform gives the difference equation shown in (2.27).

To obtain the frequency response for a given system we take the Fourier transform of the unit sample response given by

$$H(e^{j\omega}) = \sum_{k=-\infty}^{\infty} h(n)e^{-j\omega n} \tag{2.34}$$

But we have seen that the system function $H(z)$ can be obtained from the unit sample response by the Z transform as

$$H(z) = \sum_{h=-\infty}^{\infty} h(n)e^{-j\omega n} \tag{2.35}$$

Upon examination of (2.34) and (2.35) it is easily seen that the frequency response, if it exists, can be obtained by replacing the z in the system function by $e^{j\omega}$ as follows:

$$H(e^{j\omega}) = H(z)\big|_{z=e^{j\omega}} \qquad (2.36)$$

The notation $H(e^{j\omega})$ earlier given to the frequency response was selected in anticipation of this result. The relationships just developed for conversion between difference equation, system function, frequency response, and impulse response are conceptually summarized in Fig. 2.7 and illustrated in Examples 2.12 and 2.13.

EXAMPLE 2.12

Given the first-order linear constant coefficient difference equation

$$y(n) = ay(n - 1) + x(n)$$

find the impulse response of the system by first finding the system function $H(z)$ and then taking the inverse transform. Assume a causal system.

Solution. Take the $\mathcal{Z}$ transform of both sides ignoring initial conditions, collect terms, and divide $Y(z)$ by $X(z)$ to get $H(z)$ as follows:

$$Y(z) = az^{-1}Y(z) + X(z)$$
$$(1 - az^{-1})Y(z) = X(z)$$
$$H(z) = \frac{Y(z)}{X(z)} = \frac{1}{1 - az^{-1}} = \frac{z}{z - a}$$

By the causality requirement, the inverse transform of $z/(z - a)$ is a positive time exponential and with the region of convergence $|z| > |a|$, $h(n)$ becomes

$$h(n) = \mathcal{Z}^{-1}\left(\frac{z}{z - a}\right) = a^n u(n)$$

EXAMPLE 2.13

Given that $H(z) = (z + 1)/(z^2 - 2z + 3)$ represents a causal system, find a difference equation realization and the frequency response of the system.

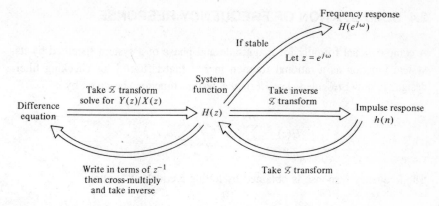

Figure 2.7 Relationships between difference equation, system function, impulse response, and frequency response for stable causal systems represented by linear, constant coefficient difference equations.

Solution. Since the system is causal, first write $H(z)$ in terms of negative powers of z

$$H(z) = \frac{Y(z)}{X(z)} = \frac{z + 1}{z^2 - 2z + 3} = \frac{z^{-1} + z^{-2}}{1 - 2z^{-1} + 3z^{-2}}$$

Now cross multiply:

$$Y(z)(1 - 2z^{-1} + 3z^{-2}) = X(z)(z^{-1} + z^{-2})$$

Taking the inverse transform yields the difference equation

$$y(n) - 2y(n - 1) + 3y(n - 2) = x(n - 1) + x(n - 2)$$

and rearranging gives the following difference equation realization:

$$y(n) = 2y(n - 1) - 3y(n - 2) + x(n - 1) + x(n - 2)$$

The frequency response $H(e^{j\omega})$ obtained by letting $z = e^{j\omega}$ becomes

$$H(e^{j\omega}) = \frac{z + 1}{z^2 - 2z + 3}\bigg|_{z = e^{j\omega}} = \frac{e^{j\omega} + 1}{e^{2j\omega} - 2e^{j\omega} + 3}$$

$$= \frac{(1 + \cos\omega) + j\sin\omega}{\cos 2\omega - 2\cos\omega + 3 + j(\sin 2\omega - 2\sin\omega)}$$

2.4 COMPUTATION OF FREQUENCY RESPONSE

A computational formula for magnitude and phase of a system specified by its system function as a rational fraction in z^{-1} that is useful for checking filter designs is now presented. Let $H(z)$, the system function, be given by

$$H(z) = \sum_{r=0}^{M} b_r z^{-r} \Big/ \sum_{k=0}^{N} a_k z^{-k}$$

The frequency response is obtained by letting z equal $e^{j\omega}$, giving

$$H(e^{j\omega}) = \sum_{r=0}^{M} b_r e^{j\omega r} \Big/ \sum_{k=0}^{N} a_k e^{-j\omega k}$$

By expanding the exponentials into sines and cosines and collecting real and imaginary parts, $H(e^{j\omega})$ is found to be

$$H(e^{j\omega}) = \frac{\displaystyle\sum_{r=0}^{M} b_r \cos r\omega - j \sum_{r=0}^{M} b_r \sin r\omega}{\displaystyle\sum_{k=0}^{N} a_k \cos k\omega - j \sum_{k=0}^{N} a_k \sin k\omega}$$

From the above, the magnitude and phase are easily seen to be

$$|H(e^{j\omega})| = \left[\frac{\left(\displaystyle\sum_{r=0}^{M} b_r \cos r\omega\right)^2 + \left(\displaystyle\sum_{r=0}^{M} b_r \sin r\omega\right)^2}{\left(\displaystyle\sum_{k=0}^{N} a_k \cos k\omega\right)^2 + \left(\displaystyle\sum_{k=0}^{N} a_k \sin k\omega\right)^2} \right]^{1/2} \tag{2.37}$$

$$\arg H(e^{j\omega}) = \tan^{-1}\left[\frac{-\displaystyle\sum_{r=0}^{N} b_r \sin r\omega}{\displaystyle\sum_{r=0}^{N} b_r \cos r\omega} \right] - \tan^{-1}\left[\frac{-\displaystyle\sum_{k=0}^{N} a_k \sin k\omega}{\displaystyle\sum_{k=0}^{N} a_k \cos k\omega} \right] \tag{2.38}$$

z **Transforms of Truncated Delayed Signals** It will be useful to have the transforms of delayed versions of a given signal that have been truncated with

n	$\cdots$ -2	-1	0	1	2	3
$\underline{x(n)}$	$\cdots$ $x(-2)$	$x(-1)$	$x(0)$	$x(1)$	$x(2)$	$x(3)$ $\cdots$
$\underline{x(n-1)}$	$\cdots$ $x(-3)$	$x(-2)$	$x(-1)$	$\dot{x}(0)$	$x(1)$	$x(2)$ $\cdots$
$\underline{x(n-k)}$	$\cdots$ $x(-(k+2))$	$x(-(k+1))$	$x(-k)$	$x(-(k-1))$	$x(-(k-2))$	$x(-(k-3))$
	$\cdots$ $\downarrow$	$\downarrow$	$\downarrow$	$\downarrow$	$\downarrow$	$\downarrow$ $\cdots$
	z^2	z^1	z^0	z^{-1}	z^{-2}	z^{-3}

Figure 2.8 Delayed versions of a sequence $x(n)$

a unit step function. In Fig. 2.8 we see various delayed versions of the signal $x(n)$ with a dotted line giving the position of truncation obtained when multiplying by a unit step.

If we define $\mathcal{Z}[x(n)u(n)]$ to be $X(z)$, then the $\mathcal{Z}$ transform of $x(n - k)u(n)$ is given by

$$\mathcal{Z}[x(n - k)u(n)] = \sum_{n=-\infty}^{\infty} x(n - k)u(n)z^{-n} = \sum_{n=0}^{\infty} x(n - k)z^{-n}$$

By letting $r = n - k$, the summation can be rewritten as

$$\mathcal{Z}[x(n - k)u(n)] = \sum_{r=-k}^{\infty} x(r)z^{-(r+k)} = \sum_{r=-k}^{-1} x(r)z^{-(r+k)} + \sum_{r=0}^{\infty} x(r)z^{-(r+k)}$$

$$= x(-k) + z^{-1}x(-(k-1)) + \cdots + z^{-(k-1)}x(-1) + z^{-k}X(z)$$

$$(2.39)$$

This result, easily verified from the last two rows of Fig. 2.8, is used in the following section to solve linear constant coefficient difference equations with inputs that are stepped into a system.

2.5 SOLUTION OF LINEAR CONSTANT COEFFICIENT DIFFERENCE EQUATIONS

In Chapter 1 a method was mentioned for solving linear constant coefficient difference equations which involved finding the particular and homogeneous solution similar in approach to the classical solution of linear constant coefficient differential equations. We shall see that the $\mathcal{Z}$ transform offers an alternative and simpler approach to solving linear constant coefficient difference equations and that the method is similar to that of solving linear constant coefficient differential equations using the Laplace transform. Suppose that we wish the solution of the following linear constant coefficient difference equation:

$$\sum_{k=0}^{N} a_k y(n - k) = \sum_{r=0}^{M} b_r x(n - r), \qquad n \geq 0 \qquad (2.40)$$

subject to the following initial conditions:

$$y(i) \qquad \text{for } i = -1, \ldots, -N$$
and $$\qquad x(i) \qquad \text{for } i = -1, \ldots, -M$$

Taking the Z transform of both sides of Eq. (2.40) and using the results shown in Eq. (2.39) for the transforms of the delayed $x(n)$'s and $y(n)$'s multiplied by a unit step we have after inserting the initial conditions and rearranging, the following

$$\sum_{k=0}^{N} a_k z^{-k} Y(z) + g(z^{-1}, y(-1), y(-2), \ldots, y(-n))$$

$$= \sum_{r=0}^{M} b_r z^{-r} X(z) + h(z^{-1}, x(-1), x(-2), \ldots, x(-M)) \qquad (2.41)$$

In Eq. (2.41) $g(\cdot)$ and $h(\cdot)$ represent polynomials in z^{-1} that are functions of the initial conditions on $y(n)$ and $x(n)$, respectively. By factoring out the $Y(z)$ and $X(z)$, rearranging, and dividing by the sum we see

$$Y(z) = \frac{\displaystyle\sum_{r=0}^{M} b_r z^{-r}}{\displaystyle\sum_{k=0}^{N} a_k z^{-k}} X(z)$$
$$+ \frac{h(z^{-1}, x(-1), x(-2), \ldots, x(-M)) - g(z^{-1}, y(-1), y(-2), \ldots, y(-N))}{\displaystyle\sum_{k=0}^{N} a_k z^{-k}}$$

The solution $y(n)$ is now obtained by taking the inverse Z transform of $Y(z)$, that is,

$$y(n) = \mathcal{Z}^{-1}\left[X(z) \sum_{r=0}^{M} b_r z^{-r} \Big/ \sum_{k=0}^{N} a_k z^{-k} \right]$$

$$+ \mathcal{Z}^{-1}\left[\mathcal{N}(z^{-1}, x(-1), x(-2), \ldots, \right.$$

$$\left. x(-m), y(-1), y(-2), \ldots, y(-N)) \Big/ \sum_{k=0}^{N} a_k z^{-k} \right] \quad (2.42)$$

where

$$\mathcal{N}(z^{-1}, x(-1), z(-2), \ldots, y(-1), y(-2), \ldots, y(-N)) = h(\cdot) - g(\cdot)$$

Therefore the procedure for solving a linear constant coefficient difference equation is to take the $\mathcal{Z}$ transform using initial conditions, factor and rearrange in the $\mathcal{Z}$ plane to get $Y(z)$, and then take the inverse $\mathcal{Z}$ transform of $Y(z)$. The following example illustrates the procedure:

EXAMPLE 2.14

Find the solution to the following linear constant coefficient difference equation:

$$y(n) - \tfrac{3}{2} y(n-1) + \tfrac{1}{2} y(n-2) = (\tfrac{1}{4})^n \qquad \text{for } n \geq 0$$

with initial conditions $y(-1) = 4$ and $y(-2) = 10$.

Solution. Taking the $\mathcal{Z}$ transform of both sides gives

$$Y(z) - \tfrac{3}{2}[y(-1) + z^{-1}Y(z)] + \tfrac{1}{2}[y(-2) + z^{-1}y(-1) + z^{-2}Y(z)] = \frac{z}{z - \tfrac{1}{4}}$$

Substituting in the initial conditions and rearranging gives

$$Y(z)(1 - \tfrac{3}{2}z^{-1} + \tfrac{1}{2}z^{-2}) = \frac{z}{z - \tfrac{1}{4}} + 1 - 2z^{-1} = \frac{2z^2 - \tfrac{9}{4}z + \tfrac{1}{2}}{z(z - \tfrac{1}{4})}$$

and dividing by $(1 - \tfrac{3}{2}z^{-1} + \tfrac{1}{2}z^{-2})$, $Y(z)$ is seen to be

$$Y(z) = \frac{z(2z^2 - \tfrac{9}{4}z + \tfrac{1}{2})}{(z - \tfrac{1}{4})(z - \tfrac{1}{2})(z - 1)}$$

By partial fraction expansion we find

$$Y(z) = \frac{\frac{1}{3}z}{z - \frac{1}{4}} + \frac{z}{z - \frac{1}{2}} + \frac{\frac{2}{3}z}{z - 1}$$

Taking the inverse $\mathcal{z}$ transform gives the solution of the difference equation as

$$y(n) = [\tfrac{1}{3}(\tfrac{1}{4})^n + (\tfrac{1}{2})^n + \tfrac{2}{3}] u(n)$$

The following example examines the steady state and transient responses for a first-order system excited by a sinusoidal input. Although the presentation is only for the first-order system, it can be shown that the relationship established for the steady state response in terms of the transfer function of the system is a general result for stable systems and sinusoidal inputs.

EXAMPLE 2.15

Given a first-order system characterized by the following difference equation

$$y(n) = Ky(n - 1) + x(n), \qquad n \geq 0$$

with initial condition $y(-1)$, find the system function $H(z)$ and the response of the system to an input $x(n)$ given by

$$x(n) = \cos \omega_0 n \, u(n)$$

Assume $|K| < 1$ in order to have a stable system.

Isolate the transient and steady state portions of the response and verify that Eq. (1.22) holds in the steady state.

Solution. The system function $H(z)$ is seen from Example 2.12 to be

$$H(z) = z/(z - K)$$

The solution of the difference equation can be found by using the $\mathcal{z}$ transform method described earlier. Taking the $\mathcal{z}$ transform of the governing difference equation and using the initial condition gives

$$Y(z) = K[y(-1) + z^{-1}Y(z)] + X(z) \qquad (2.43)$$

The transform of the input $x(n)$ is easily obtained from Table 2.1 as

$$X(z) = \frac{z^2 - z \cos \omega_0}{z^2 - 2z \cos \omega_0 + 1}$$

Substituting this $X(z)$ into (2.43) and taking the $Kz^{-1} Y(z)$ to the left side of the equation gives

$$Y(z)(1 - Kz^{-1}) = Ky(-1) + \frac{z^2 - z \cos \omega_0}{z^2 - 2z \cos \omega_0 + 1}$$

Dividing through by $(1 - Kz^{-1})$, $Y(z)$ is obtained as

$$Y(z) = \frac{Ky(-1)}{1 - Kz^{-1}} + \frac{z^2 - z \cos \omega_0}{(1 - Kz^{-1})(z^2 - 2z \cos \omega_0 + 1)}$$

$$= Y_1(z) + Y_2(z)$$

The response $y(n)$, obtained by taking the inverse transform of $Y(z)$, is composed of two terms. The first is the zero input response due to initial condition and the second is the forced response due to the input $x(n)$. The first term $Y_1(z)$ is already in a convenient form for taking the inverse; however, the second part must be expanded using a partial fraction expansion as follows:

$$\frac{Y_2(z)}{z} = \frac{z(z - \cos \omega_0)}{(z - K)(z^2 - 2z \cos \omega_0 + 1)} = \frac{A}{z - K} + \frac{B}{z - e^{j\omega_0}} + \frac{C}{z - e^{-j\omega_0}}$$

where A, B, and C are determined as

$$A = \frac{z(z - \cos \omega_0)}{z^2 - 2z \cos \omega_0 + 1}\bigg|_{z = K} = \frac{K(K - \cos \omega_0)}{K^2 - 2K \cos \omega_0 + 1}$$

$$B = \frac{z(z - \cos \omega_0)}{(z - K)(z - e^{-j\omega_0})}\bigg|_{z = e^{j\omega_0}} = \frac{e^{j\omega_0}(e^{j\omega_0} - \cos \omega_0)}{(e^{j\omega_0} - K)(e^{j\omega_0} - e^{-j\omega_0})} = \frac{e^{j\omega_0}}{2(e^{j\omega_0} - K)}$$

$$C = B^*$$

$Y(z)$ is then easily found by multiplying $Y_2(z)/z$ by z and adding the results to $Y_1(z)$ as follows:

$$Y(z) = \frac{Ky(-1)z}{(z - K)} + \frac{A}{z - K} + \frac{B}{z - e^{j\omega_0}} + \frac{B^*}{z - e^{-j\omega_0}}$$

By taking the inverse Z transform of $Y(z)$, $y(n)$ is seen to be

$$y(n) = [Ky(-1) + A]K^n u(n) + B(e^{j\omega_0})^n u(n) + B^* (e^{-j\omega_0})^n u(n)$$

The last two terms are recognized as complex conjugates and $y(n)$ can be further simplified since the sum of those terms is as follows:

$$B(e^{j\omega_0})^n u(n) + B^* (e^{-j\omega_0})^n u(n) = 2 \operatorname{Re} [Be^{j\omega_0 n}] u(n)$$

Using the system function and B derived earlier, the above may be written as

$$2 \operatorname{Re} [Be^{j\omega_0 n}] u(n) = 2 \operatorname{Re} \left(\frac{\frac{1}{2} e^{j\omega_0}}{e^{j\omega_0} - K} \cdot e^{j\omega_0 n} \right) u(n)$$

$$= \operatorname{Re} [H(z)|_{z = e^{j\omega_0}} \cdot e^{j\omega_0 n}] u(n)$$

But $H(z)$ evaluated at $z = e^{j\omega_0}$ can be written in exponential form as

$$H(z)|_{z = e^{j\omega_0}} = |H(e^{j\omega_0})| e^{j[\arg H(e^{j\omega_0})]}$$

Therefore we are finally able to write $2 \operatorname{Re} [Be^{j\omega_0 n}]$ in terms of a cosine wave as

$$2 \operatorname{Re} [Be^{j\omega_0 n}] = \operatorname{Re} [|H(e^{j\omega_0})| e^{j\omega_0 n} e^{j[\arg H(e^{j\omega_0})]}]$$

$$= |H(e^{j\omega_0})| \cos [\omega_0 n + \arg H(e^{j\omega_0})]$$

and $y(n)$ becomes

$$y(n) = \underbrace{[Ky(-1) + A]K^n u(n)}_{\text{Transient response}} + \underbrace{|H(e^{j\omega_0})| \cos [\omega_0 n + \arg H(e^{j\omega_0})] u(n)}_{\text{Steady state response}}$$

As $|K| < 1$, the first term will eventually go to zero as n approaches ∞ and thus is called the transient response. The second term is the steady state response and is identical to that of Eq. (1.22) of Chapter 1, where the system was assumed always excited.

2.6 SUMMARY

In this chapter the Z transform has been defined and used to explore the relationships that exist between the system function, impulse response, and difference equation characterizations of linear shift invariant discrete-time systems. It has been shown that the Z transform plays the same role as the Laplace transform does for analyzing systems represented by linear constant coefficient differential equations. Important results established are summarized in Table 2.3 for both linear time invariant continuous- and discrete-time systems showing the analogous structure.

TABLE 2.3   BRIEF COMPARISON OF IMPORTANT RESULTS FOR
CONTINUOUS-TIME AND DISCRETE-TIME SYSTEMS

Continuous Time

Linear time invariant system: $x_a(t)$ input; $y_a(t)$ output

Linear constant coefficient differential equation representation:

$$C_N\frac{d^N y_a(t)}{dt^N} + C_{N-1}\frac{d^{N-1}y_a(t)}{dt^{N-1}} + \cdots + C_0 y_a(t) = d_M\frac{d^M x_a(t)}{dt^M} + d_{M-1}\frac{d^{M-1}x_a(t)}{dt^{M-1}} + \cdots + d_0 x_a(t)$$

System function:

$$Y_a(s)/X_a(s) \overset{\triangle}{=} H_a(s) = \sum_{i=0}^{M} d_i s^i \Big/ \sum_{i=0}^{N} c_i s^i$$

Frequency response: $H_a(j\Omega) \overset{\triangle}{=} H_a(s)|_{s\to j\Omega}$

Impulse response: $h_a(t) \overset{\triangle}{=} \mathcal{L}^{-1}[H_a(s)]$

Output as a convolution integral:

$$y_a(t) = \int_{-\infty}^{\infty} x(\tau)h_a(t-\tau)d\tau = \int_{-\infty}^{\infty} h(\tau)x_a(t-\tau)d\tau$$

Stability: if and only if $\displaystyle\int_{-\infty}^{\infty} |h_a(t)|\, dt < \infty$ bounded input–bounded output (BIBO)

Causality: if and only if $h_a(t) = 0$ for $t < 0$

Discrete-time

Linear shift invariant system: $x(n)$ input; $y(n)$ output

Linear constant coefficient difference equation representation:

$$a_N y(n-N) + a_{N-1}y(n-(N-1)) + \cdots + a_0 y(n) = b_M x(n-M) + b_{M-1}x(n-(M-1))$$
$$+ \cdots + b_0 x(n)$$

System function:

$$Y(z)/X(z) = H(z) \overset{\triangle}{=} \sum_{i=0}^{M} b_i z^{-i} \Big/ \sum_{i=0}^{N} a_i z^{-i}$$

Frequency response: $H(e^{j\omega}) = H(z)|_{z\to e^{j\omega}}$

Unit sample response: $h(n) = \mathcal{Z}^{-1}[H(z)]$

Output as a convolution sum: $y(n) = \displaystyle\sum_{k=-\infty}^{\infty} h(k)x(n-k) = \sum_{k=-\infty}^{\infty} h(n-k)x(k)$

Stability: if and only if $\displaystyle\sum_{k=-\infty}^{\infty} |h(k)| < \infty$ (BIBO)

Causality: if and only if $h(n) = 0$ for $n < 0$

BIBLIOGRAPHY

Antoniou, Andreas. *Digital Filters: Analysis and Design*. McGraw-Hill, New York, 1979.

Jong, M. T. *Methods of Discrete Signal and System Analysis*. McGraw-Hill, New York, 1982.

Oppenheim, A. V., and R. W. Schafer. *Digital Signal Processing*. Prentice-Hall, Englewood Cliffs, NJ, 1975.

Stearns, S. D. *Digital Signal Analysis*. Hayden, Rochelle Park, NJ, 1975.

PROBLEMS

2.1 Given the following sequence: $x(n) = -3^n u(-n - 1) + u(n) - 2^n u(n)$,

   (a) Illustrate the sequence.

   (b) What is the region of convergence of the Z transform of the negative n part of the sequence?

   (c) What is the region of convergence of the Z transform of the positive and zero n part of the sequence?

   (d) What is the region of convergence of the Z transform of the entire sequence?

   (e) Find the Z transform of $x(n)$. What effect do the zeros of $X(z)$ have on the region of convergence? the poles?

   (f) If $x(n)$ is a unit sample response of a system, is the system causal? Stable?

2.2 For the pairs below, tell which one has a better chance of converging and tell why.

   (a) Discrete-time Fourier transform or Z transform.

   (b) Laplace transform or continuous time Fourier transform.

2.3 Find the Z transform of the following discrete-time signals, remembering to include the region of convergence for each:

   (a) $x(n) = 2^n u(n) + 3(\frac{1}{2})^n u(n)$

   (b) $x(n) = 3(-\frac{1}{2})^n u(n) - 2(3)^n u(-n - 1)$

   (c) $x(n) = (-\frac{1}{5})^n u(n) + 5(\frac{1}{2})^{-n} u(-n - 1)$

   (d) $x(n) = 3e^{-2n} u(n) + 2(4)^n u(-n - 1) + 5\delta(n)$

   (e) $x(n) = 3(\frac{1}{2})^n u(n) + 5(\frac{1}{4})^n u(-n - 1)$

   (f) $x(n) = 3(\frac{1}{3})^n u(-n - 1) + 5(\frac{1}{4})^n u(n)$

   (g) $x(n) = 3a^{-0.5n} u(n) + 4(2)^n u(-n - 1)$

   (h) $x(n) = \frac{1}{2}\delta(n) + \delta(n - 1) - \frac{1}{2}\delta(n - 2)$

   (i) $x(n) = 2\delta(n - 3) - 2\delta(n + 3)$

2.4 In Section 2.1.2, a number of important properties of the Z transform were presented without proof. Prove the following: (a) linearity, (b) translation, (c) multiplication by an exponential, (d) multiplication by a ramp, (e) convolution (time domain), (f) convolution (z domain), (g) initial value theorem, and (h) final value theorem.

2.5 Derive the $\mathcal{Z}$ transform of the following sequences:
(a) $\sin(\omega_0 n)\,u(n)$; (b) $\cos(\omega_0 n)u(n)$; (c) $n\,u(n)$; (d) $\sin(\omega_0 n + \theta)\,u(n)$;
(e) $\cos(\omega_0 n + \theta)\,u(n)$; (f) $\sinh(\omega_0 n)\,u(n)$; (g) $\cosh(\omega_0 n)\,u(n)$.
Check your results with the entries from Table 2.1 and Appendix C.

2.6 Find the $\mathcal{Z}$ transform of the following discrete-time signals, remembering to include the region of convergence for each one. Use the definitions of the $\mathcal{Z}$ transform, the tables, and the properties.
(a) $x(n) = u(n - 2)$
(b) $x(n) = n\,u(n - 1)$
(c) $x(n) = e^{-3n}\,u(n - 1)$
(d) $x(n) = 3 \cdot 2^n\,u(-n)$
(e) $x(n) = (n - 1)^2\,e^{-n}\,u(n - 1)$
(f) $x(n) = (n + 0.5)(\frac{1}{3})^n\,u(n)$
(g) $x(n) = 0.25u(-n - 1) + (\frac{1}{2})^n\,u(n) - n(\frac{1}{4})^n\,u(n) + (\frac{1}{8})^n\,u(n - 1)$

2.7 Find the $\mathcal{Z}$ transform of the following discrete-time signals including the region of convergence:
(a) $x(n) = 4\cos(anT)\,u(n)$
(b) $x(n) = 2\cos[0.2\pi(n - 1)]\,u(n)$
(c) $x(n) = (n - 1)\cos[\omega_0 n]\,u(n - 1)$
(d) $x(n) = e^{-3n}\sin(\pi n/6)\,u(n)$
(e) $x(n) = n^2 x(n)$, where $\mathcal{Z}[x(n)] = X(z)$ with region of convergence $a < |z| < b$

2.8 Using the partial fraction expansion method, find the inverse $\mathcal{Z}$ transform of $X(z)$ given by

$$X(z) = \frac{z}{(z + 2)(z - 3)}$$

for the following regions of convergence: (a) $|z| > 3$, (b) $2 < |z| < 3$, and (c) $|z| < 2$.

2.9 Suppose $X(z)$ is given as follows:

$$X(z) = \frac{z(z^2 - 4z + 5)}{(z - 3)(z - 1)(z - 2)}$$

Find $x(n)$ for the following regions of convergence using the partial fraction expansion method:
(a) $2 < |z| < 3$ (b) $|z| > 3$ (c) $|z| < 1$.

2.10 Using the partial fraction expansion method, determine $x(n)$ for the following:
(a) $X(z) = (z^3 + z^2)/[(z - 1)(z - 3)]$ with ROC $|z| > 3$
(b) $X(z) = (z^2 + 2)/[(z - 1)(z - 3)]$ with ROC $1 < |z| < 3$
(c) $X(z) = (z^3 - 2z^2)/[(z - \frac{1}{2})(z + \frac{1}{4})]$ with ROC $\frac{1}{4} < |z| < \frac{1}{2}$
(d) Repeat (c) for ROC $|z| > \frac{1}{2}$
(e) $X(z) = (z + 1)^2/z$ with ROC $0 < |z| < \infty$
(f) $X(z) = (z^3 + 2z^2)/[(z - \frac{1}{2})^2(z + \frac{3}{4})]$ with ROC $|z| > \frac{3}{4}$

2.11 Given $X(z) = (z^2 + z)/[(z - \frac{1}{2} + j\frac{1}{2})(z - \frac{1}{2} + j\frac{1}{2})]$ with ROC $|z| > 0.707$, find $x(n)$ in a form that contains no complex numbers.

2.12 Using the sinusoidal results of Table 2.1 directly without performing a partial fraction expansion, find the inverse $\mathbb{Z}$ transform of

$$X(z) = z^2/(z^2 - z + 1) \text{ for ROC } |z| > 1$$

2.13 Find $x(n)$ for the following $X(z)$ using the tables,

$$X(z) = \frac{2z}{z^2 - 1.414z + 1} + \frac{z}{z - \frac{1}{2}} + \frac{3z}{z - 3} + \frac{z}{(z - \frac{1}{8})^3}$$

with ROC $1 < |z| < 3$.

2.14 By division find the inverse $\mathbb{Z}$ transform of $X(z) = 2/(z + \frac{1}{3})$ for the following regions of convergence: (a) $|z| < \frac{1}{3}$ and (b) $|z| > \frac{1}{3}$.

2.15 Find $x(-2)$, $x(-1)$, $x(0)$, $x(1)$, and $x(2)$ by division for

$$X(z) = \frac{(2z^2 - 3z)}{(z - 1)(z - 2)}$$

with ROC $|z| > 2$.

2.16 Given $\mathbb{Z}[x(n)] = X(z) = (z + 3)/(z + 2)^2$ with ROC $|z| > 2$,
(a) Use the division method to evaluate $x(n)$ at $n = 0$ and $n = +1$.
(b) Check your value at $n = 0$ by using the initial value property.

2.17 Given $X(z) = z(z + 1)/(z^2 - 3z + 2)$, find $x(n)$ for the following regions of convergence using the method indicated:
(a) ROC $|z| > 2$ by partial fractions
(b) ROC $|z| < 1$ by partial fractions
(c) ROC $1 < |z| < 2$ by partial fractions
(d) ROC $|z| > 2$ by division
(e) ROC $|z| < 1$ by division

2.18 In evaluating the inverse $\mathbb{Z}$ transform by the complex inversion integral, a contour integral must be computed. Evaluate the following typical contour integrals for the contour c specified:

(a) $\oint_c \frac{1}{z^2} dz$ $\qquad c : |z| = 1$

(b) $\oint_c \frac{1}{z} dz$ $\qquad c : |z| = 3$

(c) $\oint_c \frac{z^{-2}}{z - 1} dz$ $\qquad c : |z| = 2$

(d) $\frac{1}{2\pi j} \oint_c \frac{2}{z(z - 2)} dz$ $\qquad c : |z| = 1$

(e) Repeat (b) and (c) $\qquad c : |z| = 0.5$

(f) $\frac{1}{2\pi j} \oint_c \frac{1}{z^5} dz$ $\qquad c : |z| = \frac{1}{5}$

(g) $\dfrac{1}{2\pi j}\displaystyle\oint_c \dfrac{z^3}{(z-1)^2}\,dz$ $c : |z| = 2$

(h) $\displaystyle\oint_c \dfrac{z^3}{(z-1)^3}\,dz$ $c : |z| = 3$

(i) $\displaystyle\oint_c \dfrac{(z-1)^2}{z^3}\,dz$ $c : |z| = \frac{1}{2}$

(j) $\displaystyle\oint_c \dfrac{1}{(z-1)^2(z+2)}\,dz$ $c : |z| = 3$

(k) $\displaystyle\oint_c \dfrac{1}{z^3(z+0.5)}\,dz$ $c : |z| = \frac{1}{4}$

(l) $\displaystyle\oint_c \dfrac{z-2}{z^2(z+2)}\,dz$ $c : |z| = 1$

2.19 Evaluate $x(-2)$, $x(-1)$, $x(0)$, and $x(1)$ using the complex inversion integral for
$$X(z) = 1/(z-2) \quad \text{with ROC } |z| > 2.$$

2.20 Evaluate the inverse $\mathcal{Z}$ transform of the following $X(z)$ for the region of convergence indicated by using the complex inversion integral:
   (a) $X(z) = z^2/[(z-\frac{1}{2})(z-\frac{1}{4})]$ for ROC $|z| > \frac{1}{2}$
   (b) Repeat (a) for ROC $|z| < \frac{1}{4}$
   (c) $X(z) = z^2/(z-\frac{1}{4})^2$ for ROC $|z| > \frac{1}{4}$
   (d) $X(z) = z^2/(z-\frac{1}{2})$ for ROC $|z| > \frac{1}{2}$

2.21 Find the inverse $\mathcal{Z}$ transform of the following by any method for the region of convergence indicated.

   (a) $X(z) = \dfrac{z(z+1)}{(z-1)^2(z-\frac{1}{2})}$ for ROC $|z| > 1$

   (b) $X(z) = \dfrac{z^3 + 2z^2 + 3z}{(z-\frac{1}{2})(z^2 - 2z + 4)}$ for ROC $|z| > 2$

   (c) $X(z) = \dfrac{z^{-1}}{-2z^{-2} - z^{-1} + 1}$ for ROC $1 < |z| < 2$

   (d) $X(z) = \dfrac{2}{z(z-\frac{1}{2})}$ for ROC $|z| > \frac{1}{2}$

   (e) $X(z) = (z^2 + z)/[(z-1)(z-3)]$ for ROC $1 < |z| < 3$

   (f) $X(z) = (z^2 + ze^{-3T} - 2ze^{-2T})/[z^2 - (e^{-2T} + e^{-3T})z + e^{-5T}]$
     for ROC $|z| > e^{-2T}$

   (g) $X(z) = z^2/(z-1)$ for ROC $|z| > 1$

2.22 Let $x(n) = x_1(n) * x_2(n)$, where $x_1(n) = (\frac{1}{3})^n\, u(n)$ and $x_2(n) = (\frac{1}{5})^n\, u(n)$.
   (a) Find $X(z)$ by using the convolution property for $\mathcal{Z}$ transforms.
   (b) Find $x(n)$ by taking the inverse transform of $X(z)$ using the partial fraction expansion method.

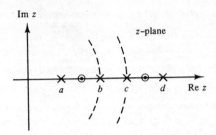

2.23 Given the pole-zero plot on the right for a system function $H(z)$ with region of convergence $b < |z| < c$ and $a < b < c < d$.

(a) What is the requirement on b and c such that the system is stable?

(b) What is the requirement on b and c such that the system is causal?

(c) If d is less than 1, is the system stable? Why?

(d) If d is less than 1, is the system causal?

(e) For $a = 1$, $b = 2$, $c = 3$, and $d = 4$, which poles contribute to the positive time part of the inverse transform?

(f) For $a = 1$, $b = 2$, $c = 3$, and $d = 4$, which poles contribute to the negative time part of the inverse transform?

2.24 In taking the inverse transform of $X_k(z) = z/(z - a)^k$ for the region of convergence $|z| > |a|$ it is useful to apply the transformation $z = 1/p$ when finding $x_k(n)$ for n less than zero. Show that $x_k(n) = 0$ for $n < 0$ by using this transformation.

2.25 Given that the $\mathcal{Z}$ transform of $x(n)u(n)$ is equal to $X(z)$, show that the transform of $x(n + r)u(n)$ for r an integer greater than zero is as follows:

$$\mathcal{Z}[x(n + r)\,u(n)] = z^r X(z) = [z^r x(0) + z^{r-1}x(1) + \cdots + zx(r - 1)]$$

2.26 A causal system is represented by the following difference equation:

$$y(n) + \tfrac{1}{4}y(n - 1) = x(n) + \tfrac{1}{2}x(n - 1).$$

(a) Find the system function $H(z)$ and give the corresponding region of convergence.

(b) Find the unit sample response of the system in analytical form.

(c) Find the frequency response $H(e^{j\omega})$ and determine its magnitude and phase.

2.27 Given a causal system represented by the following linear constant coefficient difference equation:

$$y(n) = x(n) - \tfrac{1}{2}x(n - 1) + \tfrac{3}{4}y(n - 1) - \tfrac{1}{8}y(n - 2)$$

(a) Find the system function in simplest form.

(b) Give another difference equation that realizes this system.

(c) Is the system an FIR or IIR filter?

2.28 Given a causal system with system function

$$H(z) = (z - \tfrac{1}{3})/(z - \tfrac{1}{4})$$

Find a difference equation realization of this system.

2.29 Given a causal system specified by its system function $H(z)$ as follows:

$$H(z) = z(z - 1)/[(z - \tfrac{1}{2})(z + \tfrac{1}{4})]$$

(a) Find a difference equation realization of the system.

(b) Give the analytical form of the impulse response $h(n)$.

2.30 Given $H(z) = (3z^{-2} + 2z^{-1} + 5)/(6z^{-2} + 4z^{-1} + 1)$.

(a) Find a difference equation realization if $x(n)$ is the input and $y(n)$ is the output, for a causal system with $H(z)$ as its system function.

(b) How many multiplications are required for its implementation in your form?

2.31 Given a system with system function $H(z) = z/(z^2 + 1)$ for ROC $|z| > 1$.

(a) Is this a causal system? Why or why not?

(b) Is the system BIBO stable? Why or why not?

(c) Find the unit sample response $h(n)$.

(d) Find a difference equation realization of the system. Let $x(n)$ be the input and $y(n)$ be the output.

(e) Is the system linear or nonlinear?

(f) Is the system shift invariant? Why or why not?

2.32 The impulse response of a discrete-time system that is linear and shift invariant is given by

$$h(n) = (\tfrac{1}{2})^n u(n) + (-\tfrac{1}{3})^n u(n)$$

(a) Is this a causal or noncausal system? Why?

(b) Is the system an infinite impulse response (IIR) or finite impulse response (FIR) system?

(c) Is the system BIBO stable? Why or why not?

(d) Find the Fourier transform of $h(n)$.

(e) Find the $\mathbb{Z}$ transform of $h(n)$. Be sure to include region of convergence.

(f) Give a difference equation realization of the system.

(g) Find the frequency response of the system and put in magnitude and angle form.

(h) Find the energy in the sequence $h(n)$.

2.33 Find the frequency response for the system function $H(z)$ given by

$$H(z) = \frac{1 - 2z + 3z^2}{3 - 2z + z^2}$$

Find both the magnitude and phase of the frequency response and comment on your results. Show that we also have unity magnitude at all frequencies for $H(z)$ of the following general form:

$$H(z) = (a + bz + cz^2)/(c + bz + az^2)$$

A filter with this $H(z)$ is called an all pass filter. Cascaded sections of all pass filters are also all pass and are sometimes used for adjusting the phase response of a given discrete-time filter.

2.34 A linear shift invariant system is characterized by its unit sample response

$$h(n) = e^{-n} u(n)$$

(a) Is the system causal? Why?

(b) Is the system stable? Why?

(c) Find the system function $H(z)$ and give a pole zero plot.

(d) Plot $|H(e^{j\omega})|$ for all ω.

(e) Give a difference equation realization of the system.

(f) Find a realization in terms of the convolution sum.

2.35 Given a unit sample response

$$h(n) = \begin{cases} (0.36788)^n, & 0 \le n \le 3 \\ 0, & \text{elsewhere} \end{cases}$$

(a) Find the corresponding system function $H(z)$.

(b) Find the pole zero plot for this system function.

(c) Graph the magnitude of $H(z)|_{z = e^{j\omega_0}}$ as ω_0 varies from 0 to 2π.

(d) Graph the argument of $H(z)|_{z = e^{j\omega_0}}$ as ω_0 varies from 0 to 2π.

(e) If these graphs are periodic give the period.

(f) Compare your results with those of Problem 2.34.

2.36 Solve the following difference equations for $y(n)$ using $\mathcal{Z}$ transforms and the specified initial conditions:

(a) $y(n) = \frac{1}{2}y(n - 1) + x(n)$, $n \ge 0$,
    where $x(n) = (\frac{1}{2})^n u(n)$ and $y(-1) = \frac{1}{4}$.

(b) $y(n) = \frac{1}{2}y(n - 1) + x(n)$, $n \ge 0$,
    where $x(n) = (\frac{1}{4})^n u(n)$ and $y(-1) = 1$.

(c) $y(n) = y(n - 1) - y(n - 2) + 2$, $n \ge 0$,
    where $y(-2) = 1$ and $y(-1) = 2$.

(d) $y(n) + y(n - 2) = \delta(n)$, $n \ge 0$,
    where $y(-2) = 0$ and $y(-1) = 1$.

(e) $y(n) - y(n - 1) + \frac{1}{4}y(n - 2) = x(n)$, $n \ge 0$,
    where $x(n) = 2(\frac{1}{8})^n$, $y(-1) = 2$, and $y(-2) = 4$.

2.37 Suppose the A/D–$H(z)$–D/A system shown uses a sampling rate of 20,000 samples/sec. Let $x_a(t) = 10 \cos (2\pi \cdot 500t) u(t)$.

(a) Give another input signal $x_1(t)$ which gives the same samples $x(n)$.

(b) Give another input signal $x_2(t)$ which gives the same frequency (digital) for $x(n)$ but has a phase reversal.

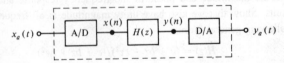

(c) What is the highest-frequency input signal that will not be aliased?

(d) If $H(z) = 1/(z - 0.5)$, find the steady state output $y_a(t)$ for the input $x_a(t)$ given above.

(e) If $H(z) = 1/(z - 0.05)$, find the steady state output $y_a(t)$ for an input $x_a(t) = 2u(t)$.

(f) Plot the magnitude of the frequency response $|H(e^{j\omega})|$ for the discrete-time filter characterized by the $H(z)$ given in (e).

(g) Plot the equivalent analog frequency response for the A/D–$H(z)$–D/A structure using the $H(z)$ of (e). Repeat with a sampling rate of 40,000 samples per second.

2.38 A digital filter $H(z)$ is connected in the A/D–$H(z)$–D/A structure with input to the configuration $x_a(t)$ and output $y_a(t)$ as shown. For $H(z) = (1 + z^{-1})^2$ and $T = 10^{-3}$ sec,

(a) Give a difference equation realization for the digital filter specified by $H(z)$.

(b) Does $H(z)$ represent an IIR or FIR filter? Why?

(c) Suppose $x_a(t) = 5 \cos 10\pi t \, u(t)$. Find the steady state output $y_a(t)$.

2.39 Let $x_a(t) = 25 \cos 100\pi t \, u(t)$. Define the sequence $x(n)$ by $x(n) = x_a(nT)$, where T is the time between samples. Assume a sampling rate of 200 samples/sec.

(a) Find T.

(b) Is $x(n)$ periodic? If so, what is the period; if not, why not?

(c) What is the digital frequency of $x(n)$?

(d) Find an $x_a(t)$ different in frequency from the one specified above that gives the same sequence $x(n)$.

(e) Suppose $x_a(t)$ given above is the input to a filter of the A/D–$H(z)$–D/A structure, with a causal filter specified by $H(z) = 2/(z + \frac{1}{2})$. Find the steady state response $y_{ss}(n)$ and the corresponding steady state response for $y_a(t)$.

2.40 Given the A/D–$H(z)$–D/A structure with $H(z) = 1 + z^{-1}$ and also the input $x_a(t) = 5 \cos (20\pi t)$.

(a) What is $x(n)$ if the sampling rate of the A/D and D/A is 200 samples/sec?

(b) What is the digital frequency of $x(n)$?

(c) Find the $y_{ss}(n)$, the steady state output to the $x(n)$ in (a).

(d) Give the steady state output for $y_a(t)$ for the input $x_a(t)$.

(e) Give an analog input of a different frequency which will give the same steady state output as that for $5 \cos (20\pi t)$.

(f) What type of analog filter would be needed to prevent aliasing? Give a respectable cutoff frequency for such a filter?

(g) Find the frequency response for the filter represented by $H(z)$. Plot magnitude and phase.

(h) Plot the magnitude of the equivalent analog frequency response for the A/D–$H(z)$–D/A structure.

Chapter 3

Analog Filter Design

3.0 INTRODUCTION

A very important approach to the design of digital filters is to apply a transformation to an existing analog filter. For this method, to be presented in Chapter 4, it is necessary to have a base or catalog of analog filters that can serve as the prototypes for the transformation. In this chapter design procedures and tables for the analog Butterworth, Chebyshev, and elliptic filters are presented to establish that base.

These analog procedures normally begin with a specification of the frequency response for the filter describing how the filter reacts in the steady state to sinusoidal inputs. If an input sinusoid is not attenuated or attenuated less than a specified tolerance as it goes through the system, it is said to be in a passband of the filter; whereas, if it is attenuated more than a specified value it is said to be stopped and within the stopband of the filter. Input sinusoids with neither a little nor a large amount of attenuation are said to be in the transition band. A typical frequency response is shown in Fig. 3.1 showing the passband, transition band, and stopband. The filter with this type of frequency response is called a *low-pass filter* as it passes all frequencies less than a certain value Ω_c, called the cutoff frequency, and attenuates or stops all frequencies past Ω_r, the stopband critical frequency. Other important basic types of filters are the high-pass (HP), bandpass (BP), and bandstop (BS) filters, whose frequency responses are shown

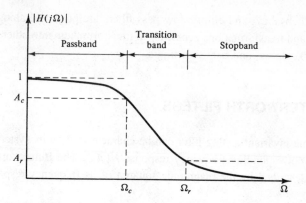

Figure 3.1 A typical required frequency response for a low-pass filter design

in Fig. 3.2. Also shown are the frequency responses for the ideal LP, HP, BP, and BS filters which exhibit no transition bands.

More complex filters can be obtained by placing these four basic types in various parallel and cascade configurations.

It is known that the low-pass, high-pass, bandpass, and bandstop filters can be obtained from a normalized low-pass filter via specific transformations in the s-plane. Therefore, prime consideration will be given to low-pass filter design. In particular, the properties and design procedures for the analog But-

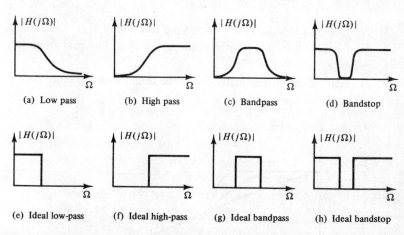

Figure 3.2 Basic types of frequency responses: (a) low pass, (b) high pass, (c) bandpass, (d) bandstop, (e) ideal low pass, (f) ideal high pass, (g) ideal bandpass, (h) ideal bandstop.

terworth, Chebyshev, and elliptic low-pass filters are presented along with the procedures and transformations necessary to transform them into other low-pass, high-pass, bandpass, and bandstop filters.

3.1 BUTTERWORTH FILTERS

A linear time invariant analog filter can be characterized by its system function $H(s)$ or its corresponding frequency response $H(j\Omega)$. The Butterworth filter of order n is described by the magnitude squared of its frequency response given below:

$$|H_n(j\Omega)|^2 = 1/[1 + (\Omega/\Omega_c)^{2n}] \qquad (3.1)$$

In Fig. 3.3 the magnitude squared frequency response of the Butterworth filter is shown for several different values of n.

The following properties are easily determined:

1. $|H_n(j\Omega)|^2 \big|_{\Omega=0} = 1$ for all n.
2. $|H_n(j\Omega)|^2 \big|_{\Omega=\Omega_c} = \frac{1}{2}$ for all finite n.
   This implies that
   $|H_n(j\Omega)|\big|_{\Omega=\Omega_c} = 0.707$
   and $20 \log |H_n(j\Omega)|\big|_{\Omega=\Omega_c} = -3.0103$.

3. $|H_n(j\Omega)|^2$ is a monotonically decreasing function of Ω.
4. As n gets larger, $|H_n(j\Omega)|^2$ approaches an ideal low-pass frequency response.
5. $|H_n(j\Omega)|^2$ is called maximally flat at the origin since all order derivatives exist and are zero.

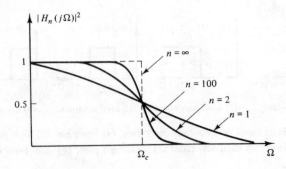

Figure 3.3 The magnitude squared frequency response for a Butterworth filter.

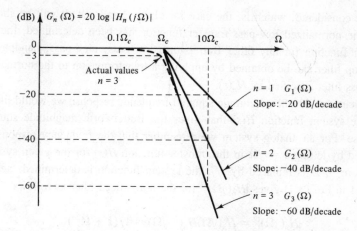

Figure 3.4 Filter gain plot for analog Butterworth filters of various orders n.

It is convenient in many cases to look at the frequency response in decibels, that is, plot $20 \log |H(j\Omega)|$ versus Ω. Figure 3.4 is a straight line approximation of the frequency response in decibels for the Butterworth filter. The straight lines are determined for $\Omega \ll \Omega_c$ and $\Omega \gg \Omega_c$ by approximating $20 \log |H_n(j\Omega)|$. We have

$$G_n(\Omega) = 20 \log |H_n(j\Omega)| = 10 \log |H_n(j\Omega)|^2$$
$$= 10 \log |1/[1 + (\Omega/\Omega_c)^{2n}]| = -10 \log [1 + (\Omega/\Omega_c)^{2n}] \quad (3.2)$$

Since the $(\Omega/\Omega_c)^{2n}$ is approximately zero for $\Omega \ll \Omega_c$, we have

$$G_n(\Omega) \approx -10 \log |1 + 0| = 0$$

For $\Omega \gg \Omega_c$ the 1 in the summation becomes insignificant compared to $(\Omega/\Omega_c)^{2n}$, so

$$G_n(\Omega) \approx -10 \log |(\Omega/\Omega_c)^{2n}| = -20n \log |\Omega/\Omega_c| \quad (3.3)$$

Therefore, the approximate gain for $\Omega \ll \Omega_c$ is 0 dB while for $\Omega \gg \Omega_c$ the slope of $G_n(\Omega)$ is $-20n$ dB/decade. For $\Omega = \Omega_c$ the actual value of $G_n(\Omega)$ is approximately -3 dB, as seen earlier, and the actual values for Ω around Ω_c can be calculated using (3.2). These values are shown by dotted lines in Fig. 3.4 for $n = 3$.

In most of the work to follow, the normalized low-pass Butterworth filter will be considered, which is, the case for $\Omega_c = 1$ rad/sec. It will be shown, once the normalized low-pass transfer function has been determined, that the transfer function for any other Butterworth low-pass, high-pass, bandpass or bandstop filter can be obtained by applying a transformation to the normalized low-pass filter specified by $H_n(s)$.

Starting with the magnitude squared frequency response we would like to find the system function $H(s)$ that gives the Butterworth magnitude squared response. For an analog system we remember that the frequency response is obtained by letting $s = j\Omega$ in the transfer function $H(s)$ for the given system. Therefore, if Ω is replaced by s/j, the system function is determined. Setting $\Omega_c = 1$ in Eq. (3.1) gives $|H_n(j\Omega)|^2$ for the normalized filter as follows:

$$|H_n(j\Omega)|^2 = H_n(j\Omega)H_n(-j\Omega) = 1/(1 + \Omega^{2n})$$

$$H_n(s) \cdot H_n(-s) = \frac{1}{1 + (s/j)^{2n}} \tag{3.4}$$

The poles of $H_n(s) \cdot H_n(-s)$ are given by the roots of the denominator, i.e.,

$$1 + (s/j)^{2n} = 0$$
or
$$s^{2n} = -1(j)^{2n} = (-1)^{n+1} \tag{3.5}$$

The roots of the above equation can be identified for the cases when n is odd and even. For n odd, the poles of $H_n(s)H_n(-s)$ become the $2n$th roots of 1, while for n even, the poles are the $2n$th roots of -1. That is,

for n odd: $\quad s_k = 1\underline{/k\pi/n}, \qquad k = 0, 1, 2, \ldots, 2n - 1$

for n even: $\quad s_k = 1\underline{/\pi/2n + k\pi/n}, \qquad k = 0, 1, 2, \ldots, 2n - 1$

These poles for n odd and even are illustrated in Fig. 3.5. For n odd we have a pole of $H_n(s) \cdot H(-s)$ at $s = 1$, and then poles equally spaced on the unit circle π/n in angle, while for n even the first pole is at $1\underline{/\pi/2n}$ and the remaining poles equally spaced around the unit circle by π/n.

If we wish the filter $H_n(s)$ to be a stable and causal filter, the poles of $H_n(s)$ are selected to be those in the left half plane and $H_n(s)$ can be written in the following form:

$$H_n(s) = \frac{1}{\prod_{\substack{\text{LHP} \\ \text{poles}}} (s - s_k)} = \frac{1}{B_n(s)} \tag{3.6}$$

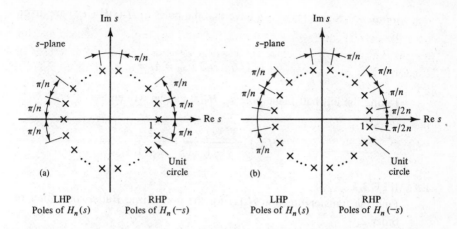

Figure 3.5 Poles of $H_n(s)H(-s)$ for a low-pass Butterworth filter with a 1-rad cutoff frequency. (a) n odd. (b) n even.

where s_k are all the left half plane poles of $H_n(s) H_n(-s)$. The denominator, $B_n(s)$, can be shown to be a Butterworth polynomial of order n. Table 3.1a gives the first five Butterworth polynomials in a real factored form.

The following examples find the first and second order Butterworth polynomials and corresponding filter transfer functions $H_1(s)$ and $H_2(s)$.

EXAMPLE 3.1

Find the transfer function $H_1(s)$ for the normalized Butterworth filter of order 1.

TABLE 3.1a BUTTERWORTH POLYNOMIALS AND
NORMALIZED LOW-PASS BUTTERWORTH
FILTERS

Order n	Butterworth polynomial $B_n(s)$
1	$s + 1$
2	$s^2 + \sqrt{2}s + 1$
3	$(s^2 + s + 1)(s + 1)$
4	$(s^2 + 0.76536s + 1)(s^2 + 1.84776s + 1)$
5	$(s + 1)(s^2 + 0.6180s + 1)(s^2 + 1.6180s + 1)$

Normalized low-pass Butterworth filters

$$H_n(s) = \frac{1}{B_n(s)}$$

Solution. Since $n = 1$ we have that the poles of $H_1(s)H_1(-s)$ are given by

$$s_1 = 1\underline{/0} \quad \text{and} \quad s_2 = 1\underline{/\pi}$$

Taking the left half plane pole s_2, $H_1(s)$ can be written as

$$H_1(s) = \frac{1}{s - (-1)} = \frac{1}{s + 1}$$

EXAMPLE 3.2

Find the transfer function $H_2(s)$ for the normalized Butterworth filter of order 2.

Solution. Since $n = 2$ we have the poles of $H_2(s) H_2(-s)$ given by

$$s_k = 1\underline{/\pi/4 + k\pi/2}, \quad k = 0, 1, 2, 3$$

These poles are shown in Fig. 3.6, and using the left-half plane poles we can express the transfer function as follows

$$H_2(s) = \frac{1}{(s - s_2)(s - s_3)}$$

$$= \frac{1}{[s - (-0.707 - 0.707j)][s - (-0.707 + 0.707j)]}$$

$$= \frac{1}{s^2 + \sqrt{2}s + 1}$$

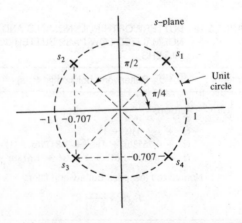

Figure 3.6 Poles of $H(s) H(-s)$ for a normalized Butterworth filter of order 2.

TABLE 3.1b BUTTERWORTH POLYNOMIALS IN STANDARD AND FACTORED FORMS

Standard form

$$B_n(s) = a_n s^n + a_{n-1} s^{n-1} + \cdots + a_1 s + a_0$$

a_8	a_7	a_6	a_5	a_4	a_3	a_2	a_1	a_0	n
							1	1	1
						1	$\sqrt{2}$	1	2
					1	2	2	1	3
				1	2.613	3.414	2.613	1	4
			1	3.236	5.236	5.236	3.236	1	5
		1	3.864	7.464	9.141	7.464	3.864	1	6
	1	4.494	10.103	14.606	14.606	10.103	4.494	1	7
1	5.126	13.138	21.848	25.691	21.848	13.138	5.126	1	8

Factored form

$$B_n(s)$$

$B_n(s)$	n
$s + 1$	1
$s^2 + \sqrt{2}s + 1$	2
$(s^2 + s + 1)(s + 1)$	3
$(s^2 + 0.7653s + 1)(s^2 + 1.8477s + 1)$	4
$(s + 1)(s^2 + 0.6180s + 1)(s^2 + 1.6180s + 1)$	5
$(s^2 + 0.5176s + 1)(s^2 + \sqrt{2}s + 1)(s^2 + 1.9318s + 1)$	6
$(s + 1)(s^2 + 0.4450s + 1)(s^2 + 1.2456s + 1)(s^2 + 1.8022s + 1)$	7
$(s^2 + 0.3986s + 1)(s^2 + 1.1110s + 1)(s^2 + 1.6630s + 1)(s^2 + 1.9622s + 1)$	8

Butterworth filter

$$H_n(s) = \frac{1}{a_n s^n + a_{n-1} s^{n-1} + \cdots + a_1 s + 1} = \frac{1}{B_n(s)}$$

Source: Kuo, Franklin F. *Network Analysis and Synthesis*. Wiley, New York, 1966, p. 372. Reprinted with permission.

Extended tables of Butterworth polynomials in both polynomial and factored form are given in Table 3.1b. The factored forms will prove to be useful in obtaining digital filters, since they provide convenient cascade realizations.

To obtain Butterworth filters with cutoff frequencies other than 1 rad/sec it is convenient to use the 1 rad/sec Butterworth filters as prototypes and apply analog-to-analog transformations.

3.1.1 Analog-to-Analog Transformations

In the previous section emphasis was placed on the one radian low-pass Butterworth filters with the stipulation that other non-normalized Butterworth filters could be derived by transformational methods. The transformational method, however, is not limited in its application to Butterworth filters or normalized filters. In the following discussion a normalized low-pass filter will be used as the prototype filter for the purpose of illustration.

If we replace s of $H(s)$, the system function for a normalized low-pass filter, by s/Ω_u, we get a new transfer function $H'(s)$, given by

$$H'(s) = H(s)|_{s \to s/\Omega_u} = H(s/\Omega_u) \tag{3.7}$$

If we evaluate the magnitude of the transfer function $H'(s)$ at $s = j\Omega$ to get the frequency response we have

$$|H'(j\Omega)| = |H(j\Omega/\Omega_u)| \tag{3.8}$$

At the value of $\Omega = \Omega_u$ we have

$$|H'(j\Omega_u)| = |H(j\Omega_u/\Omega_u)| = |H(j1)|$$

TABLE 3.2 ANALOG-TO-ANALOG TRANSFORMATION

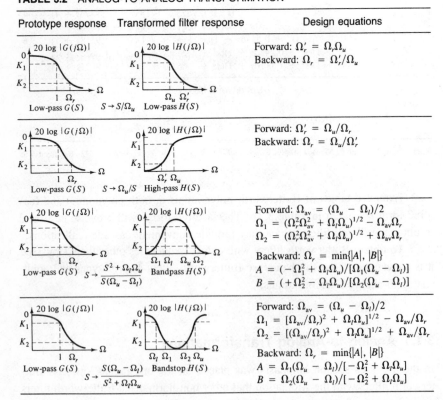

Prototype response	Transformed filter response	Design equations
Low-pass $G(S)$	$S \to S/\Omega_u$ Low-pass $H(S)$	Forward: $\Omega'_r = \Omega_r\Omega_u$ Backward: $\Omega_r = \Omega'_r/\Omega_u$
Low-pass $G(S)$	$S \to \Omega_u/S$ High-pass $H(S)$	Forward: $\Omega'_r = \Omega_u/\Omega_r$ Backward: $\Omega_r = \Omega_u/\Omega'_r$
Low-pass $G(S)$	$S \to \dfrac{S^2 + \Omega_l\Omega_u}{S(\Omega_u - \Omega_l)}$ Bandpass $H(S)$	Forward: $\Omega_{av} = (\Omega_u - \Omega_l)/2$ $\Omega_1 = (\Omega_r^2\Omega_{av}^2 + \Omega_l\Omega_u)^{1/2} - \Omega_{av}\Omega_r$ $\Omega_2 = (\Omega_r^2\Omega_{av}^2 + \Omega_l\Omega_u)^{1/2} + \Omega_{av}\Omega_r$ Backward: $\Omega_r = \min\{|A|, |B|\}$ $A = (-\Omega_1^2 + \Omega_l\Omega_u)/[\Omega_1(\Omega_u - \Omega_l)]$ $B = (+\Omega_2^2 - \Omega_l\Omega_u)/[\Omega_2(\Omega_u - \Omega_l)]$
Low-pass $G(S)$	$S \to \dfrac{S(\Omega_u - \Omega_l)}{S^2 + \Omega_l\Omega_u}$ Bandstop $H(S)$	Forward: $\Omega_{av} = (\Omega_u - \Omega_l)/2$ $\Omega_1 = [\Omega_{av}/\Omega_r)^2 + \Omega_l\Omega_u]^{1/2} - \Omega_{av}/\Omega_r$ $\Omega_2 = [(\Omega_{av}/\Omega_r)^2 + \Omega_l\Omega_u]^{1/2} + \Omega_{av}/\Omega_r$ Backward: $\Omega_r = \min\{|A|, |B|\}$ $A = \Omega_1(\Omega_u - \Omega_l)/[-\Omega_1^2 + \Omega_l\Omega_u]$ $B = \Omega_2(\Omega_u - \Omega_l)/[-\Omega_2^2 + \Omega_l\Omega_u]$

That is, the frequency response for the new transfer function evaluated at $\Omega = \Omega_u$ is equal to the value of the normalized transfer function at $\Omega = 1$. In a sense we have moved the cutoff frequency from 1 rad/sec to Ω_u rad/sec and thus have a scaling of the frequency axis. Similar transformations can be defined for taking low-pass transfer functions to high-pass, bandpass and bandstop transfer functions. Table 3.2 gives these transformations along with design equations for both forward and backward development. For example, if the transformation $s \rightarrow s/\Omega_u$ is applied to the low-pass structure as shown at the top of Table 3.2, the critical frequency Ω_r will be transformed (forward) into Ω'_r, which is Ω_r time Ω_u as seen under the design equation column. In a similar fashion, if Ω'_r is the desired critical frequency of the transformed filter, the backward equation gives the value of Ω_r that must be used such that going through the transformation $s \rightarrow s/\Omega_u$ results in the required Ω'_r. We have Ω_r equals Ω'_r/Ω_u. Procedures will now be given for the design of non-normalized low-pass and bandpass filters, and Table 3.2 provides both forward and backward design formulas for high-pass and bandstop filter designs if desired.

3.1.2 Design of Low-Pass Butterworth Filters

The filter requirements are normally given in terms of a set of critical frequencies, say Ω_1, Ω_2 and gains K_1 and K_2. A common set of conditions for the low-pass response given in Fig. 3.7 are

$$0 \geq 20 \log |H(j\Omega)| \geq K_1 \qquad \text{for all } \Omega \leq \Omega_1 \qquad (3.9)$$

$$20 \log |H(j\Omega)| \leq K_2 \qquad \text{for all } \Omega \geq \Omega_2 \qquad (3.10)$$

As seen from Eq. (3.1), the Butterworth LP frequency response is characterized by only two parameters, n, the order of the filter, and Ω_c, the cutoff frequency. If we replace $H(j\Omega)$ in Eqs. (3.9) and (3.10) by the Butterworth

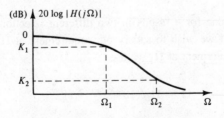

Figure 3.7 Desired filter requirements in dB form for a low-pass filter.

magnitude squared function [Eq. (3.1)] and consider that the equalities hold, n and Ω_c must satisfy the following:

$$10 \log \left[\frac{1}{1 + (\Omega_1/\Omega_c)^{2n}} \right] = K_1 \tag{3.11}$$

$$10 \log \left[\frac{1}{1 + (\Omega_2/\Omega_c)^{2n}} \right] = K_2 \tag{3.12}$$

Dividing both sides of the above equations by 10, taking the antilog and simplifying yields

$$\left(\frac{\Omega_1}{\Omega_c} \right)^{2n} = 10^{-K_1/10} - 1, \qquad \left(\frac{\Omega_2}{\Omega_c} \right)^{2n} = 10^{-K_2/10} - 1 \tag{3.13}$$

Dividing to cancel Ω_c we have the following implicit equation relating $\Omega_1, \Omega_2, K_1, K_2$, and n:

$$\left(\frac{\Omega_1}{\Omega_2} \right)^{2n} = \frac{10^{-K_1/10} - 1}{10^{-K_2/10} - 1} \tag{3.14}$$

A simple closed form answer for n is easily obtained from this expression and is given by

$$n = \frac{\log_{10}[(10^{-K_1/10} - 1)/(10^{-K_2/10} - 1)]}{2 \log_{10}(\Omega_1/\Omega_2)} \tag{3.15}$$

If n is an integer we use that value; otherwise we use the next larger integer. Defining $\lceil \cdot \rceil$ to indicate this operation we have

$$n = \left\lceil \frac{\log_{10}[(10^{-K_1/10} - 1)/(10^{-K_2/10} - 1)]}{2 \log_{10}(\Omega_1/\Omega_2)} \right\rceil \tag{3.16}$$

Using this value for n results in two different selections for Ω_c as seen from Eq. (3.13). If we wish to satisfy our requirement at Ω_1 exactly and do better than our requirement at Ω_2 we use

$$\Omega_c = \Omega_1/(10^{-K_1/10} - 1)^{1/2n} \tag{3.17}$$

while if we wish to satisfy our requirement at Ω_2 and exceed our requirement at Ω_1 we use

$$\Omega_c = \Omega_2/(10^{-K_2/10} - 1)^{1/2n} \qquad (3.18)$$

Choosing a value of Ω_c in between these gives an $H(j\Omega)$ that exceeds both requirements. An example of the design of a low-pass Butterworth filter using the procedure just described is now given.

EXAMPLE 3.3

Design an analog Butterworth filter that has a -2 dB or better cutoff frequency of 20 rad/sec and at least 10 dB of attenuation at 30 rad/sec.

Solution. The critical requirements are

$$\Omega_1 = 20, \qquad K_1 = -2, \qquad \Omega_2 = 30, \qquad K_2 = -10$$

Substituting these requirements into Eq. (3.16) gives

$$n = \left\lceil \frac{\log_{10}[(10^{0.2} - 1)/(10^1 - 1)]}{2\log_{10}(20/30)} \right\rceil = 3.3709 = 4$$

Using this value of n in (3.17) to exactly satisfy the -2 dB requirement gives

$$\Omega_c = 20/(10^{0.2} - 1)^{1/8} = 21.3868$$

The normalized low-pass Butterworth filter ($\Omega_c = 1$) for $n = 4$, can be found from Table 3.1b as

$$H_4(s) = \frac{1}{(s^2 + 0.76536s + 1)(s^2 + 1.84776s + 1)}$$

Applying a low-pass to low-pass transformation, $s \to s/\Omega_c$, with $\Omega_c = 21.3868$ gives the desired transfer function as follows:

$$H(s) = H_4(s)\big|_{s \to s/21.3868}$$

$$= \frac{1}{\left[\left(\dfrac{s}{21.3868}\right)^2 + 0.76536\left(\dfrac{s}{21.3868}\right) + 1\right]}$$

$$\times \frac{1}{\left[\left(\dfrac{s}{21.3868}\right)^2 + 1.84776\left(\dfrac{s}{21.3868}\right) + 1\right]}$$

$$= \frac{0.209210 \times 10^6}{(s^2 + 16.3686s + 457.394)(s^2 + 39.5176s + 457.394)}$$

3.1.3 Design of Bandpass Butterworth Filters

The design of a bandpass filter is also based on applying a transformation to a low-pass normalized Butterworth filter of the proper order. The typical filter requirements shown in Fig. 3.8 can be written as

$$20 \log |H(j\Omega)| \leq K_2, \qquad \Omega \leq \Omega_1$$
$$0 \leq 20 \log |H(j\Omega)| \leq K_1, \qquad \Omega \leq \Omega \leq \Omega_u$$
$$20 \log |H(j\Omega)| \leq K_2, \qquad \Omega \geq \Omega_2$$

If $H_{\text{LP}}(s)$ represents a unit bandwidth low-pass filter with critical radian frequency Ω_r, then from Table 3.2 a bandpass filter with transfer function $H_{\text{BP}}(s)$ is given by

$$H_{\text{BP}}(s) = H_{\text{LP}}(s)\big|_{s \to (s^2 + \Omega_l \Omega_u)/[s(\Omega_u - \Omega_l)]}$$

For the bandpass filter to satisfy the K_2 requirement at Ω_1 we must have equality within the transformation, that is,

$$j\Omega_r = [(j\Omega_1)^2 + \Omega_l\Omega_u]/[j\Omega_1(\Omega_u - \Omega_l)]$$

Solving the above equation for Ω_r and a similar equation to satisfy the K_2 requirement at Ω_2 gives

$$\Omega_r = (\Omega_1^2 - \Omega_l\Omega_u)/[\Omega_1(\Omega_u - \Omega_l)]$$
$$\Omega_r = (\Omega_2^2 - \Omega_l\Omega_u)/[\Omega_2(\Omega_u - \Omega_l)]$$

Depending upon the size of Ω_1 and Ω_2 with respect to the product $\Omega_l\Omega_u$, the Ω_r's above could correspond to either positive or negative frequency values. Also, in most cases the Ω_r resulting from these equations will not be equal and

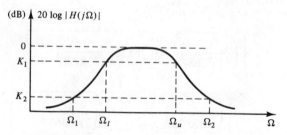

Figure 3.8 Typical bandpass requirements.

we must choose the most restrictive value, thus giving a filter that exceeds our less restrictive requirement. Therefore the selection of Ω_r becomes that given in the backward design equations for the low-pass to bandpass transformation part of Table 3.2:

$$\Omega_r = \min\{|A|,|B|\} \tag{3.19a}$$
where
$$A = (-\Omega_1^2 + \Omega_l\Omega_u)/[\Omega_1(\Omega_u - \Omega_l)] \tag{3.19b}$$
$$B = (\Omega_2^2 - \Omega_l\Omega_u)/[\Omega_2(\Omega_u - \Omega_l)] \tag{3.19c}$$

The procedure for the design of a bandpass filter, $H_{BP}(s)$, to satisfy the given set of specifications is composed of two steps. First, design a low-pass filter $H_{LP}(s)$ with the Ω_r determined above and second, apply the low-pass to bandpass transformation using the desired Ω_u and Ω_l. This procedure is illustrated in the following example.

EXAMPLE 3.4

Design an analog bandpass filter with the following characteristics:

(a) a -3.0103-dB upper and lower cutoff frequency of 50 Hz and 20 kHz
(b) a stopband attenuation of at least 20 dB at 20 Hz and 45 kHz
(c) a monotonic frequency response

Solution. The monotonic requirement can be satisfied with a Butterworth filter. From the specifications above we can identify the following critical frequencies:

$$\Omega_1 = 2\pi(20) = 125.663 \text{ rad/sec}$$
$$\Omega_2 = 2\pi(45) \cdot 10^3 = 2.82743 \times 10^5 \text{ rad/sec}$$
$$\Omega_u = 2\pi(20) \cdot 10^3 = 1.25663 \times 10^5 \text{ rad/sec}$$
$$\Omega_l = 2\pi(50) = 314.159 \text{ rad/sec}$$

Also the low-pass prototype must satisfy

$$0 \geqq 20 \log|H_{LP}(j1)| \geqq -3.0103 \text{ dB}$$
$$20 \log|H_{LP}(j\Omega_r)| \leqq -20 \text{ dB}$$

Using the Ω_1, Ω_2, Ω_u, and Ω_l in Eqs. (3.19a) and (3.19b) gives

$$A = 2.5053$$
$$B = 2.2545$$

The most critical value Ω_r is the minimum of the two, that is,

$$\Omega_r = 2.2545$$

The low-pass Butterworth filter of order n can then be easily calculated from Eq. (3.16) as

$$n = \left\lceil \left[\log \left(\frac{10^{0.30102} - 1}{10^2 - 1} \right) \right] \Big/ \left[2 \log \frac{1}{2.2545} \right] \right\rceil = \lceil 2.829 \rceil = 3$$

From the Butterworth Table 3.1b and $n = 3$ we have the low-pass prototype as

$$H_{LP} = \frac{1}{s^3 + 2s^2 + 2s + 1}$$

The required analog-to-analog transformation is determined from Ω_u and Ω_l as

$$s \to \frac{s^2 + \Omega_l\Omega_u}{s(\Omega_u - \Omega_l)} = \frac{s^2 + 3.94784 \times 10^7}{s(1.25349 \times 10^5)}$$

$H_{BP}(s)$ then is finally seen to be

$$H_{BP}(s) = \frac{1}{\left\{ \left[\dfrac{s^2 + 3.94784 \times 10^7}{s(1.25349 \times 10^5)} \right]^3 + 2\left[\dfrac{s^2 + 3.94784 \times 10^7}{s(1.25349 \times 10^5)} \right]^2 \right.}$$

$$\left. + \, 2\, \frac{s^2 + 3.947845 \times 10^7}{s(1.25349 \times 10^5)} + 1 \right\}$$

$$= \frac{1.969530 \times 10^{15} s^3}{(s^6 + 2.5069909 \times 10^5 s^5 + 3.15434 \times 10^{10} s^4}$$
$$+ \, 1.9893 \times 10^{15} s^3 + 1.245285 \times 10^{18} s^2$$
$$+ \, 3.9072593 \times 10^{20} s + 6.15289108 \times 10^{22})$$

3.2 CHEBYSHEV FILTERS

There are two types of Chebyshev filters, one containing a ripple in the passband (type 1) and the other containing a ripple in the stopband (type 2). A type 1 low-pass normalized (unit bandswith) Chebyshev filter with a ripple in the pass-band is characterized by the following magnitude squared frequency response:

$$|H_n(j\Omega)|^2 = \frac{1}{1 + \epsilon^2 T_n^2(\Omega)} \tag{3.20}$$

where $T_n(\Omega)$ is the nth order Chebyshev polynomial. The Chebyshev polynomials can be generated and thus defined from the following recursive formula:

$$T_n(x) = 2xT_{n-1}(x) - T_{n-2}(x), \qquad n > 2 \tag{3.21}$$

with $T_0(x) = 1$ and $T_1(x) = x$. A list of the first ten Chebyshev polynomials is given in Table 3.3 for reference.

In Fig. 3.9 we see the plot of $T_5(\Omega)$ and the corresponding magnitude squared plot for the type 1 Chebyshev filter. It is noticed that for $n = 5$, the Chebyshev polynomial oscillates between plus and minus one within the interval $x = -1$ to 1, while outside that interval it grows toward plus and minus infinity. This same type of oscillation takes place for every Chebyshev polynomial and causes the equal magnitude ripple in the $|H_n(j\Omega)|^2$ as shown in Fig. 3.9(b). The amplitude of $|H_n(j\Omega)|^2$ oscillates between 1 and $1/(1 + \epsilon^2)$ as Ω goes from -1 to $+1$, but the periods of the oscillations are not equal. The $|H_n(j\Omega)|^2$ heads to zero outside -1 to 1 because the magnitude of the Chebyshev polynomial heads toward ∞ as Ω goes away from -1 negatively or $+1$ positively.

It can easily be seen from the list of Chebyshev polynomials that $T_n(x)$ at $x = 0$ is 1 when n is even and zero when n is odd, resulting in $|H_n(j\Omega)|^2$ to be $1/(1 + \epsilon^2)$ at $\Omega = 0$ for n even and 1 at $\Omega = 0$ for n odd. The two general

TABLE 3.3 THE FIRST TEN CHEBYSHEV POLYNOMIALS

n	$T_n(x)$
0	1
1	x
2	$2x^2 - 1$
3	$4x^3 - 3x$
4	$8x^4 - 8x^2 + 1$
5	$16x^5 - 20x^3 + 5x$
6	$32x^6 - 48x^4 + 18x^2 - 1$
7	$64x^7 - 112x^5 + 56x^3 - 7x$
8	$128x^8 - 256x^6 + 160x^4 - 32x^2 + 1$
9	$256x^9 - 576x^7 + 432x^5 - 120x^3 + 9x$
10	$512x^{10} - 1280x^8 + 1120x^6 - 400x^4 + 50x^2 - 1$

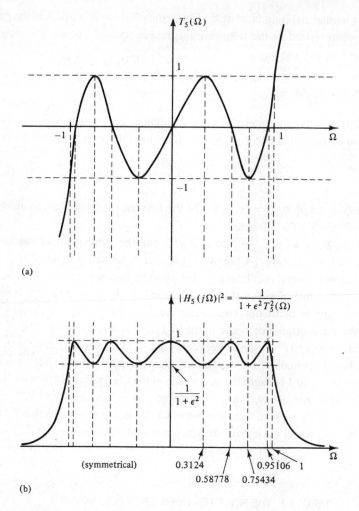

Figure 3.9 (a) Plot of the fifth-order Chebyshev polynomial $T_5(\Omega)$; (b) plot of corresponding $|H_5(j\Omega)|^2$ for the type 1 Chebyshev filter.

shapes of magnitude squared frequency response of the type 1 Chebyshev filter for n odd and even are given in Fig. 3.10.

The following properties are easily observable from Fig. 3.10 and Eq. (3.20).

(1) The magnitude squared frequency response oscillates between 1 and $1/(1 + \epsilon^2)$ within the passband, the so-called equiripple, and has a value of $1/(1 + \epsilon^2)$ at $\Omega = 1$, the so-called cutoff frequency.

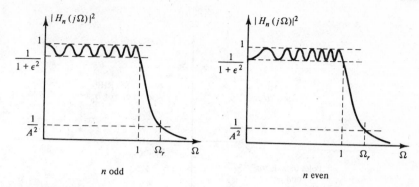

Figure 3.10 Magnitude squared frequency responses for the normalized type 1 Chebyshev filter of odd and even n.

(2) The magnitude squared frequency response $|H_n(j\Omega)|^2$ is monotonic outside the passband, including both transition band and the stopband. The stopband begins at Ω_r with magnitude squared frequency response at value $1/A^2$.

To obtain the causal stable transfer function $H_n(s)$ that corresponds to the Chebyshev magnitude squared function given in Eq. (3.20), we must find the poles of $H_n(s)H_n(-s)$ and select the left half plane poles for $H_n(s)$. The poles of $H_n(s) \cdot H_n(-s)$ are obtained by finding the roots of

$$1 + \epsilon^2 T_n^2(s/j) = 0 \qquad (3.22)$$

It can be shown [10] that the roots of (3.22), that is, the poles of $H_n(s) \cdot H_n(-s)$ fall on an ellipse as shown in Fig. 3.11(a).

If $s_k = \sigma_k + j\Omega_k$ represents a pole, then σ_k and Ω_k were shown in Weinberg [10] to satisfy

$$\frac{\sigma_k^2}{a^2} + \frac{\Omega_k^2}{b^2} = 1 \qquad (3.23)$$

where a, b, σ_k, and Ω_k are given by

$$a = \tfrac{1}{2}\{[1 + (1 + \epsilon^2)^{1/2}]/\epsilon\}^{1/n} - \tfrac{1}{2}\{[1 + (1 + \epsilon^2)^{1/2}]/\epsilon\}^{-1/n} \qquad (3.24)$$
$$b = \tfrac{1}{2}\{[1 + (1 + \epsilon^2)^{1/2}]/\epsilon\}^{1/n} + \tfrac{1}{2}\{[1 + (1 + \epsilon^2)^{1/2}]\epsilon/\}^{-1/n}$$
$$\sigma_k = -a\sin\left[(2k - 1)\pi/2n\right]$$
$$\Omega_k = b\cos\left[(2k - 1)\pi/2n\right] \qquad k = 1, 2, \ldots, 2n \qquad (3.25)$$

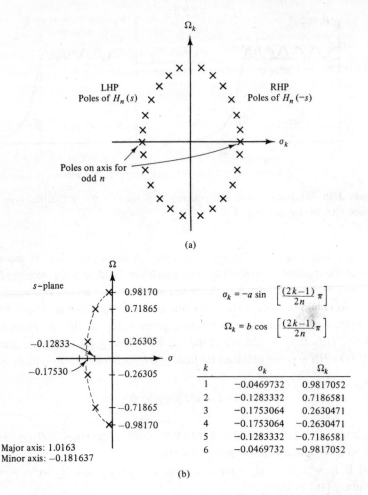

(a)

$$\sigma_k = -a \sin \left[\frac{(2k-1)}{2n} \pi \right]$$

$$\Omega_k = b \cos \left[\frac{(2k-1)}{2n} \pi \right]$$

k	σ_k	Ω_k
1	−0.0469732	0.9817052
2	−0.1283332	0.7186581
3	−0.1753064	0.2630471
4	−0.1753064	−0.2630471
5	−0.1283332	−0.7186581
6	−0.0469732	−0.9817052

(b)

Figure 3.11   (a) Poles of $H_n(s)H_n(-s)$ for the normalized low-pass Chebyshev filter. (b) Poles of $H(s)$ for $n = 6$, $\epsilon = 0.7647831$.

Using the left half plane poles only, $H_n(s)$ can be written in the following form,

$$H_n(s) = \frac{K}{\displaystyle\prod_{\substack{LHP \\ poles}} (s - s_k)} = \frac{K}{V_n(s)} \tag{3.26}$$

where K is a normalizing factor whose value makes $H(0)$ equal to 1 for n odd and $1/(1 + \epsilon^2)^{1/2}$ for n even. In Eq. (3.26), $V_n(s)$ is a polynomial in s given by $V_n(s) = s^n + b_{n-1}s^{n-1} + \cdots + b_1 s + b_0$. The normalizing constant K is easily seen for n odd and even to be

$$K = V_n(0) = b_0, \qquad n \text{ odd}$$
$$K = V_n(0)/(1 + \epsilon^2)^{1/2}, \qquad n \text{ even}$$

Two different forms of tables have been established for $V_n(s)$ for various n and given values of ϵ. Table 3.4 gives the $V_n(s)$ in polynomial form for $n = 1$ to 10 and ϵ corresponding to $\frac{1}{2}$, 1, 2, and 3 dB ripples, while Table 3.5 gives the zeros [poles of $H_n(s)$] for the same n and ϵ. As for the case of the Butterworth filter, the table with the poles will be the most useful. In Fig. 3.11(b), the pole locations of $H_n(s)$ for $n = 6$ and $\epsilon = 0.7647831$ are shown. Note that the poles are not equally spaced but do exhibit symmetry with respect to the σ axis.

We see that the normalized Chebyshev low-pass filter has only two parameters, ϵ and n. Normal filter design proceeds from a set of specifications which include acceptable ripple, critical frequencies, and stopband attenuation. The acceptable ripple determines the ϵ, while the stopband attenuation and ϵ determine the order n as discussed in the following section.

Selection of n For low-pass filter design we are usually interested not only in the cutoff frequency but in the stopband attenuation. Usual specifications might be that the magnitude squared frequency response is less than a certain value $1/A^2$ at a frequency Ω_r in the stopband as seen in Fig. 3.10. It has been shown [10] that the n that satisfies a specified ripple characterized by ϵ and a stopband gain of $1/A$ at a particular Ω_r is given by

$$n = \left\lceil \frac{\log_{10}[g + (g^2 - 1)^{1/2}]}{\log_{10}[\Omega_r + (\Omega_r^2 - 1)^{1/2}]} \right\rceil \qquad (3.27)$$

where
$$A = 1/|H_n(j\Omega_r)|$$
$$g = [(A^2 - 1)/\epsilon^2]^{1/2} \qquad (3.28)$$

The following examples use (3.27) and (3.28) in the designs of normalized and unnormalized Chebyshev low-pass filters.

EXAMPLE 3.5

Design a low-pass 1-rad/sec bandwidth Chebyshev filter with the following characteristics:

(a) Acceptable passband ripple of 2 dB.

TABLE 3.4 POLYNOMIALS $V_n(s)$ USED IN CHEBYSHEV FILTER DESIGN FOR $\frac{1}{2}$-, 1-, 2-, and 3-dB RIPPLES.

Chebyshev filter $H_n(s) = \dfrac{K_n}{V_n(s)}$ $K_n = \begin{cases} b_0/(1+\epsilon^2)^{1/2} & \text{for } n \text{ even} \\ b_0 & \text{for } n \text{ odd} \end{cases}$

$$V_n(s) = s^n + b_{n-1}s^{n-1} + \cdots + b_1 s + b_0$$

n	b_0	b_1	b_2	b_3	b_4	b_5	b_6	b_7	b_8	b_9
\multicolumn{11}{c}{a. $\frac{1}{2}$ dB Ripple(ϵ = 0.3493114, ϵ^2 = 0.1220184)}										
1	2.8627752									
2	1.5162026	1.4256245								
3	0.7156938	1.5348954	1.2529130							
4	0.3790506	1.0254553	1.7168662	1.1973856						
5	0.1789234	0.7525181	1.3095747	1.9373675	1.1724909					
6	0.0947626	0.4323669	1.1718613	1.5897635	2.1718446	1.1591761				
7	0.0447309	0.2820722	0.7556511	1.6479029	1.8694079	2.4126510	1.1512176			
8	0.0236907	0.1525444	0.5735604	1.1485894	2.1840154	2.1492173	2.6567498	1.1460801		
9	0.0111827	0.0941198	0.3408193	0.9836199	1.6113880	2.7814990	2.4293297	2.9027337	1.1425705	
10	0.0059227	0.0492855	0.2372688	0.6269689	1.5274307	2.1442372	3.4409268	2.7097415	3.1498757	1.1400664
\multicolumn{11}{c}{b. 1-dB Ripple (ϵ = 0.5088471, ϵ^2 = 0.2589254)}										
1	1.9652267									
2	1.1025103	1.0977343								
3	0.4913067	1.2384092	0.9883412							
4	0.2756276	0.7426194	1.4539248	0.9528114						
5	0.1228267	0.5805342	0.9743961	1.6888160	0.9368201					
6	0.0689069	0.3070808	0.9393461	1.2021409	1.9308256	0.9282510				
7	0.0307066	0.2136712	0.5486192	1.3575440	1.4287950	2.1760778	0.9231228			
8	0.0172267	0.1073447	0.4478257	0.8468243	1.8369024	1.6551557	2.4230264	0.9198113		
9	0.0076767	0.0706048	0.2441864	0.7863109	1.2016071	2.3781188	1.8814798	2.6709468	0.9175476	
10	0.0043067	0.0344971	0.1824512	0.4553892	1.2444914	1.6129856	2.9815094	2.1078524	2.9194657	0.9159320

140

n	b_0	b_1	b_2	b_3	b_4	b_5	b_6	b_7	b_8	b_9
				c. 2-dB Ripple ($\epsilon = 0.7647831$ $\epsilon^2 = 0.5848932$)						
1	1.3075603									
2	0.6367681	0.8038164								
3	0.3268901	1.0221903	0.7378216							
4	0.2057651	0.5167981	1.2564819	0.7162150						
5	0.0817225	0.4593491	0.6934770	1.4995433	0.7064606					
6	0.0514413	0.2102706	0.7714618	0.8670149	1.7458587	0.7012257				
7	0.0204228	0.1660920	0.3825056	1.1444390	1.0392203	1.9935272	0.6978929			
8	0.0128603	0.0729373	0.3587043	0.5982214	1.5795807	1.2117121	2.2422529	0.6960646		
9	0.0051076	0.0543756	0.1684473	0.6444677	0.8568648	2.0767479	1.3837464	2.4912897	0.6946793	
10	0.0032151	0.0233347	0.1440057	0.3177560	1.0389104	1.158287	2.6362507	1.5557424	2.7406032	0.6936904
				d. 3-dB Ripple ($\epsilon = 0.9976283$ $\epsilon^2 = 0.9952623$)						
1	1.0023773									
2	0.7079478	0.6448996								
3	0.2505943	0.9283480	0.5972404							
4	0.1769869	0.4047679	1.1691176	0.5815799						
5	0.0626391	0.4079421	0.5488626	1.4149847	0.5744296					
6	0.0442467	0.1634299	0.6990977	0.6906098	1.6628481	0.5706979				
7	0.0156621	0.1461530	0.3000167	1.0518448	0.8314411	1.9115507	0.5684201			
8	0.0110617	0.0564813	0.3207646	0.4718990	1.4666990	0.9719473	2.1607148	0.5669476		
9	0.0039154	0.0475900	0.1313851	0.5834984	0.6789075	1.9438443	1.1122863	2.4101346	0.5659234	
10	0.0027654	0.0180313	0.1277560	0.2492043	0.9499208	0.9210659	2.4834205	1.2526467	2.6597378	0.5652218

Source: Weinberg, Louis. Network Analysis and Synthesis. McGraw-Hill, New York, 1962. Reprinted with permission of the author.

TABLE 3.5 ZEROS OF POLYNOMIAL $V_n(s)$ DERIVED FROM THE CHEBYSHEV APPROXIMATION FOR $\tfrac{1}{2}$-, 1-, 2-, and 3-dB RIPPLES

Chebyshev filter $H_n(s) = \dfrac{K_n}{V_n(s)}$ $K_n = \begin{cases} b_0 & \text{for } n \text{ odd} \\ b_0/(1+\epsilon^2)^{1/2} & \text{for } n \text{ even} \end{cases}$

n = 1	n = 2	n = 3	n = 4	n = 5	n = 6	n = 7	n = 8	n = 9	n = 10
a. 1/2-dB Ripple ($\epsilon = 0.3493114$, $\epsilon^2 = 0.1220184$)									
-2.8627752	-0.7128122 ±j1.0040425	-0.6264565	-0.1753531 ±j1.0162529	-0.3623196	-0.0776501 ±j1.0084608	-0.2561700	-0.0436201 ±j1.0050021	-0.1984053	-0.0278994 ±j1.0032732
		-0.3132282 ±j1.0219275	-0.4233398 ±j0.4209457	-0.1119629 ±j1.0115574	-0.2121440 ±j0.7382446	-0.0570032 ±j1.0064085	-0.1242195 ±j0.8519996	-0.0344527 ±j1.0040040	-0.0809672 ±j0.9050658
				-0.2931227 ±j0.6251768	-0.2897940 ±j0.2702162	-0.1597194 ±j0.8070770	-0.1859076 ±j0.5692879	-0.0992026 ±j0.8829063	-0.1261094 ±j0.7182643
						-0.2308012 ±j0.4478939	-0.2192929 ±j0.1999073	-0.1519873 ±j0.6553170	-0.1589072 ±j0.4611541
								-0.1864400 ±j0.3486869	-0.1761499 ±j0.1589029
b. 1-dB Ripple ($\epsilon = 0.5088471$, $\epsilon^2 = 0.2589254$)									
-1.9652267	-0.5488672 ±j0.8951286	-0.4941706	-0.1395360 ±j0.9833792	-0.2894933	-0.0621810 ±j0.9934115	-0.2054141	-0.0350082 ±j0.9964513	-0.1593305	-0.0224144 ±j0.9977755
		-0.2470853 ±j0.9659987	-0.3368697 ±j0.4073290	-0.0894584 ±j0.9901071	-0.1698817 ±j0.7272275	-0.0457089 ±j0.9952839	-0.0996950 ±j0.8447506	-0.0276674 ±j0.9972297	-0.0650493 ±j0.9001063
				-0.2342050 ±j0.6119198	-0.2320627 ±j0.2661837	-0.1280736 ±j0.7981557	-0.1492041 ±j0.564443	-0.0796652 ±j0.8769490	-0.1013166 ±j0.7143234
						-0.1850717 ±j0.4429430	-0.1759983 ±j0.1982065	-0.1220542 ±j0.6508954	-0.1276664 ±j0.4586271
								-0.1497217 ±j0.3463342	-0.1415193 ±j0.1580321

c. 2-dB Ripple ($\epsilon = 0.7647831$, $\epsilon^2 = 0.5848932$)

$n = 1$	$n = 2$	$n = 3$	$n = 4$	$n = 5$	$n = 6$	$n = 7$	$n = 8$	$n = 9$	$n = 10$
−1.3075603	−0.4019082 ±j0.6893750	−0.3689108	−0.1048872 ±j0.9579530	−0.2183083	−0.0469732 ±j0.9817052	−0.1552958	−0.0264924 ±j0.9897870	−0.1206298	−0.0169758 ±j0.9934868
		−0.1844554 ±j0.9230771	−0.2532202 ±j0.3967971	−0.0674610 ±j0.9734557	−0.1283332 ±j0.7186581	−0.0345566 ±j0.9866139	−0.0754439 ±j0.8391009	−0.0209471 ±j0.919471	−0.0767332 ±j0.7112580
				−0.1766151 ±j0.6016287	−0.1753064 ±j0.2630471	−0.0968253 ±j0.7912029	−0.1129098 ±j0.5606693	−0.0603149 ±j0.8723036	−0.0492657 ±j0.8962374
						−0.1399167 ±j0.4390845	−0.1331862 ±j0.1968809	−0.0924078 ±j0.6474475	−0.0966894 ±j0.4566558
								−0.1133549 ±j0.3444996	−0.1071810 ±j0.1573528

d. 3-dB Ripple ($\epsilon = 0.9976283$, $\epsilon^2 = 0.9952623$)

$n = 1$	$n = 2$	$n = 3$	$n = 4$	$n = 5$	$n = 6$	$n = 7$	$n = 8$	$n = 9$	$n = 10$
1.0023773	−0.3224498 ±j0.7771576	−0.2986202	−0.0851704 ±j0.9464844	−0.1775085	−0.0382295 ±j0.9764060	−0.1264854	−0.0215782 ±j0.9867664	−0.0982716	−0.0138320 ±j0.9915418
		−0.1493101 ±j0.9038144	−0.2056195 ±j0.3920467	−0.0548531 ±j0.9659238	−0.1044450 ±j0.7147788	−0.0281456 ±j0.9826957	−0.0614494 ±j0.8365401	−0.0170647 ±j0.9895516	−0.0401419 ±j0.8944827
				−0.1436074 ±j0.5969738	−0.1426745 ±j0.2616272	−0.0788623 ±j0.7880608	−0.0919655 ±j0.5589582	−0.0491358 ±j0.8701971	−0.0625225 ±j0.7098655
						−0.1139594 ±j0.4373407	−0.1084807 ±j0.1962800	−0.0752804 ±j0.6458839	−0.0787829 ±j0.4557617
								−0.0923451 ±j0.3436677	−0.0873316 ±j0.1570448

Source: Weinberg, Louis. *Network Analysis and Synthesis.* McGraw-Hill, New York, 1962. Reprinted by permission of author.

(b) Cutoff radian frequency of 1 rad/sec.

(c) Stopband attenuation of 20 dB or greater beyond 1.3 rad/sec.

Solution. To satisfy the ripple and stopband attenuation requirements, we see from the magnitude squared plot of Fig. 3.10 that

$$20 \log |H_n(j1)| = 20 \log [1/(1 + \epsilon^2)]^{1/2} = 10 \log [1/(1 + \epsilon^2)] = -2$$
$$20 \log |H_n(j1.3)| = 20 \log |(1/A^2)^{1/2}| = 20 \log (1/A) = -20$$

The above equations are easily solved giving an $\epsilon = 0.76478$ and an $A = 10$. Using (3.27) and (3.28), the corresponding g and n are determined as follows:

$$g = [(100 - 1)/(0.76478)^2]^{1/2} = 13.01$$
$$n = \left\lceil \frac{\log_{10}\{13.01 + [(13.01)^2 - 1]^{1/2}\}}{\log_{10}\{1.3 + [(1.3)^2 - 1]^{1/2}\}} \right\rceil = \lceil 4.3 \rceil = 5$$

Using the 2-dB ripple part of Table 3.3 for $n = 5$, and the fact that $H(0) = 1$ since n is odd, we have the desired Chebyshev unit bandwidth low-pass filter as

$$H_5(s) = K/(s^5 + b_4 s^3 + \cdots + b_1 s + b_0)$$
$$= \frac{0.08172}{s^5 + 0.70646 s^4 + 1.4995 s^3 + 0.6934 s^2 + 0.459349 s + 0.08172}$$

An expression for $H_5(s)$ in terms of quadratic factors obtained by using the poles from Table 3.4 is as follows.

$$H_5(s) = \frac{0.08172}{(s + 0.218303)[s - (-0.06746 + 0.97345j)][s - (-0.06746 - 0.97345j)]}$$
$$\times \frac{1}{[s - (-0.1766151 + 0.6016j)][s - (-0.1766151 - 0.6016j)]}$$

This type of quadratic form proves useful in realizing digital filters by cascading as will be shown in Chapter 4. Multiplying out the pairs of linear factors with complex conjugate pairs results in

$$H_5(s) = \frac{0.08172}{(s + 0.218303)(s^2 + 0.134922 s + 0.95215)(s^2 + 0.35323 s + 0.393115)}$$

EXAMPLE 3.6

Design a Chevbyshev low-pass filter to satisfy the following specifications: (a) acceptable passband ripple of 2 dB, (b) cutoff frequency of 40 rad/sec, and (c) stopband attenuation of 20 dB or more at 52 rad/sec.

Solution. The general approach is to first change the requirements to those of a low-pass unit bandwidth prototype, design such a low-pass filter, and then apply a low-pass to low-pass transformation to that prototype.

From Table 3.2 we see that if s is replaced by s/Ω_u, a 1-rad/sec LPF with critical frequency Ω_r is transformed to an LPF with cutoff Ω_u and a critical frequency Ω_r'. The required Ω_r for the normalized filter is obtained from the backward design equation of Table 3.2 as

$$\Omega_r = \Omega_r'/\Omega_u = 52/40 = 1.3$$

Therefore we need to design a normalized low-pass Chevyshev filter with 2-dB cutoff at 1 rad/sec and a 20-dB attenuation at 1.3 rad/sec and then apply the transformation $s \rightarrow s/40$. As it happens, just such a low-pass filter was designed in Example 3.5. So, by using the $H_5(s)$ of Example 3.5, $H_d(s)$ can be written as

$$H_d(s) = H_5(s)\big|_{s \rightarrow s/40}$$

$$= \frac{0.08172}{(s/40 + 0.218303)(s^2/40^2 + 0.134922s/40 + 0.95215)}$$

$$\times \frac{1}{s^2/40^2 + 0.35323s/40 + 0.393115)}$$

$$= \frac{8.366 \times 10^6}{(s + 8.73212)(s^2 + 5.3969s + 1523.44)(s^2 + 14.1292s + 628.984)}$$

The designs of Chebyshev high-pass, bandpass, and band-elimination filters are easily accomplished in a parallel manner. First, the backward design formulas of Table 3.2 are used to obtain low-pass filter requirements; second, a Chebyshev filter is found to satisfy those requirements; and third, a transformation is performed to get the required $H_d(s)$.

It is worth noting at this time that after the $H_d(s)$ has been found for the analog filter, a synthesis procedure must be undertaken to obtain the implementation in physical components, i.e., resistors, capacitors, inductors, etc. We shall see that once the digital filter transfer function $H(z)$ has been found, synthesis is automatic, with the only inaccuracies resulting from quantization of the coefficients and finite word length arithmetic. In a sense the discrete-time system is its own model. The design procedures for digital filters and special structures to imitate analog filters are explored in Chapter 4.

3.3 ELLIPTIC FILTERS

We have seen that by accepting a ripple in the passband, a low-pass Chebyshev type 1 filter has a sharper cutoff, that is, a narrower transition band than a Butterworth filter of the same order n. By accepting both a passband and stopband ripple it seems reasonable that even a narrower transition width might be possible for a given n. The low-pass elliptic filter provides such a smaller transition width and is optimum in the sense that no other filter of the same order has a narrower transition width for a given passband ripple and stopband attenuation. The mathematical development of the elliptic filter, although interesting, is quite complex, involving elliptic integrals, and will not be presented. Antoniou [1], Guillemin [5], and Storer [9] provide the mathematical development for those so motivated.

The magnitude squared frequency response of the normalized low-pass elliptic filter of order n is defined by

$$|H_n(j\Omega)|^2 = \frac{1}{1 + \epsilon^2 R_n^2(\Omega)} \qquad (3.29)$$

where $R_n(\Omega)$ is a Chevyshev rational function of Ω determined from the specified ripple characteristics. Typical plots of the magnitude squared frequency response for odd and even order elliptic filters are shown in Fig. 3.12.

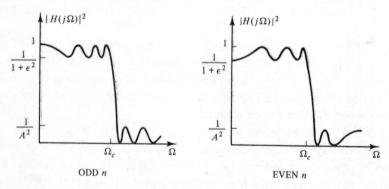

Figure 3.12 Magnitude squared frequency response of elliptic low-pass filters of odd and even orders.

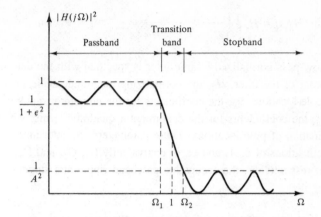

Figure 3.13 Magnitude squared frequency response of a normalized low-pass elliptic filter.

As for the Chebyshev and Butterworth filters, it is convenient to use normalized low-pass filters as a basis and obtain all other low-pass, high-pass, bandpass and bandstop filters via the proper transformation of those normalized low-pass filters. The magnitude squared frequency response for the normalized low-pass elliptic filter with radian cutoff frequency Ω_1 and critical stopband radian frequency Ω_2 is shown in Fig. 3.13.

Whereas the Chebyshev normalized filter had a Ω_1 equal to 1, it is convenient to use elliptic filters where the magnitude squared frequency response at $\Omega = 1$ is not either the passband ripple or the stopband gain. The $\Omega = 1$ turns out to be the geometric mean of Ω_2 and Ω_1, that is

$$(\Omega_1 \Omega_2)^{1/2} = 1 \tag{3.30}$$

A parameter Ω_r representing the sharpness of the transition region is defined as the ratio of Ω_2 and Ω_2 as follows:

$$\Omega_r = \Omega_2 / \Omega_1 \tag{3.31}$$

Thus a large value of Ω_r indicates a large transition band, while a small value of Ω_r means a small transition band. The filter transfer function $H_n(s)$ can be found by taking the $|H_n(j\Omega)|^2$ of Eq. (3.29), replacing $j\Omega$ by S, factoring it into left half and right half plane poles and zeros, and then taking only the left half plane poles and zeros. The transfer function $H_n(s)$, for the normalized low-pass elliptic filter is then given for odd and even n by

$$H_n(s) = \frac{H_0}{(s + s_0)} \prod_{i=1}^{(n-1)/2} \frac{s^2 + A_{0i}}{s^2 + B_{1i}s + B_{0i}}, \qquad \text{odd } n \tag{3.32}$$

$$H_n(s) = H_0 \prod_{i=1}^{n/2} \frac{s^2 + A_{0i}}{s^2 + B_{1i}s + B_{0i}}, \qquad \text{even } n \qquad (3.33)$$

The low-pass normalized elliptic filter is specified with the determination of n, the order of the filter; H_0, the normalizing magnitude scale factor; $-s_0$, the single pole location; A_{0i}, the coefficients for the numerator quadratic terms; and B_{1i}, B_{0i}, the coefficients for the denominator quadratic terms, which allow the determination of pole locations. These parameters are determined from the design specifications of ϵ, A, and Ω_r or equivalently G_1, G_2, and Ω_r, where G_1 and G_2 are, respectively, the acceptable passband ripple and stopband gain in dB:

$$G_1 = 20 \log [1/(1 + \epsilon^2)^{1/2}] = 20 \log |H_n(j\Omega_1)|$$
$$G_2 = 20 \log (1/A^2) = 20 \log |H_n(j\Omega_2)|$$

Because of the additional degree of freedom, a simple table like those for the Chebyshev and Butterworth filters is not possible; however, by restricting the passband ripples and stopband gains to be among a given set, tables have been generated. Table 3.6 represents such a table for stopband gains of -20, -30, -40, -50, -60, and -70 dB and passband ripples of 0.5, 1, 2, and 3 dB. By fixing the G_1 and G_2 in this fashion the Ω_r requirement will not be satisfied exactly; however, an Ω_r can be selected that exceeds the requirements. Various other choices for fixing the parameters exist. A common procedure, [1, 6] involves a sequential evaluation of a number of equations to select an n that will satisfy a prescribed G_1 and Ω_r while allowing the stopband gain to float as long as it is less than the required G_2.

3.3.1 Determination of n for Normalized Elliptic Filters

The design of a low-pass normalized elliptic filter to satisfy the specifications G_1, G_2, and Ω_r using Table 3.6 is straightforward, requiring minimal work. First locate the portion of the table corresponding to G_1 and G_2. Second, look down the Ω_r column for that portion of the table until a value less than the determined Ω_r value is found. Finally, the filter order is the n to the far left of that row, and the coefficients for the filter are found in all rows corresponding to that n. The values of Ω_1 and Ω_2 for the resulting filter are obtained as follows from the table value of Ω_r:

$$\Omega_1 = 1/(\Omega_r)^{1/2}, \qquad \Omega_2 = (\Omega_r)^{1/2} \qquad (3.34)$$

TABLE 3.6 COEFFICIENTS OF THE TRANSFER FUNCTION $H_N(s)$ FOR SELECTED NORMALIZED LOW-PASS ELLIPTIC FILTERS WITH GIVEN PASSBAND RIPPLES AND STOPBAND GAINS

$$H_N(s) = \frac{H_0}{D_N(s)} \prod_{i=1}^{r} \frac{s^2 + A_{0i}}{s^2 + B_{1i}s + B_{0i}} \qquad D_N(s) = \begin{cases} s + s_0, & N \text{ odd} \\ 1, & N \text{ even} \end{cases} \quad r = \left\lfloor \frac{N}{2} \right\rfloor$$

(a) Passband ripple: 0.5 dB; stopband gains: -20, -30, -40, -50, -60, -70 dB

Passband ripple = 0.5 dB; stopband gain = -20 dB

N	i	A_{0i}	B_{0i}	B_{1i}	H_0/s_0	Ω_r
2	1	5.33789	0.566660	0.809390	0.100220E + 000	2.76261
3	1	1.75640	0.808321	0.359160	0.306214E + 000 0.667292	1.42189
4	1	4.38105	0.611195	0.931959	0.100219E + 000	1.13188
	2	1.21841	0.927132	0.136543		
5	1	1.65076	0.827787	0.412816	0.303895E + 000 0.667292	1.04465
	2	1.07211	0.973640	0.049395		
	1	4.36790	0.611899	0.933855	0.100218E + 000	1.01553
6	2	1.19243	0.934830	0.156221		
	3	1.02486	0.990620	0.017576		
	1	1.64918	0.828092	0.413652	0.303861E + 000 0.667292	1.00545
7	2	1.06401	0.976479	0.056384		
	3	1.00870	0.996681	0.006219		
	1	4.36811	0.611846	0.933864	0.100192E + 000	1.00192
8	2	1.19207	0.934928	0.156548		
	3	1.02213	0.991634	0.020051		
	4	1.00306	0.998827	0.002197		
	1	1.64927	0.828047	0.413695	0.303786E + 000 0.667292	1.00068
9	2	1.06390	0.976512	0.056505		
	3	1.00775	0.997041	0.007093		
	4	1.00108	0.999586	0.000775		

Passband ripple = 0.5 dB; stopband gain = -30 dB

N	i	A_{0i}	B_{0i}	B_{1i}	H_0/s_0	Ω_r
2	1	9.51248	0.318702	0.639007	0.316294E − 001	4.80880
3	1	2.46997	0.597384	0.382044	0.121878E + 000 0.511761	1.92322
4	1	6.46603	0.398996	0.822201	0.316297E − 001	1.32446
	2	1.47114	0.798764	0.191032		
5	1	2.14490	0.648724	0.480774	0.118807E + 000 0.511761	1.12912
	2	1.18132	0.907216	0.088080		
	1	6.38228	0.402050	0.828822	0.316296E − 001	1.05394
6	2	1.38680	0.826821	0.237025		
	3	1.07474	0.958727	0.039181		
	1	2.13439	0.650591	0.484325	0.118701E + 000 0.511761	1.02299
7	2	1.15171	0.920785	0.108419		
	3	1.03168	0.981925	0.017159		
	1	6.37941	0.402154	0.829052	0.316289E − 001	1.00989
8	2	1.38394	0.827819	0.238663		
	3	1.06301	0.964898	0.048043		
	4	1.01359	0.992137	0.007464		
	1	2.13409	0.650636	0.484454	0.118683E + 000 0.511761	1.00427
9	2	1.15070	0.921256	0.109144		
	3	1.02680	0.984652	0.021005		
	4	1.00586	0.996589	0.003238		

(continued overleaf)

TABLE 3.6 (continued)

		Passband ripple = 0.5 dB; stopband gain = −40 dB				
N	i	A_{0i}	B_{0i}	B_{1i}	H_0/s_0	Ω_r
2	1	16.91940	0.179222	0.486687	0.100001E − 001	8.48925
3	1	3.55131	0.422730	0.352622	0.476460E − 001 0.417693	2.71147
4	1	9.15630	0.274972	0.710817	0.100001E − 001	1.62842
	2	1.84784	0.651784	0.213089		
5	1	2.77710	0.502279	0.488797	0.450049E − 001	1.27264
	2	1.35383	0.808559	0.117164	0.417693	
6	1	8.84271	0.281790	0.725407	0.100001E − 001	1.12697
	2	1.64666	0.706755	0.288016		
	3	1.16198	0.899904	0.061260		
7	1	2.73666	0.507485	0.497501	0.448511E − 001	1.06110
	2	1.27960	0.841245	0.155925	0.417693	
	3	1.07732	0.949005	0.031209		
8	1	8.82410	0.282208	0.726296	0.100001E − 001	1.02987
	2	1.63523	0.710223	0.292713		
	3	1.13002	0.917716	0.080817		
	4	1.03765	0.974360	0.015692		
9	1	2.73426	0.507796	0.498030	0.448397E − 001	1.01471
	2	1.27524	0.843263	0.158322	0.417693	
	3	1.06253	0.958272	0.040984		
	4	1.01851	0.987192	0.007839		

		Passband ripple = 0.5 dB; stopband gain = −50 dB				
N	i	A_{0i}	B_{0i}	B_{1i}	H_0/s_0	Ω_r
2	1	30.08815	0.100784	0.366736	0.316224E − 002	15.06069
3	1	5.16223	0.293107	0.306184	0.184313E − 001 0.353804	3.90430
4	1	12.64431	0.196244	0.613021	0.316227E − 002	2.06924
	2	2.37957	0.513565	0.211829		
5	1	3.57098	0.389887	0.467413	0.166582E − 001	1.48469
	2	1.59948	0.694940	0.132909	0.353804	
6	1	11.78364	0.207299	0.637856	0.316226E − 002	1.24101
	2	1.97593	0.593298	0.311703		
	3	1.29224	0.819447	0.078667		
7	1	3.45793	0.399761	0.483182	0.165115E − 001	1.12543
	2	1.44837	0.750659	0.191107	0.353804	
	3	1.15057	0.896718	0.045000		
8	1	11.70976	0.208315	0.640111	0.316227E − 002	1.06683
	2	1.94346	0.600989	0.321208		
	3	1.22440	0.854532	0.111494		
	4	1.07981	0.942055	0.025247		
9	1	3.44799	0.400656	0.484610	0.164979E − 001	1.03606
	2	1.43554	0.755869	0.196528	0.353804	
	3	1.11717	0.917494	0.063236		
	4	1.84294	0.967841	0.014012		

					Passband ripple = 0.5 dB; stopband gain = −60 dB	
N	i	A_{0i}	B_{0i}	B_{1i}	H_0/s_0	Ω_r
2	1	53.50591	0.056674	0.275431	0.999962E−003	26.76230
3	1	7.54288	0.201322	0.258673	0.709340E−002 0.307274	5.67937
4	1	17.20286	0.143036	0.529149	0.999995E−003	2.68325
	2	3.11024	0.396229	0.197979		
5	1	4.56247	0.304405	0.433499	0.607279E−002 0.307274	1.77664
	2	1.93110	0.581652	0.137383		
	1	15.26368	0.157911	0.565141	0.999998E−003	1.40138
6	2	2.38314	0.493638	0.315921		
	3	1.47091	0.727066	0.089646		
	1	4.30526	0.319374	0.457276	0.596437E−002	1.21984
7	2	1.66031	0.659035	0.212758	0.307274	
	3	1.25499	0.828773	0.056241		
	1	15.04762	0.159784	0.569572	0.999987E−003	1.12427
8	2	2.30879	0.506752	0.331430		
	3	1.34713	0.781898	0.136095		
	4	1.14323	0.895182	0.034429		
	1	4.27529	0.321225	0.460193	0.595147E−002	1.07151
9	2	1.63040	0.669021	0.222400	0.307274	
	3	1.19187	0.864916	0.084292		
	4	1.08212	0.936789	0.020762		

					Passband ripple = 0.5 dB; stopband gain = −70 dB	
N	i	A_{0i}	B_{0i}	B_{1i}	H_0/s_0	Ω_r
2	1	95.15055	0.031870	0.206642	0.316202E−003	47.58053
3	1	11.04815	0.137677	0.215816	0.272333E−002 0.271702	8.30124
4	1	23.20338	0.105540	0.457272	0.316228E−003	3.52135
	2	4.10055	0.301978	0.178887		
5	1	5.80087	0.238928	0.395497	0.219218E−002 0.271702	2.16377
	2	2.36605	0.478023	0.134253		
	1	19.37406	0.123292	0.504132	0.316225E−003	1.61413
6	2	2.88118	0.408976	0.307897		
	3	1.70453	0.632060	0.094687		
	1	5.29170	0.258622	0.427326	0.212501E−002	1.34764
7	2	1.91963	0.572328	0.222883	0.271702	
	3	1.39386	0.751192	0.064034		
	1	18.86032	0.126183	0.511481	0.316226E−003	1.20471
8	2	2.73400	0.427892	0.329879		
	3	1.49984	0.705734	0.153367		
	4	1.23028	0.836406	0.042104		
	1	5.21875	0.261736	0.432290	0.211499E−002	1.12338
9	2	1.85998	0.588119	0.237504	0.271702	
	3	1.28762	0.804333	0.101979		
	4	1.13819	0.894399	0.027178		

(continued overleaf)

TABLE 3.6   (continued)

(b) Passband ripple: 1 dB; stopband gains: -20, -30, -40, -50, -60, -70 dB

Passband ripple $= 1$ dB; stopband gain $= -20$ dB

N	i	A_{0i}	B_{0i}	B_{1i}	H_0/s_0	Ω_r
2	1	4.42342	0.497233	0.676727	0.100185E+000	2.32474
3	1	1.58565	0.790229	0.282927	0.281080E+000 0.565168	1.30797
4	1	3.81475	0.536633	0.768217	0.100185E+000	1.09029
	2	1.15956	0.926578	0.099029		
5	1	1.51852	0.808049	0.318242	0.279829E+000	1.02826
	2	1.04886	0.975703	0.032771	0.565168	
	1	3.80873	0.537071	0.769217	0.100184E+000	1.00902
6	2	1.14367	0.933194	0.110761		
	3	1.01550	0.992101	0.010654		
	1	1.51782	0.808242	0.318627	0.279814E+000	1.00290
7	2	1.04421	0.977941	0.036573	0.565168	
	3	1.00497	0.997446	0.003444		
	1	3.80866	0.537076	0.769228	0.100185E+000	1.00093
8	2	1.14350	0.933266	0.110888		
	3	1.01405	0.992834	0.011881		
	4	1.00160	0.999176	0.001111		
	1	1.51812	0.808099	0.318714	0.279611E+000	1.00030
9	2	1.04421	0.977936	0.036644	0.565168	
	3	1.00451	0.997680	0.003845		
	4	1.00052	0.999734	0.000359		

Passband ripple $= 1$ dB; stopband gain $= -30$ dB

N	i	A_{0i}	B_{0i}	B_{1i}	H_0/s_0	Ω_r
2	1	7.88158	0.279699	0.538632	0.316284E−001	4.00423
3	1	2.20293	0.586596	0.311981	0.113223E+000 0.430700	1.73254
4	1	5.72672	0.348111	0.682880	0.316286E−001	1.25040
	2	1.37628	0.803477	0.148347		
5	1	1.96934	0.633529	0.383990	0.111180E+000	1.09554
	2	1.13918	0.914405	0.064612	0.430700	
	1	5.67743	0.350184	0.687075	0.316278E-001	1.03799
6	2	1.31626	0.828467	0.179733		
	3	1.05468	0.964084	0.027111		
	1	1.96322	0.634884	0.386055	0.111123E+000	1.01536
7	2	1.11870	0.925872	0.077672	0.430700	
	3	1.02200	0.985162	0.011201		
	1	5.67614	0.350237	0.687189	0.316265E−001	1.00625
8	2	1.31463	0.829171	0.180623		
	3	1.04690	0.969003	0.032478		
	4	1.00893	0.993908	0.004598		
	1	1.96311	0.634904	0.386117	0.111109E+000	1.00255
9	2	1.11814	0.926187	0.078043	0.430700	
	3	1.01891	0.987211	0.013399		
	4	1.00364	0.997505	0.001883		

			Passband ripple = 1 dB; stopband gain = − 40 dB			
N	i	A_{0i}	B_{0i}	B_{1i}	H_0/s_0	Ω_r
2	1	14.01843	0.157290	0.411245	0.100001E − 001	7.04488
3	1	3.14896	0.416088	0.292413	0.445207E − 001 0.349732	2.41619
4	1	8.20047	0.238719	0.591849	0.100002E − 001	1.51549
	2	1.70946	0.658897	0.171046		
5	1	2.55417	0.489991	0.396962	0.425507E − 001 0.349732	1.21868
	2	1.28995	0.819650	0.090443		
6	1	7.98807	0.243668	0.601988	0.100001E − 001	1.09887
	2	1.55441	0.709739	0.225936		
	3	1.12899	0.909504	0.045382		
7	1	2.52698	0.494109	0.402679	0.424523E − 001 0.349732	1.04600
	2	1.23338	0.848902	0.117614		
	3	1.05957	0.955761	0.022185		
8	1	7.97731	0.243925	0.602513	0.999982E − 002	1.02168
	2	1.54681	0.712467	0.228870		
	3	1.10516	0.924805	0.058531		
	4	1.02799	0.978647	0.010708		
9	1	2.52561	0.494317	0.402976	0.424450E − 001 0.349732	1.01029
	2	1.23053	0.850439	0.119046		
	3	1.04885	0.963393	0.028495		
	4	1.01325	0.989756	0.005137		

			Passband ripple = 1 dB; stopband gain = − 50 dB			
N	i	A_{0i}	B_{0i}	B_{1i}	H_0/s_0	Ω_r
2	1	24.92930	0.088451	0.310124	0.316222E − 002	12.48471
3	1	4.56481	0.288843	0.255657	0.172697E − 001 0.295023	3.46061
4	1	11.40195	0.169815	0.510902	0.316227E − 002	1.90819
	2	2.18639	0.520870	0.173001		
5	1	3.29186	0.379847	0.382947	0.158547E − 001 0.295023	1.40723
	2	1.51047	0.707746	0.105563		
6	1	10.76686	0.178188	0.529085	0.316228E − 002	1.19891
	2	1.86027	0.596487	0.249239		
	3	1.24463	0.832197	0.060612		
7	1	3.20886	0.388081	0.394060	0.157487E − 001 0.295023	1.10127
	2	1.38863	0.759531	0.148534		
	3	1.12337	0.906935	0.033616		
8	1	10.71789	0.178872	0.530555	0.316226E − 002	1.05264
	2	1.83653	0.602956	0.255692		
	3	1.19050	0.863861	0.084091		
	4	1.06383	0.949367	0.018289		
9	1	3.20232	0.388746	0.394956	0.157403E − 001 0.295023	1.02767
	2	1.37933	0.763830	0.152090		
	3	1.09715	0.925100	0.046265		
	4	1.03346	0.972738	0.009847		

(continued overleaf)

TABLE 3.6 (continued)

N	i	A_{0i}	B_{0i}	B_{1i}	H_0/s_0	Ω_r
			Passband ripple = 1 dB; stopband gain = −60 dB			
2	1	44.33121	0.049740	0.232972	0.999985E−003	22.17688
3	1	6.66138	0.198506	0.216668	0.665490E−002 0.255373	5.02121
4	1	.15.57508	0.123525	0.441175	0.999992E−003	2.46079
	2	2.84628	0.402671	0.163210		
5	1	4.21402	0.296236	0.356928	0.580460E−002	1.67161
	2	1.81233	0.594376	0.110961	0.255373	
	1	14.06238	0.135110	0.468386	0.999998E−003	1.34354
6	2	2.24081	0.496567	0.255540		
	3	1.40684	0.741381	0.070756		
	1	4.01386	0.309159	0.374430	0.572054E−002	1.18547
7	2	1.58609	0.668120	0.168476	0.255373	
	3	1.21724	0.841598	0.043338		
	1	13.90612	0.136441	0.471452	0.999997E−003	1.10308
8	2	2.18302	0.508082	0.266640		
	3	1.30382	0.792531	0.105321		
	4	1.12001	0.905337	0.025900		
	1	3.99239	0.310618	0.376394	0.571125E−002	1.05822
9	2	1.56295	0.676792	0.175193	0.255373	
	3	1.16520	0.874493	0.063713		
	4	1.06755	0.944252	0.015252		

N	i	A_{0i}	B_{0i}	B_{1i}	H_0/s_0	Ω_r
			Passband ripple = 1 dB; stopband gain = −70 dB			
2	1	78.83301	0.027971	0.174802	0.316225E−003	39.42285
3	1	9.75117	0.135787	0.181032	0.255651E−002 0.225201	7.33052
4	1	21.05659	0.091036	0.381317	0.316225E−003	3.21902
	2	3.74387	0.307249	0.148233		
5	1	5.36546	0.232326	0.326592	0.210113E−002	2.02567
	2	2.21128	0.489620	0.109539	0.225201	
	1	17.95466	0.105129	0.417498	0.316226E−003	1.53842
6	2	2.70749	0.411512	0.250860		
	3	1.62167	0.646546	0.075892		
	1	4.95302	0.249732	0.350692	0.204637E−002	1.30200
7	2	1.82935	0.581019	0.178653	0.225201	
	3	1.34448	0.765478	0.050359		
	1	17.56029	0.107259	0.422786	0.316225E−003	1.17576
8	2	2.58820	0.428608	0.267153		
	3	1.44651	0.716853	0.120739		
	4	1.19910	0.848748	0.032478		
	1	4.89742	0.252292	0.354195	0.203877E−002	1.10451
9	2	1.78108	0.595224	0.189259	0.225201	
	3	1.25395	0.815222	0.078793		
	4	1.11791	0.904223	0.020566		

(c) Passband ripple: 2 dB; stopband gains: -20, -30, -40, -50, -60, -70 dB

Passband ripple $= 2$ dB; stopband gain $= -20$ dB

N	i	A_{0i}	B_{0i}	B_{1i}	H_0/s_0	Ω_r
2	1	3.60961	0.454891	0.537326	0.100103E + 000	1.94332
3	1	1.42939	0.793180	0.204089	0.254443E + 000 0.458898	1.20808
4	1	3.25882	0.486218	0.597266	0.100102E + 000	1.05569
	2	1.10765	0.935564	0.063585		
5	1	1.39116	0.807316	0.223995	0.253878E + 000 0.458898	1.01567
	2	1.02976	0.981070	0.018680		
6	1	3.25657	0.486437	0.597679	0.100102E + 000	1.00447
	2	1.09913	0.940286	0.069417		
	3	1.00845	0.994532	0.005396		
7	1	1.39096	0.807384	0.224146	0.253839E + 000 0.458898	1.00128
	2	1.02749	0.982483	0.020362		
	3	1.00242	0.998428	0.001551		
8	1	3.25736	0.486314	0.597662	0.100040E + 000	1.00037
	2	1.09913	0.940275	0.069488		
	3	1.00782	0.994938	0.005883		
	4	1.00069	0.999548	0.000446		
9	1	1.39157	0.807065	0.224311	0.253438E + 000 0.458898	1.00011
	2	1.02755	0.982437	0.020419		
	3	1.00224	0.998540	0.001696		
	4	1.00020	0.999870	0.000129		

Passband ripple $= 2$ dB; stopband gain $= -30$ dB

N	i	A_{0i}	B_{0i}	B_{1i}	H_0/s_0	Ω_r
2	1	6.42917	0.255975	0.433953	0.316259E − 001	3.29235
3	1	1.95290	0.593773	0.238474	0.104183E + 000 0.345928	1.55690
4	1	4.99348	0.312154	0.537291	0.316258E − 001	1.18280
	2	1.28752	0.820026	0.105666		
5	1	1.79533	0.634366	0.285641	0.102947E + 000 0.345928	1.06594
	2	1.10083	0.927270	0.042702		
6	1	4.96765	0.313430	0.539562	0.316259E − 001	1.02460
	2	1.24776	0.840500	0.124605		
	3	1.03723	0.971673	0.016632		
7	1	1.79218	0.635240	0.286656	0.102920E + 000 0.345928	1.00929
	2	1.08781	0.935985	0.050009		
	3	1.01401	0.989124	0.006386		
8	1	4.96715	0.313454	0.539608	0.316253E − 001	1.00353
	2	1.24694	0.840933	0.125007		
	3	1.03254	0.975137	0.019424		
	4	1.00531	0.995847	0.002438		
9	1	1.79226	0.635210	0.286689	0.102890E + 000 0.345928	1.00134
	2	1.08756	0.936150	0.050173		
	3	1.01227	0.990460	0.007452		
	4	1.00202	0.998417	0.000930		

(continued overleaf)

TABLE 3.6 (continued)

		Passband ripple = 2 dB; stopband gain = −40 dB				
N	i	A_{0i}	B_{0i}	B_{1i}	H_0/s_0	Ω_r
2	1	11.43468	0.143955	0.332767	0.100001E−001	5.76107
3	1	2.76901	0.423046	0.229015	0.412982E−001	2.13923
					0.278675	
4	1	7.25202	0.212344	0.467290	0.100001E−001	1.40842
	2	1.57676	0.677934	0.127954		
5	1	2.33100	0.490174	0.302683	0.399132E−001	1.16811
	2	1.22913	0.838222	0.064276	0.278675	
	1	7.11859	0.215746	0.473545	0.100000E−001	1.07316
6	2	1.46261	0.722961	0.164483		
	3	1.09825	0.922931	0.030620		
	1	2.31409	0.493238	0.305999	0.398550E−001	1.03262
7	2	1.18820	0.862873	0.081414	0.278675	
	3	1.04352	0.964199	0.014224		
	1	7.11297	0.215892	0.473812	0.999999E−002	1.01470
8	2	1.45794	0.724945	0.166087		
	3	1.08158	0.935151	0.038502		
	4	1.01955	0.983564	0.006530		
	1	2.31340	0.493363	0.306141	0.398506E−001	1.00665
9	2	1.18649	0.863937	0.082158	0.278675	
	3	1.03626	0.969979	0.017824		
	4	1.00884	0.992494	0.002982		

		Passband ripple = 2 dB; stopband gain = −50 dB				
N	i	A_{0i}	B_{0i}	B_{1i}	H_0/s_0	Ω_r
2	1	20.33444	0.080952	0.251283	0.316225E−002	10.19181
3	1	3.99860	0.294311	0.202406	0.160814E−001	3.04137
					0.233674	
4	1	10.17228	0.150255	0.403833	0.316226E−002	1.75285
	2	1.99900	0.538766	0.132731		
5	1	3.01161	0.379339	0.295684	0.149963E−001	1.33243
	2	1.42380	0.728758	0.078074	0.233674	
	1	9.72751	0.156399	0.415895	0.316226E−002	1.15868
6	2	1.74395	0.608865	0.186369		
	3	1.19866	0.850107	0.043146		
	1	2.95385	0.385982	0.302761	0.149248E−001	1.07859
7	2	1.32899	0.775193	0.107118	0.233674	
	3	1.09753	0.919969	0.023037		
	1	9.69746	0.156833	0.416742	0.316217E−002	1.03963
8	2	1.72753	0.614050	0.190306		
	3	1.15715	0.877392	0.058421		
	4	1.04897	0.958048	0.012076		
	1	2.94989	0.386446	0.303257	0.149195E−001	1.02017
9	2	1.32263	0.778532	0.109203	0.233674	
	3	1.07785	0.935006	0.030969		
	4	1.02487	0.978220	0.006269		

		Passband ripple = 2 dB; stopband gain = −60 dB				
N	i	A_{0i}	B_{0i}	B_{1i}	H_0/s_0	Ω_r
2	1	36.16031	0.045523	0.188850	0.999994E − 003	18.09398
3	1	5.82458	0.202473	0.172384	0.620820E − 002 0.201327	4.39729
4	1	13.96818	0.108944	0.348857	0.999992E − 003	2.24440
	2	2.58877	0.417854	0.126933		
5	1	3.86405	0.295378	0.277502	0.551997E − 002	1.56860
	2	1.69531	0.615100	0.083998	0.201327	
	1	12.83274	0.117802	0.367723	0.999990E − 003	1.28693
6	2	2.09697	0.507386	0.194121		
	3	1.34376	0.761617	0.052028		
	1	3.71433	0.306340	0.289299	0.545771E − 002	1.15215
7	2	1.51120	0.683872	0.124582	0.201327	
	3	1.18037	0.858228	0.030942		
	1	12.72577	0.118715	0.369634	0.999991E − 003	1.08281
8	2	2.05385	0.517202	0.201368		
	3	1.26045	0.807913	0.075699		
	4	1.09763	0.917695	0.017963		
	1	3.69977	0.307452	0.290490	0.545154E − 002	1.04572
9	2	1.49403	0.691114	0.128833	0.201327	
	3	1.13880	0.887078	0.044501		
	4	1.05373	0.952885	0.010283		

		Passband ripple = 2 dB; stopband gain = −70 dB				
N	i	A_{0i}	B_{0i}	B_{1i}	H_0/s_0	Ω_r
2	1	64.30347	0.025599	0.141715	0.316223E − 003	32.15951
3	1	8.51907	0.138568	0.144359	0.238690E − 002 0.176885	6.40894
4	1	18.94147	0.080138	0.301571	0.316227E − 003	2.92363
	2	3.39482	0.319442	0.116149		
5	1	4.92838	0.231379	0.254932	0.200503E − 002	1.88906
	2	2.05780	0.508450	0.084092	0.176885	
	1	16.50253	0.091214	0.327385	0.316224E − 003	1.46330
6	2	2.53159	0.420680	0.192438		
	3	1.53918	0.667139	0.056962		
	1	4.60436	0.246641	0.271782	0.196188E − 002	1.25686
7	2	1.73771	0.595918	0.134255	0.176885	
	3	1.29542	0.784230	0.036926		
	1	16.21208	0.092748	0.330857	0.316227E − 003	1.14735
8	2	2.43787	0.435853	0.203560		
	3	1.39254	0.732913	0.088741		
	4	1.16836	0.864042	0.023267		
	1	4.56376	0.248701	0.274035	0.195631E − 002	1.08620
9	2	1.69990	0.608362	0.141317	0.176885	
	3	1.22011	0.829611	0.056613		
	4	1.09811	0.915850	0.014401		

(continued overleaf)

TABLE 3.6 (continued)

(d) Passband ripple: 3 dB; stopband gains: −20, −30, −40, −50, −60, −70 dB

Passband ripple = 3 dB; stopband gain = −20 dB

N	i	A_{0i}	B_{0i}	B_{1i}	H_0/s_0	Ω_r
2	1	3.16206	0.446651	0.451221	0.999995E-001	1.73915
3	1	1.34231	0.807114	0.157389	0.237355E+000 0.394899	1.15516
4	1	2.92756	0.472487	0.493658	0.999995E−001	1.03853
	2	1.08010	0.945318	0.044619		
5	1	1.31717	0.818329	0.170012	0.237048E+000 0.394899	1.00996
	2	1.02041	0.985364	0.011943		
6	1	2.92670	0.472578	0.493859	0.999824E−001	1.00260
	2	1.07480	0.948718	0.047988		
	3	1.00531	0.996140	0.003149		
7	1	1.31710	0.818355	0.170085	0.237014E+000 0.394899	1.00068
	2	1.01909	0.986290	0.012829		
	3	1.00139	0.998987	0.000827		
8	1	2.92702	0.472526	0.493855	0.999560E−001 1.00018	
	2	1.07480	0.948717	0.048015		
	3	1.00497	0.996385	0.003382		
	4	1.00036	0.999734	0.000217		
9	1	1.31754	0.818103	0.170205	0.236721E+000 0.394899	1.00005
	2	1.01913	0.986257	0.012860		
	3	1.00131	0.999048	0.000891		
	4	1.00010	0.999930	0.000058		

Passband ripple = 3 dB; stopband gain = −30 dB

N	i	A_{0i}	B_{0i}	B_{1i}	H_0/s_0	Ω_r
2	1	5.62939	0.251455	0.369762	0.316227E−001	2.90352
3	1	1.80966	0.609296	0.193943	0.984438E−001 0.294542	1.45814
4	1	4.55072	0.300629	0.449156	0.316227E−001	1.14542
	2	1.23703	0.836431	0.081201		
5	1	1.69070	0.645119	0.227931	0.975896E−001 0.294542	1.05020
	2	1.07973	0.937581	0.030987		
6	1	4.53463	0.301525	0.450566	0.316224E−001	1.01783
	2	1.20754	0.853567	0.094050		
	3	1.02807	0.977008	0.011414		
7	1	1.68878	0.645734	0.228523	0.975702E−001 0.294542	1.00640
	2	1.07039	0.944451	0.035677		
	3	1.01004	0.991640	0.004150		
8	1	4.53448	0.301532	0.450586	0.316200E−001	1.00231
	2	1.20705	0.853853	0.094273		
	3	1.02485	0.979584	0.013112		
	4	1.00361	0.996975	0.001502		
9	1	1.68889	0.645692	0.228545	0.975407E−001 0.294542	1.00083
	2	1.07025	0.944549	0.035766		
	3	1.00890	0.992580	0.004765		
	4	1.00130	0.998906	0.000543		

			Passband ripple = 3 dB; stopband gain = − 40 dB			
N	i	A_{0i}	B_{0i}	B_{1i}	H_0/s_0	Ω_r
2	1	10.01180	0.141419	0.284766	$0.999991E-002$	5.05584
3	1	2.54917	0.436001	0.190366	$0.392774E-001$ 0.235639	1.98022
4	1	6.67811	0.203053	0.391813	$0.999996E-002$	1.34663
	2	1.49919	0.696465	0.102519		
5	1	2.19519	0.498382	0.246744	$0.382019E-001$	1.13934
	2	1.19395	0.853367	0.049526	0.235639	
	1	6.58231	0.205696	0.396137	$0.999996E-002$	1.05892
6	2	1.40720	0.737064	0.129340		
	3	1.08088	0.932780	0.022704		
	1	2.18313	0.500844	0.248944	$0.381626E-001$	1.02545
7	2	1.16144	0.874707	0.061632	0.235639	
	3	1.03473	0.969917	0.010161		
	1	6.57883	0.205793	0.396297	$0.999973E-002$	1.01109
8	2	1.40391	0.738612	0.130362		
	3	1.06788	0.942934	0.028071		
	4	1.01510	0.986681	0.004499		
	1	2.18274	0.500923	0.249027	$0.381575E-001$	1.00485
9	2	1.16026	0.875502	0.062090	0.235639	
	3	1.02925	0.974534	0.012526		
	4	1.00660	0.994131	0.001982		

			Passband ripple = 3 dB; stopband gain = − 50 dB			
N	i	A_{0i}	B_{0i}	B_{1i}	H_0/s_0	Ω_r
2	1	17.80401	0.079527	0.215323	$0.316224E-002$	8.93009
3	1	3.66965	0.303970	0.169877	$0.153424E-001$ 0.196605	2.79862
4	1	9.42948	0.143008	0.338892	$0.316227E-002$	1.66148
	2	1.88811	0.556103	0.108630		
5	1	2.84042	0.385226	0.243575	$0.144422E-001$	1.28850
	2	1.37246	0.746126	0.062138	0.196605	
	1	9.08300	0.148050	0.347729	$0.316225E-002$	1.13534
6	2	1.67291	0.621998	0.149783		
	3	1.17170	0.863608	0.033383		
	1	2.79564	0.390934	0.248624	$0.143880E-001$	1.06567
7	2	1.29287	0.788544	0.083790	0.196605	
	3	1.08263	0.929154	0.017340		
	1	9.06166	0.148372	0.348291	$0.316224E-002$	1.03239
8	2	1.66023	0.626392	0.152510		
	3	1.13726	0.887785	0.044465		
	4	1.04059	0.963840	0.008851		
	1	2.79284	0.391297	0.248944	$0.143845E-001'$	1.01611
9	2	1.28801	0.791298	0.085194	0.196605	
	3	1.06658	0.942090	0.022947		
	4	1.02015	0.981709	0.004477		

(continued overleaf)

TABLE 3.6 (continued)

		Passband ripple = 3 dB; stopband gain = −60 dB				
N	i	A_{0i}	B_{0i}	B_{1i}	H_0/s_0	Ω_r
2	1	31.66062	0.044721	0.161891	0.999982E − 003	15.84610
3	1	5.33757	0.209329	0.145315	0.593160E − 002 0.168753	4.03471
4	1	12.99985	0.103391	0.292827	0.999993E − 003	2.11596
	2	2.43545	0.432543	0.105070		
5	1	3.65012	0.299596	0.229883	0.533723E − 002 0.168753	1.50711
	2	1.62514	0.632389	0.068114		
6	1	12.06873	0.110910	0.307139	0.999998E − 003	1.25325
	2	2.00866	0.518843	0.157996		
	3	1.30601	0.777153	0.041293		
7	1	3.52768	0.309409	0.238672	0.528678E − 002 0.168753	1.13252
	2	1.46533	0.697339	0.099388		
	3	1.15851	0.870254	0.024042		
8	1	11.98665	0.111628	0.308485	0.100000E − 002	1.07105
	2	1.97347	0.527579	0.163278		
	3	1.23409	0.819865	0.059153		
	4	1.08453	0.926218	0.013672		
9	1	3.51659	0.310331	0.239495	0.528205E − 002 0.168753	1.03859
	2	1.45140	0.703663	0.102420		
	3	1.12295	0.896257	0.034067		
	4	1.04578	0.958601	0.007671		

		Passband ripple = 3 dB; stopband gain = −70 dB				
N	i	A_{0i}	B_{0i}	B_{1i}	H_0/s_0	Ω_r
2	1	56.30123	0.025149	0.121502	0.316224E − 003	28.15950
3	1	7.80139	0.143329	0.121935	0.228214E − 002 0.147835	5.87251
4	1	17.66937	0.075925	0.253151	0.316227E − 003	2.74749
	2	3.18636	0.331232	0.096742		
5	1	4.66128	0.234458	0.211860	0.194364E − 002 0.147835	1.80680
	2	1.96513	0.524255	0.068958		
6	1	15.60039	0.085536	0.273180	0.316226E − 003	1.41799
	2	2.42333	0.430381	0.157837		
	3	1.48926	0.683143	0.045939		
7	1	4.38669	0.248487	0.224767	0.190706E − 002 0.147835	1.22975
	2	1.68126	0.608671	0.108451		
	3	1.26583	0.798034	0.029282		
8	1	15.36540	0.086788	0.275721	0.316225E − 003	1.13043
	2	2.34394	0.444323	0.166219		
	3	1.35940	0.745475	0.070539		
	4	1.14995	0.874828	0.018148		
9	1	4.35404	0.250271	0.226394	0.190259E − 002 0.147835	1.07541
	2	1.64930	0.619957	0.113684		
	3	1.19948	0.840239	0.044277		
	4	1.08638	0.923761	0.011054		

EXAMPLE 3.7

Find the transfer function $H(s)$ for a normalized elliptic filter that will satisfy the following specifications: $G_1 = -0.5$ dB, $G_2 = -30$ dB, and $\Omega_r = 1.21$.

Solution. From Table 3.6(a) with passband ripple of -0.5 dB and a stopband gain of -30 dB the smallest value of n that satisfies the Ω_r requirement is $n = 5$, giving an Ω_r of 1.12912. Using the rows for n equals 5 the coefficients A_{0i}, B_{1i}, B_{0i} (for $i = 1, 2, \ldots, 5$), S_0, and H_0 are found, and upon substitution into Eq. (3.32) gives the transfer function $H_5(s)$ as

$$H_5(s) = \frac{0.118807(s^2 + 2.14490)(s^2 + 1.18132)}{(s + 0.511761)(s^2 + 0.480774s + 0.648724)(s^2 + 0.088080s + 0.907216)}$$

The resulting Ω_1 and Ω_2 are easily calculated from Eq. (3.34) as 0.941087 and 1.06260. A plot of 20 log $|H_5(j\Omega)|$, given in Fig. 3.14, shows that the G_1 and G_2 requirements are exactly satisfied while the Ω_r requirement is exceeded. The apparent discontinuity in the plot results because a different vertical scale was used in the passband and stopband to more clearly show the ripple structure.

The design of an unnormalized elliptic low-pass filter satisfying a G_1 dB ripple, cutoff at Ω_1', and a G_2 dB gain at critical stopband frequency Ω_2' can be

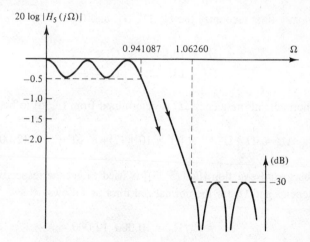

Figure 3.14 Frequency response for the elliptic filter designed in Example 3.7.

obtained by applying a LP → LP transformation of a suitable normalized elliptic filter as follows:

$$H(s) = H_{LP}(s)\big|_{s \to s/\Omega_0} \tag{3.35}$$

For the elliptic filter, Ω_0 is selected to be the geometric mean of Ω_1' and Ω_2' written as follows:

$$\Omega_0 = (\Omega_1' \Omega_2')^{1/2} \tag{3.36}$$

The corresponding low-pass filter requirements Ω_1, Ω_2, and Ω_r are obtained as follows:

$$\Omega_1 = \Omega_1'/\Omega_0 \tag{3.37a}$$
$$\Omega_2 = \Omega_2'/\Omega_0 \tag{3.37b}$$
$$\Omega_r = \Omega_2/\Omega_1 = \Omega_2'/\Omega_1' \tag{3.37c}$$

This procedure is illustrated in the following example.

EXAMPLE 3.8

Find the transfer function for an elliptic low-pass filter with -2-dB cutoff value at 10,000 rad/sec and a stopband attenuation of 40 dB for all Ω past 14,400 rad/sec.

Solution. First recognize the Ω_1', Ω_2', G_1, and G_2 as

$$\Omega_1' = 10{,}000, \qquad G_1 = -2$$
$$\Omega_2' = 14{,}400, \qquad G_2 = -40$$

The normalizing frequency Ω_0 is determined from (3.36) to be

$$\Omega_0 = (\Omega_2' \cdot \Omega_1')^{1/2} = [(1 \times 10^4)(1.44 \times 10^4)]^{1/2} = 12{,}000$$

The backward equation [Eq. (3.37)] is used to get the respective critical frequencies and Ω_r for the normalized filter as follows

$$\Omega_1 = \Omega_1'/\Omega_0 = 10{,}000/12{,}000 = \tfrac{5}{6}$$
$$\Omega_2 = \Omega_2'/\Omega_0 = 14{,}400/12{,}000 = \tfrac{6}{5}$$
$$\Omega_r = \Omega_2/\Omega_1 = (\tfrac{6}{5})/(\tfrac{5}{6}) = 1.44$$

From the 2-dB and stopband gain of -40 dB part of Table 3.6(c) it is seen that an $n = 4$ gives an $\Omega_r = 1.40842$, thus satisfying the Ω_r requirement of 1.44. From the $n = 4$ rows and Eq. (3.33) the normalized low-pass elliptic filter becomes

$$H_{LP}(s) = \frac{0.01(s^2 + 7.25202)(s^2 + 1.57676)}{(s^2 + 0.467290s + 0.212344)(s^2 + 0.127954s + 0.677934)}$$

The required elliptic filter with transfer function $H(s)$ is finally obtained from Eq. (3.35), using $\Omega_0 = 12000$ as follows

$$H(s) = H_{LP}(s)|_{s \to s/12000}$$

$$= \frac{0.01\left[\left(\frac{s}{12000}\right)^2 + 7.25202\right]\left[\left(\frac{s}{12000}\right)^2 + 1.57676\right]}{\left[\left(\frac{s}{12000}\right)^2 + 0.467290\frac{s}{12000} + 0.212344\right]}$$

$$\times \frac{1}{\left[\left(\frac{s}{12000}\right)^2 + \frac{0.127954s}{12000} + 0.677930\right]}$$

Simplifying this expression gives the desired transfer function as

$$H(s) = \frac{0.01(s^2 + 1.04429 \times 10^9)(s^2 + 2.27053 \times 10^8)}{(s^2 + 5607.48s + 30{,}577{,}536)(s^2 + 1535.448s + 97{,}622{,}496)}$$

3.4 GENERAL FILTER FORMS

From Eqs. (3.20) and (3.29) it is seen that the magnitude squared frequency response, $|H(j\Omega)|^2$, for both Chebyshev and elliptic low-pass filters are of the form

$$|H(j\Omega)|^2 = \frac{1}{1 + \epsilon^2 F^2(\Omega)}$$

The $F(\Omega)$ is a Chebyshev polynomial for a Chebyshev filter and a Chebyshev rational function for the elliptic filter. If ϵ is set to 1 and $F(\Omega)$ is selected as powers of Ω the Butterworth filter is described. The $F(\Omega)$'s are shown in

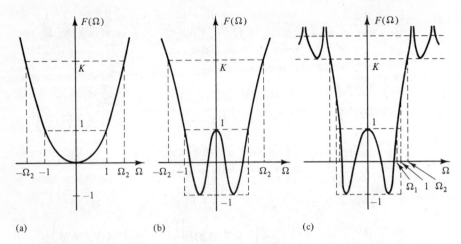

Figure 3.15 $F(\Omega)$ for (a) Butterworth, (b) Chebyshev, and (c) elliptic filters.

Fig. 3.15 for the normalized low-pass Butterworth, Chebyshev, and elliptic filters.

What made these $F(\Omega)$'s good for low-pass design? First of all $|F(\Omega)|$ < 1 for all $|\Omega|$ < some Ω_1. For both the Butterworth and Chebyshev filter Ω_1 was 1. Second $|F(\Omega)|$ is greater than or equal to some large value K for all $|\Omega|$ greater than Ω_2. These selections in turn make

$$\frac{1}{1 + \epsilon^2} < |H(j\Omega)|^2 \leq 1 \qquad \text{for } |\Omega| < \Omega_1$$

$$|H(j\Omega)|^2 < \frac{1}{1 + \epsilon^2 K} \qquad \text{for } |\Omega| > \Omega_2$$

Therefore it seems reasonable that $F(\Omega)$'s that satisfy these requirements give a normalized low-pass filter. By selecting $F(\Omega)$'s from a family of functions a family of low-pass filters similar to those for Butterworth, Chebyshev, and elliptic filters could be defined. Although they are not presented here, other families of filters generated in this fashion are described in the literature. For further information consult Antoniou [1].

3.5 SUMMARY

In this chapter design formulas, procedures, and tables have been presented for the normalized Butterworth, Chebyshev, and elliptic low-pass filters. Other types of filters, including unnormalized low-pass, high-pass, bandpass, and bandstop filters, were in turn designed by reflecting the filter requirements to a normalized filter and then applying the proper type of transformation (LP→LP, LP→HP, LP→BP, LP→BS) to such a designed low-pass filter. The design procedures involved simple tabular look up for both the Butterworth and Chebyshev filters once the order of the filter was calculated from a given set of formulas. For the design of elliptic filters a tabular look up was also possible but with some loss of flexibility since the table contained only prescribed values of passband ripple, and stopband gain.

In general, to satisfy the same set of specifications, the order of the resulting elliptic filter was smaller than or equal to that for a Chebyshev filter, which was smaller than or equal to that for a Butterworth filter. The analog filters designed in this chapter can serve as a catalog or base from which digital filters can be designed by a transformational procedure. This procedure is fully described in the following chapter.

REFERENCES

1. Antoniou, Andreas. *Digital Filters: Analysis and Design*. McGraw-Hill, New York, 1979.
2. Cadzow, James A. *Discrete-time Systems*. Prentice-Hall, Englewood Cliffs, NJ, 1973.
3. Christian, E., and E. Eisenmann. *Filter Design Tables and Graphs*. Wiley, New York, 1966.
4. Daniels, R. W. *Approximation Methods for Electronics Filter Design*. McGraw-Hill, New York, 1974.
5. Guillemin, E. A. *Synthesis of Passive Networks*. Wiley, New York, 1957.
6. Johnson, D. E. *Introduction to Filter Theory*. Prentice-Hall, Englewood Cliffs, NJ, 1976.
7. Johnson, D. E., J. R. Johnson, and M. D. Kashefi. "Ultraspherical Rational Filters," *IEEE Transactions on Circuit Theory* CT-20 (September 1973), 569–599.
8. Kuo, Franklin F. *Network Analysis and Synthesis*. Wiley, New York, 1966.
9. Storer, J. E., *Passive Network Synthesis*. McGraw-Hill, New York, 1962.
10. Weinberg, Louis, *Network Analysis and Synthesis*. McGraw-Hill, New York, 1962.

PROBLEMS

3.1 The system function $H_5(s)$ represents a 1 rad/sec fifth-order normalized Butterworth filter.

(a) Give $H_5(s)$ in both the polynomial and quadratic factored forms.

(b) What is the gain $|H_5(j\Omega)|$ at $\Omega = 1$ rad/sec? What is the gain in decibels?

(c) Repeat (a) and (b) for a Chebyshev type I filter with $\epsilon = 0.7647831$.

3.2 Obtain the transfer function $H_9(s)$ for a normalized Butterworth filter of order 9 in the quadratic factored form.

3.3 Verify the entry in the Chebyshev Table 3.4 for 1 dB ripple and $n = 5$.

3.4 Given that $G(s) = 1/(s^2 + \sqrt{2}s + 1)$ represents a normalized second-order low-pass Butterworth filter,

   (a) Plot 20 log $|G(j\Omega)|$ for Ω from 0 to 100. At what radian frequency Ω is the magnitude down 3 dB? down 20 dB?

   (b) Apply the LP to HP transformation, $s \to 10/s$, to the $G(s)$ to obtain a new filter $H(s)$ and plot 20 log $|H(j\Omega)|$ for Ω from 0 to 100. Does the new filter perform as expected? At what radian frequency is the filter magnitude down 3 dB? down 20 dB?

   (c) The transformation $s \to 5s/(s^2 + 50)$ is applied to the $G(s)$ given, resulting in a new filter $H(s)$. What type of filter results and what are the critical frequencies? Plot 20 log $|H(j\Omega)|$ to verify your conclusions.

3.5 Design both (a) a Butterworth and (b) a Chebyshev analog low-pass filter that have a -3-dB cutoff frequency of 100 rad/sec and a stopband attenuation of 25 dB or greater for all radian frequencies past 250 rad/sec. Plot 20 log $|H_a(j\Omega)|$ for your filters and show that you satisfy the requirements at the critical frequencies.

3.6 What is the order n of an analog Chebyshev low-pass filter that has a pass band from 0 to 200 Hz with acceptable ripple of 1 dB and a monotonic stop band that is down at least 40 dB at and beyond 250 Hz? Repeat for a Butterworth filter and compare the n's. What conclusion can you make?

3.7 We wish to design an analog low-pass filter which will have a -1-dB cutoff frequency at 75 Hz and have greater than 20 dB of attenuation for all frequencies greater than 150 Hz. Find $H_a(s)$ that will satisfy those requirements and plot 20 log $|H_a(j\Omega)|$ and arg $H(j\Omega)$ for

   (a) Butterworth (maximally flat) filter.

   (b) Chebyshev type I (equiripple passband) filter.

   (c) Elliptic filter.

   Compare the orders of the filters and comment on your results.

3.8 Design an analog bandpass filter to satisfy the following specifications: (S1) A -3-dB upper and lower cutoff frequency of 100 Hz and 3.8 kHz; (S2) stopband attenuation of 20 dB at 20 Hz and 8 kHz; (S3) no ripple within both stopband and passband. Check your design by plotting 20 log $|H_a(j\Omega)|$ and arg $H_a(j\Omega)$.

3.9 Repeat Problem 3.8 for -3 dB upper and lower cutoff of 50 Hz and 20 kHz with a stopband attenuation of 20 dB at 10 Hz and 60 kHz.

3.10 Design (a) a Butterworth and (b) a Chebyshev analog high-pass filter that will pass all signals of radian frequencies greater than 200 rad/sec with no more than 2 dB of attenuation and have a stopband attenuation of greater than 20 dB for all Ω less than 100 rad/sec.

Chapter 4

Digital Filter Design

4.0 DISCRETE-TIME FILTERS

A discrete-time filter takes a discrete-time input sequence $x(n)$ and produces a discrete-time output sequence $y(n)$. A special class of linear discrete-time shift-invariant system can be characterized by a unit sample response $h(n)$, a system function $H(z)$, or a difference equation realization. If such a system has the following difference equation representation:

$$\sum_{k=0}^{N} a_k y(n-k) = \sum_{k=0}^{M} b_k x(n-k) \qquad (4.1)$$

then, as has been shown earlier, the corresponding system function, $H(z)$, is given by

$$H(z) = \frac{\displaystyle\sum_{k=0}^{M} b_k z^{-k}}{\displaystyle\sum_{k=0}^{N} a_k z^{-k}} \qquad (4.2)$$

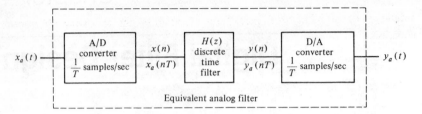

Figure 4.1 Simulation of an analog filter.

If the system is BIBO stable, its frequency response is obained by letting $z = e^{j\omega}$ as follows

$$H(e^{j\omega}) = \frac{\sum_{k=0}^{M} b_k e^{-j\omega k}}{\sum_{k=0}^{N} a_k e^{-j\omega k}} \tag{4.3}$$

A filter may be required to have a given frequency response, or a specific response to an impulse, step, or ramp, or simulate a continuous analog system. To simulate an analog filter the discrete-time filter is used in the analog-to-digital–$H(z)$–digital-to-analog structure shown in Fig. 4.1. The analog input signal, $x_a(t)$, is converted to a sequence $x(n)$ by the analog-to-digital converter with a specified sample rate $1/T$, T being the time between samples. This sequence $x(n)$ is then passed through the digital filter $H(z)$ giving an output sequence $y(n)$ which is changed to the continuous output $y_a(t)$ by a digital-to-analog converter.

The analog-to-digital (A/D) converter can be thought of roughly as a sampler and coder while the digital-to-analog (D/A) converter in many cases represents a decoder and holder followed by a low-pass filter. The next few sections will look at several different techiques for designing $H(z)$, including (a) the numerical solution to a differential equation, (b) the bilinear transformation method, (c) the digital-to-digital transformation approach, and (d) the impulse invariant approach.

4.1 DESIGN BY USING NUMERICAL SOLUTIONS OF DIFFERENTIAL EQUATIONS

We shall begin with the assumption that we would like to design a digital filter $H(z)$ which, when used in the A/D–$H(z)$–D/A structure shown in Fig. 4.1, simulates a continuous-time linear filter specified by the following differential equation:

$$\sum_{k=0}^{N} c_k \frac{d^k y_a(t)}{dt^k} = \sum_{k=0}^{M} d_k \frac{d^k x_a(t)}{dt^k} \tag{4.4}$$

This filter has input $x_a(t)$ and output $y_a(t)$ and can be characterized by its system function $H_a(s)$, which is easily seen, by taking the Laplace transform of (4.4) and ignoring initial conditions, to be

$$H_a(s) = \frac{\displaystyle\sum_{k=0}^{M} d_k s^k}{\displaystyle\sum_{k=0}^{N} c_k s^k} \tag{4.5}$$

Suppose that we approximate the derivatives by backward differences. The first backward difference $\nabla^{(1)}[\cdot]$ is defined by

$$\nabla^{(1)}[y(n)] = [y(n) - y(n-1)]/T \tag{4.6}$$

Higher-order backward differences are found by applying the first backward difference repeatedly, as follows:

$$\nabla^{(k)}[y(n)] = \nabla^{(1)}[\nabla^{(k-1)}[y(n)]] \tag{4.7}$$

Using the kth-order differences as approximations to the derivatives in Eq. (4.4) we have

$$\sum_{k=0}^{N} c_k \nabla^{(k)}[y_a(nT)] \cong \sum_{k=0}^{M} d_k \nabla^{(k)}[x_a(nT)] \tag{4.8}$$

The above equation represents a numerical approach for obtaining $y_a(nT)$, the sampled version of $y_a(t)$, which is the solution of the original differential equation for a given input signal $x_a(t)$. The $\mathcal{Z}$ transform of the first- and kth-order differences are given below:

$$\begin{aligned}
\mathcal{Z}(\nabla^{(1)}[y(n)]) &= \mathcal{Z}([y(n) - y(n-1)]/T) = Y(z)(1 - z^{-1})/T \\
\mathcal{Z}(\nabla^{(k)}[y(n)]) &= Y(z)[(1 - z^{-1})/T]^k
\end{aligned} \tag{4.9}$$

Letting $x(n)$ and $y(n)$ represent $x_a(nT)$ and $y_a(nT)$, respectively, the $\mathcal{Z}$ transform of both sides of Eq. (4.8), assuming equality, gives by linearity of the $\mathcal{Z}$ transform the following

$$\sum_{k=0}^{N} c_k \mathcal{Z}(\nabla^{(k)}[y(n)]) = \sum_{k=0}^{M} d_k \mathcal{Z}(\nabla^{(k)}[x(n)]) \qquad (4.10)$$

Using Eq. (4.9) we can write Eq. (4.10) as

$$\sum_{k=0}^{N} c_k[(1 - z^{-1})/T]^k Y(z) = \sum_{k=0}^{M} d_k[(1 - z^{-1})/T]^k X(z) \qquad (4.11)$$

From Eq. (4.11) the system function representing the numerical solution of the differential equation is easily seen to be

$$H(z) = \frac{Y(z)}{X(z)} = \frac{\displaystyle\sum_{k=0}^{M} d_k[(1 - z^{-1})/T]^k}{\displaystyle\sum_{k=0}^{N} c_k[(1 - z^{-1})/T]^k} \qquad (4.12)$$

Examining Eq. (4.5) we see that $H(z)$ can be obtained by replacing s in $H_a(s)$ by $(1 - z^{-1})/T$, that is,

$$H(z) = H_a(s)\big|_{s \to (1 - z^{-1})/T} \qquad (4.13)$$

Therefore, replacing derivatives by backward differences corresponds to mappings from the s-plane to the z-plane and z-plane to the s-plane given by

$$s = \frac{1 - z^{-1}}{T}, \qquad z = \frac{1}{1 - sT} \qquad (4.14)$$

As the frequency response for the analog system is obtained by letting $s = j\Omega$, it is of interest to look at the image in the z-plane of the $j\Omega$ axis of the s-plane, which from (4.14) is

$$z = \frac{1}{1 - j\Omega T} = \frac{1}{1 + \Omega^2 T^2} + j\frac{\Omega T}{1 + \Omega^2 T^2} = x + jy \qquad (4.15)$$

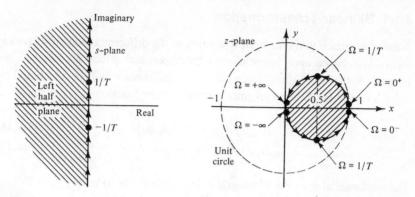

Figure 4.2 The image in the z-plane of the $j\Omega$ axis of the s-plane for the mapping $z = 1/(1 - ST)$.

It is easy to see that x, the real part of z, and y, the imaginary part of z, are related by

$$x^2 + y^2 = x \tag{4.16}$$

Completing the square in Eq. (4.16) gives the following equation:

$$(x - \tfrac{1}{2})^2 + y^2 = \tfrac{1}{4} \tag{4.17}$$

Thus, the image in the z-plane of the $j\Omega$ axis of the s-plane is a circle of radius $\tfrac{1}{2}$, as shown in Fig. 4.2.

It can be shown that the left half of the s-plane transforms to the area inside the circle given by Eq. (4.17); therefore the transformation $s = (1 - z^{-1})/T$ will take poles in the left half s-plane inside the unit circle in the z-plane. This means that an application of the transformation to a stable analog filter $H_a(s)$ gives a stable digital filter—certainly a desirable property. However, since the frequency response of the digital filter is obtained by evaluating $H(z)$ on the unit circle, $z = e^{j\omega}$, the shape of the equivalent frequency response of the A/D–$H(z)$–D/A structure would not be similar to that of $H_a(s)$. To preserve the shape of the frequency response we would like to have the transformation from analog filter to digital filter take the $j\Omega$ axis of the s-plane into the unit circle in the z-plane. The following section will give such a transformation.

4.1.1 Bilinear Transformation

Another approach to the numerical solution of the differential equation given in Eq. (4.4) is based upon application of the trapezoidal rule to approximate the integral. Rather than show a general proof, the method will be illustrated by using a first-order differential equation and transfer function $H_a(s)$ given by

$$a_1 y_a'(t) + a_0 y_a(t) = b_0 x(t) \tag{4.18}$$

$$H_a(s) = b_0/(a_1 s + a_0) \tag{4.19}$$

The fundamental theorem of integral calculus allows us to write

$$y_a(t) = \int_{t_0}^{t} y_a'(t)\, dt + y_a(t_0) \tag{4.20}$$

Since Eq. (4.20) holds for any t and any t_0, we let $t = nT$ and $t_0 = (n - 1)T$ to get

$$y_a(nT) = \int_{(n-1)T}^{nT} y_a'(t)\, dt + y_a[(n - 1)T] \tag{4.21}$$

Using the trapezoidal rule to approximate the integral and assuming equality a recursive relationship for determining $y_a(nT)$ can be found from Eq. (4.21) as follows:

$$y_a(nT) = y_a[(n-1)T] + (T/2)\{y_a'(nT) + y_a'[(n - 1)T]\} \tag{4.22}$$

The derivatives in (4.22) can be evaluated from (4.18), giving the following difference equation for the approximate numerical solution of the differential equation:

$$y_a(nT) = y_a[(n - 1)T] + T/2 \left\{ \frac{-a_0}{a_1} y_a(nT) + \frac{b_0}{a_1} x(nT) \right.$$
$$\left. - \frac{a_0}{a_1} y_a[(n - 1)T] + \frac{b_0}{a_1} x[(n - 1)T] \right\} \tag{4.23}$$

Taking the $\mathcal{Z}$ transform of (4.23), with $X(z)$ and $Y(z)$ the $\mathcal{Z}$ transforms of $x_a(nT)$ and $y_a(nT)$, respectively, and rearranging to get $Y(z)$ over $X(z)$, the system function can be shown to be

$$H(z) = \frac{Y(z)}{X(z)} = \frac{b_0}{a_1 \dfrac{2}{T}\left(\dfrac{1 - z^{-1}}{1 + z^{-1}}\right) + a_0} \tag{4.24}$$

By comparing Eqs. (4.19) and (4.24) we see that $H(z)$ can be obtained by replacing s in $H_a(s)$ by $(2/T)(1 - z^{-1})/(1 + z^{-1})$, as follows:

$$H(z) = H_a(s)\big|_{s \to [(2/T)(1 - z^{-1})]/(1 + z^{-1})]} \tag{4.25}$$

By using a state variable representation of the general differential equation given in Eq. (4.4) and the basic approach just discussed, it can be shown that the transformation derived above for a first-order differential equation holds in the general case. This bilinear transformation is characterized by the following:

$$s = \frac{2(1 - z^{-1})}{T(1 + z^{-1})}, \qquad z = \frac{1 + sT/2}{1 - sT/2} \tag{4.26}$$

The image in the z-plane of the $j\Omega$ axis from the s-plane is shown in Fig. 4.3. It can be shown in a fashion similar to the previous section that the bilinear transformation given in Eq. (4.26) has the following properties.

(1) The entire $j\Omega$ axis of the s-plane goes into the unit circle of the z-plane.

(2) The left half side of the s-plane is transformed inside the unit circle.

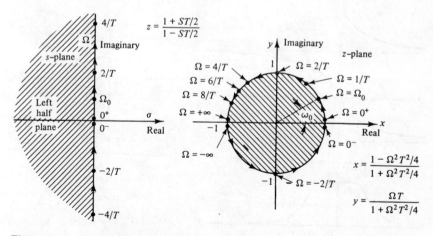

Figure 4.3 The image in the z-plane of the $j\Omega$ axis of the s-plane for the mapping $z = [1 + ST/2]/[1 - ST/2]$.

Therefore a stable analog filter would be transformed into a stable digital filter. While the frequency response of the analog filter and digital filter have the same amplitudes, there is a nonlinear relationship between corresponding digital and analog frequencies. Since this relationship is important for the design of digital filters using the bilinear transformation, it will be investigated in more detail.

Letting $z = e^{j\omega}$ and $s = j\Omega$ in the first part of Eq. (4.26) gives the following relationship:

$$j\Omega = \frac{2(1 - e^{-j\omega})}{T(1 + e^{-j\omega})} \qquad (4.27)$$

Dividing through by j and recognizing the tangent function on the right-hand side of Eq. (4.27), we have

$$\Omega = \frac{2}{T}\tan(\omega/2) \qquad (4.28)$$

The inverse relationship is found from (4.28) to be

$$\omega = 2\tan^{-1}(\Omega T/2) \qquad (4.29)$$

These relationships are shown in Fig. 4.4. We see that if the bilinear transformation is applied to an $H_a(s)$ with critical frequency Ω_c, the digital filter will

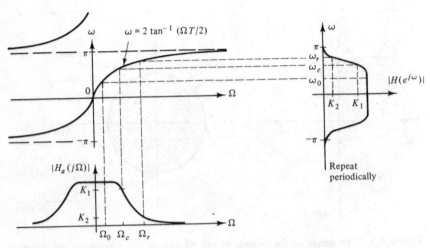

Figure 4.4 Relationship between analog and digital frequencies when applying the bilinear transformation.

have a critical frequency ω_c given by $\omega_c = 2 \tan^{-1}(\Omega_c T/2)$. If the resulting $H(z)$ is used in an A/D–$H(z)$–D/A structure, the equivalent critical frequency becomes

$$\Omega_{c\,eq} = \frac{2}{T} \tan^{-1}(\Omega_c T/2)$$

which will give Ω_c only if $\Omega_c \cdot T/2$ is so small that $\tan^{-1}(\Omega T/2)$ is approximately equal to $\Omega T/2$. This warping of the critical frequency will be compensated for in the design procedure using the bilinear transformation by prewarping.

4.1.2 Design of Digital Filters Using the Bilinear Transformation

The design of a digital filter to satisfy a set of digital specifications is required. The specifications for a digital filter usually take the form of a set of critical frequencies $\{\omega_1, \omega_2, \ldots, \omega_N\}$ and a corresponding set of magnitude requirements $\{K_1, K_2, \ldots, K_N\}$. When an analog filter is used as the prototype for the bilinear transformation method we have seen that the relationship between digital and analog frequencies is nonlinear and governed by Eqs. (4.28) and (4.29). Therefore, to get the proper digital frequency, we must design an analog filter with analog critical frequencies Ω_i': $i = 1, 2, \ldots, N$ given by

$$\Omega_i' = \frac{2}{T} \tan(\omega_i/2), \qquad i = 1, 2, \ldots, N \tag{4.30}$$

This operation will be referred to as prewarping. The analog magnitude requirements are not changed and remain the same as the corresponding digital requirements. An analog filter $H_a(s)$ is then designed to satisfy the prewarped specifications given by $\Omega_1', \Omega_2', \ldots, \Omega_n'$ and $K_1, K_2, \ldots, K_N$ and the bilinear transform is applied to that $H_a(s)$, i.e.,

$$H(z) = H_a(s)\big|_{s = (2/T)[(1-z^{-1})/(1+z^{-1})]}$$

As the T in the Ω_i' and the T in the bilinear transform cancel in the procedure described above for low-pass filter design, it is convenient to just use T equal to 1 in both places. This is easily seen since if the Ω_i' comes from an analog-to-analog transformation of a unit radian frequency, we have $s \rightarrow s/\Omega_i'$, and when the bilinear transformation $s \rightarrow 2(1 - z^{-1})/[T(1 - z^{-1})]$ is used, the cascade of transformations is given by

$$s \rightarrow \frac{2}{T} \frac{(1 - z^{-1})}{(1 + z^{-1})\Omega_i'} = \frac{2(1 - z^{-1})}{T(1 + z^{-1})(2/T)\tan(\omega_i/2)} = \frac{(1 - z^{-1})}{(1 + z^{-1})\tan(\omega_i/2)} \quad (4.31)$$

Therefore, the transformation given in Eq. (4.31) does not have a T in the expression.

The procedure for the design of a digital filter using the bilinear transformation is summarized in Fig. 4.5 and consists of (Step 1) prewarping the digital specifications, (Step 2) designing an analog filter to meet the prewarped specifications, and (Step 3) applying the bilinear transformation. T was arbitrarily set to 1 for the purpose of illustration but can be set equal to any value, since, as we have seen, it cancels in the design.

The following example illustrates the procedure just described for the design of a low-pass digital filter using the bilinear transformation method.

EXAMPLE 4.1

Design and realize a digital low-pass filter using the bilinear transformation method to satisfy the following characteristics: (a) monotonic stopband and passband; (b) -3.01-dB cutoff frequency of 0.5π rad; (c) magnitude down at least 15 dB at 0.75π rad. The required frequency response is shown in Fig. 4.6.

Solution. The design procedure is that of using the bilinear transformation on an analog prototype and consists of the following three steps:

Step 1. Prewarp the critical digital frequencies $\omega_1 = 0.5\pi$ and $\omega_2 = 0.75\pi$ using $T = 1$ sec to get

$$\Omega_1' = \frac{2}{T}\tan\frac{\omega_1}{2} = 2\tan(0.5\pi/2) = 2.000$$

$$\Omega_2' = \frac{2}{T}\tan\frac{\omega_2}{2} = 2\tan(0.75\pi/2) = 4.8282$$

Step 2. Design an analog low-pass filter with critical frequencies Ω_1' and Ω_2' that satisfy

$$0 \geq 20\log|H_a(j\Omega_1')| \geq -3.01 \text{ dB} = K_1$$

and

$$20\log|H_a(j\Omega_2')| \leq -15 \text{ dB} = K_2$$

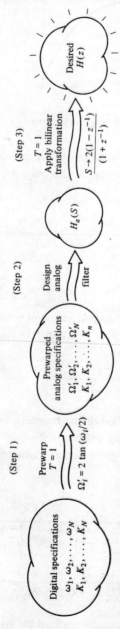

Figure 4.5 Procedure for the design of a digital filter using the bilinear transformation method.

(Step 1)

Prewarp
$T = 1$

$\Omega_i' = 2 \tan(\omega_i/2)$

Digital specifications
$\omega_1, \omega_2, \ldots, \omega_N$
$K_1, K_2, \ldots, K_N$

Prewarped analog specifications
$\Omega_1', \Omega_2', \ldots, \Omega_N', K_n$
$K_1, K_2, \ldots, K_n$

(Step 2)

Design
analog
filter

$H_a(S)$

(Step 3)

$T = 1$
Apply bilinear
transformation

$$S \to \frac{2(1 - z^{-1})}{(1 + z^{-1})}$$

Desired
$H(z)$

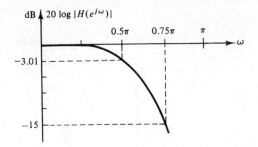

Figure 4.6 Required frequency response for Example 4.1.

A Butterworth filter is used to satisfy the monotonic property and has an order n and critical frequency Ω_c determined by Eqs. (3.16) and (3.17) as follows:

$$n = \left\lceil \frac{\log_{10}[(10^{+3.01/10} - 1)/(10^{15/10} - 1)]}{2\log_{10}[2/4.8282]} \right\rceil = \lceil 1.9412 \rceil = 2$$

$$\Omega_c = \frac{2.000}{(10^{3.01/10} - 1)^{1/4}} = 2$$

Therefore the required prewarped analog filter using the Butterworth Table 3.1b and the low-pass to low-pass transformation from Table 3.2 is

$$H_a(s) = \frac{1}{s^2 + \sqrt{2}s + 1} \Bigg|_{s \to s/2} = \frac{4}{s^2 + 2\sqrt{2}s + 4}$$

Step 3. Applying the bilinear transformation ($T = 1$) to $H_a(s)$ will take the prewarped analog filter to a digital filter with system function $H(z)$ that will satisfy the given digital requirements:

$$H(z) = H_a(s)\big|_{s \to [2(1 - z^{-1})/(1 + z^{-1})]}$$

$$= \frac{4}{\left[\dfrac{2(1 - z^{-1})}{(1 + z^{-1})}\right]^2 + 2\sqrt{2}\left[\dfrac{2(1 - z^{-1})}{(1 + z^{-1})}\right] + 4}$$

$$= \frac{1 + 2z^{-1} + z^{-2}}{3.4142135 + 0.5857865z^{-2}}$$

This digital filter can be realized by specification of a difference equation obtained from the transfer function $H(z)$ given by

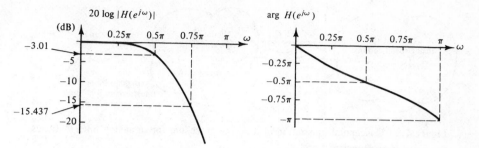

Figure 4.7 Frequency response for digital filter designed in Example 4.1.

$$H(z) = \frac{Y(z)}{X(z)} = \frac{1 + 2z^{-1} + z^{-2}}{3.4142135 + 0.5857865z^{-2}}$$

Cross multiplying gives

$$Y(z)[3.4142135 + 0.5857865z^{-2}] = X(z)[1 + 2z^{-1} + z^{-2}]$$

and taking the inverse $\mathcal{Z}$ transform we find

$$3.4142135y(n) + 0.5857865y(n - 2) = x(n) + 2x(n - 1) + x(n - 2)$$

By rearranging and scaling, $y(n)$ can be realized by the following difference equation:

$$y(n) = 0.2928932[x(n) + 2x(n - 1) + x(n - 2)] - 0.1715729y(n - 2)$$

As a check of the filter design the frequency response of the designed filter with system function $H(z)$ should be obtained. The frequency response is obtained by plotting $20 \log |H(e^{j\omega})|$ versus ω. Figure 4.7 shows that the frequency response for the $H(z)$ given in Step 3 satisfies the -3.01-dB requirement at 0.5π rad and exceeds the -15-dB requirement at 0.75π. Finding the frequency response for a designed filter is a moment of truth and should be a standard step in any digital filter design.

4.2 ANALOG DESIGN USING DIGITAL FILTERS

In many cases we are required to simulate an analog filter using the A/D–$H(z)$–D/A structure given in Fig. 4.1. We will usually be given a set of analog requirements

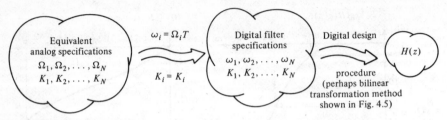

Figure 4.8   Conceptual illustration of a design procedure for an analog filter that uses the A/D–$H(z)$–D/A structure.

with critical frequencies $\Omega_1, \Omega_2, \ldots, \Omega_N$ and corresponding decibel frequency response magnitudes $K_1, K_2, \ldots, K_N$. The sampling rate $1/T$ of the A/D converter will be specified or can be determined from the input signals under consideration. The general approach for the design is to first convert the analog requirements to digital requirements and then design the digital filter using the technique discussed in the previous section. This procedure is conceptually illustrated in Fig. 4.8. The conversion of the analog specifications to digital specifications is through the formula $\omega_i = \Omega_i T$. To show that this is true, suppose the input to the equivalent analog filter is

$$x_a(t) = \sin \Omega_i t$$

The output of the A/D converter with sampling rate $1/T$ becomes

$$x(n) = x_a(nT) = \sin(\Omega_i T \cdot n) = \sin \omega_i n$$

Thus, the magnitude of the discrete-time sinusoid signal is the same as the continuous time sinusoid, while the digital frequency ω_k is given in terms of the analog frequency Ω_i by

$$\omega_i = \Omega_i T \tag{4.32}$$

Therefore, the specifications for the digital filter become $\omega_1, \omega_2, \ldots, \omega_N$ with corresponding frequency response magnitudes $K_1, K_2, \ldots, K_N$. To complete the procedure we design a digital filter to satisfy these requirements. The following example illustrates the procedure.

EXAMPLE 4.2

Design a digital filter $H(z)$ that when used in an A/D–$H(z)$–D/A structure gives an equivalent low-pass analog filter with (a) -3.01-dB cutoff frequency of 500 Hz, (b) monotonic stop and passband, (c) magnitude of frequency response down at least 15 dB at 750 Hz, (d) sample rate of 2000 samples/sec.

Solution. The analog specifications become

$$\Omega_1 = 2\pi f_1 = 2\pi \cdot 500 = \pi \cdot 10^3 \, \text{rad/sec}, \qquad K_1 = -3.01 \, \text{dB}$$
$$\Omega_2 = 2\pi f_2 = 2\pi \cdot 750 = 1.5 \cdot \pi \cdot 10^3 \, \text{rad/sec}, \qquad K_2 = -15 \, \text{dB}$$

and the corresponding digital specifications using Eq. (4.32) become

$$\omega_1 = \Omega_1 T = \pi \cdot 10^3 \cdot (1/2000) = 0.5\pi \, \text{rad}, \qquad K_1 = -3.01 \, \text{dB}$$
$$\omega_2 = \Omega_2 T = 1.5\pi \cdot 10^3 \cdot (1/2000) = 0.75\pi \, \text{rad} \qquad K_2 = -15 \, \text{dB}$$

The next step in the procedure is to design a digital filter given by $H(z)$ that satisfies these digital specifications. In Example 4.1, a digital filter was designed with these same specifications. (What a stroke of luck!) Therefore, $H(z)$, the system function for the digital filter in the A/D–$H(z)$–D/A structure is

$$H(z) = (1 + 2z^{-1} + z^{-2})/(3.4142135 + 0.5857865z^{-2})$$

4.3 DESIGN OF DIGITAL FILTERS USING DIGITAL-TO-DIGITAL TRANSFORMATIONS

We have seen in Chapter 3 that one method for the design of analog filters relied on applying a transformation to an analog low-pass filter with a unit bandwidth. The transformations given in Table 3.2 replace the s in the low-pass prototype with some function of s, say $f(s)$. It was shown that we could obtain low-pass, high-pass, bandpass, and bandstop filters by selecting the proper $f(s)$. The low-pass unit bandwidth analog Chebyshev and Butterworth filters of various orders n and ripple ϵ served as families of prototypes to use with the proper transformation.

Similarly, a set of transformations can be found that take a low-pass digital filter and turn it into a high-pass, bandpass, bandstop, or another low-pass digital filter. These transformations are given in Table 4.1. The transformations all take the form of replacing the z^{-1} in $H(z)$ by $g(z^{-1})$, some function of z^{-1}. The conceptual diagram for the digital filter design by using a digital-to-digital transformation is shown in Fig. 4.9. Starting with a set of digital specifications and

TABLE 4.1 DIGITAL-TO-DIGITAL TRANSFORMATIONS

Type		Transformation	Design formulas
From	To		
Low pass → Low pass		$z^{-1} \rightarrow \dfrac{z^{-1} - \alpha}{1 - \alpha z^{-1}}$	$\alpha = \dfrac{\sin\left[(\theta_p - \omega_p)/2\right]}{\sin\left[(\theta_p + \omega_p)/2\right]}$
Low pass → High pass		$z^{-1} \rightarrow -\dfrac{z^{-1} + \alpha}{1 + \alpha z^{-1}}$	$\alpha = -\dfrac{\cos\left[(\omega_p + \theta_p)/2\right]}{\cos\left[(\omega_p - \theta_p)/2\right]}$
Low pass → Bandpass		$z^{-1} \rightarrow -\dfrac{z^{-2} - \dfrac{2\alpha k}{k+1}z^{-1} + \dfrac{k-1}{k+1}}{\dfrac{k-1}{k+1}z^{-2} - \dfrac{2\alpha k}{k+1}z^{-1} + 1}$	$\alpha = \dfrac{\cos\left[(\omega_1 + \omega_2)/2\right]}{\cos\left[(\omega_2 - \omega_1)/2\right]}$ $k = \cot\left[(\omega_2 - \omega_1)/2\right]\tan(\theta_p/2)$
Low pass → Bandstop		$z^{-1} \rightarrow \dfrac{z^{-2} - \dfrac{2\alpha}{1+k}z^{-1} + \dfrac{1-k}{1+k}}{\dfrac{1-k}{1+k}z^{-2} - \dfrac{2\alpha}{1+k}z^{-1} + 1}$	$\alpha = \dfrac{\cos\left[(\omega_2 + \omega_1)/2\right]}{\cos\left[(\omega_2 - \omega_1)/2\right]}$ $k = \tan\left[(\omega_2 - \omega_1)/2\right]\tan(\theta_p/2)$

(Each row of the Type column contains filter magnitude response sketches: $20\log|H(e^{j\omega})|$ vs ω with passband edge θ_p and level K_1 on the "From" side, and the corresponding transformed response on the "To" side with edges ω_p, or ω_1, ω_2, and level K_1.)

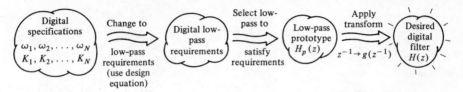

Figure 4.9 Conceptual diagram for the design of digital filters using digital-to-digital transformations.

using the inverse of the design equations given in Table 4.1, a set of low-pass digital requirements can be established. A low-pass digital prototype filter $H_p(z)$ is then selected to satisfy these requirements and the proper digital-to-digital transformation is applied to give the desired $H(z)$.

To make the digital-to-digital transformation procedure feasible for digital filter design, a catalog of low-pass digital filters of the Chebyshev and Butterworth type should be developed. The following two examples give two approaches for developing the Butterworth catalog, which is shown in Table 4.2 for reference.

EXAMPLE 4.3

We desire the design of a unit bandwidth 3-dB digital Butterworth filter of order one by using the conventional bilinear transformation procedure given in Fig. 4.8.

Solution.   (a) Prewarp the $\omega_1 = 1$ rad requirement to get

$$\Omega' = 2\tan(\omega/2) = 2\tan(\tfrac{1}{2}) = 1.092605$$

(b) Use the $n = 1$ analog Butterworth filter as a prototype applying a low-pass to low-pass transformation to get $H_p(s)$, that is,

$$H_p(s) = \frac{1}{s + 1}\bigg|_{s=(s/1.092605)} = \frac{1}{0.9152438s + 1}$$

(c) Go through the bilinear transformation

$$H(z) = H_p(s)\big|_{s\rightarrow[2(1-z^{-1})/1+z^{-1})]} = \frac{1}{0.9152438 \cdot 2(1 - z^{-1})/(1 + z^{-1}) + 1}$$

$$= \frac{1 + z^{-1}}{2.8305 - 0.83052z^{-1}}$$

TABLE 4.2 TABLE OF NORMALIZED BUTTERWORTH DIGITAL FILTERS OF ORDER 1–8

$$H_{B1}(z) = \frac{0.353296(1 + z^{-1})}{1 - 0.293408z^{-1}}$$

$$H_{B2}(z) = \frac{0.144106(1 + z^{-1})^2}{1 - 0.677496z^{-1} + 0.253921z^{-2}}$$

$$H_{B3}(z) = \frac{5.71568 \times 10^{-2}(1 + z^{-1})^3}{(1 - 0.760595z^{-1} + 0.407722z^{-2})(1 - 0.293408z^{-1})}$$

$$H_{B4}(z) = \frac{2.24832 \times 10^{-2}(1 + z^{-1})^4}{(1 - 0.817391z^{-1} + 0.512840z^{-2})(1 - 0.607963z^{-1} + 0.125228z^{-2})}$$

$$H_{B5}(z) = \frac{8.81325 \times 10^{-3}(1 + z^{-1})^5}{(1 - 0.857603z^{-1} + 0.587265z^{-2})(1 - 0.642925z^{-1} + 0.189935z^{-2})(1 - 0.293408z^{-1})}$$

$$H_{B6}(z) = \frac{3.44859 \times 10^{-3}(1 + z^{-1})^6}{(1 - 0.887350z^{-1} + 0.642321z^{-2})(1 - 0.677491z^{-1} + 0.253911z^{-2})(1 - 0.596097z^{-1} + 0.103267z^{-2})}$$

$$H_{B7}(z) = \frac{1.34804 \times 10^{-3}(1 + z^{-1})^7}{(1 - 0.910178z^{-1} + 0.684572z^{-2})(1 - 0.708757z^{-1} + 0.311778z^{-2})(1 - 0.614630z^{-1} + 0.137566z^{-2})(1 - 0.293408z^{-1})}$$

$$H_{B8}(z) = \frac{5.26606 \times 10^{-4}(1 + z^{-1})^8}{(1 - 0.928225z^{-1} + 0.717973z^{-2})(1 - 0.736365z^{-1} + 0.362876z^{-2})(1 - 0.635778z^{-1} + 0.176708z^{-2})(1 - 0.592014z^{-1} + 0.095709z^{-2})}$$

EXAMPLE 4.4

In this example a 3-dB unit bandwidth digital Butterworth filter of order one is designed by using a low-pass to low-pass digital-to-digital transformation of a low-pass digital filter obtained by using the bilinear transformation directly on a 3-dB unit bandwidth analog filter without prewarping.

(a) Apply the bilinear transformation directly to $H_a(s)$, the Butterworth filter of order one, to get $H_1(z)$:

$$H_1(z) = \frac{1}{s + 1}\Bigg|_{s \to [2(1 - z^{-1})/(1 + z^{-1})]} = \frac{1 + z^{-1}}{3 - z^{-1}}$$

(b) Because no prewarping was done, the critical frequency for the digital filter is warped by formula Eq. (4.29). Using $T = 1$ gives

$$\omega_c = 2\tan^{-1}(\Omega T/2) = 2\tan^{-1}(1 \cdot \tfrac{1}{2}) = 0.9272952$$

(c) From Table 4.1 we see that a digital frequency of critical frequency θ_p can be transformed to another digital filter with cutoff ω_p by replacing z^{-1} by $(z^{-1} - \alpha)/(1 - z^{-1}\alpha)$. Letting ω_p equal one and θ_p equal 0.9272952, α is determined as follows:

$$\alpha = \frac{\sin[(\theta_p - \omega_p)/2]}{\sin[(\theta_p + \omega_p)/2]} = \frac{\sin[0.9272952 - 1)/2]}{\sin[(0.9272952 + 1)/2]} = -0.04425$$

Therefore, the desired filter $H(z)$ is easily shown to be the same answer as in the previous example as follows:

$$H(z) = H_1(z)\big|_{z^{-1} \to (z^{-1} + 0.04425)/(1 + 0.04425z^{-1})}$$

$$= \frac{1 + (z^{-1} + 0.04425)/(1 + 0.04425z^{-1})}{3 - (z^{-1} + 0.04425)/(1 + 0.04425z^{-1})}$$

$$= \frac{1 + z^{-1}}{2.8305 - 0.8305z^{-1}}$$

Unit bandwidth low-pass digital Butterworth and Chebyshev filters of higher orders can be generated as above forming a family of transfer functions that can be used for digital filter design using the digital-to-digital transformation method. Table 4.2 is such a Butterworth table.

Determination of n for Digital Butterworth Filter   The design requirements for a digital low-pass filter are usually given in terms of a cutoff frequency ω_1 and stopband frequency ω_2 with their corresponding gains K_1 and K_2 as follows:

$$0 \geq 20 \log |H(e^{j\omega_1})| \geq K_1, \qquad 20 \log |H(e^{j\omega_2})| \leq K_2$$

Using these requirements it is necessary to determine the order of the digital unit bandwidth low-pass Butterworth prototype and corresponding critical frequency ω_p. The formula for the order of the digital filter can be obtained by using the prewarped digital frequencies Ω_1' and Ω_2' in the standard formula for the analog Butterworth filter, as follows:

$$n = \left\lceil \frac{\log_{10}[(10^{-K_1/10} - 1)/(10^{-K_2/10} - 1)]}{2 \log_{10}\{[\tan(\omega_1/2)]/[\tan(\omega_2/2)]\}} \right\rceil \qquad (4.33)$$

The corresponding value of ω_p to exactly satisfy the K_1 requirement must also be determined. Using the 1-rad analog Butterworth filter of order n, the value of Ω_p of frequency in rad/sec that satisfies the gain K_1 requirement is seen to be

$$\Omega_p = (10^{-K_1/10} - 1)^{-1/2n} 2 \tan(\omega_1/2) \qquad (4.34)$$

and the corresponding radian frequency in the digital domain after the bilinear transformation becomes

$$\omega_p = 2 \tan^{-1}[(10^{-K_1/10} - 1)^{-1/2n} \tan(\omega_1/2)] \qquad (4.35)$$

Therefore, the required digital low-pass filter is determined from $H_{Bn}(z)$, the digital Butterworth filter, as

$$H(z) = H_{Bn}(z)\big|_{z^{-1} \to (z^{-1} - \alpha)/(1 - \alpha z^{-1})}$$

where
$$\alpha = \frac{\sin[(\theta_p - \omega_1)/2]}{\sin[(\theta_p + \omega_1)/2]}$$

Table 4.2 gives $H_{Bn}(z)$ for values of n from 1 to 8. An example of the digital-to-digital transformation design procedure is now presented.

EXAMPLE 4.5

   Using the digital-to-digital transformation method and Table 4.2, find the system function $H(z)$ for a low-pass digital filter that satisfies the following set of requirements: (a) monotonic stop and passbands; (b) -3.0102 dB cutoff digital frequency of 0.5π; (c) attenuation at and past 0.75π is at least 15 dB (see Figure 4.6).

Solution. Because of the monotonic requirement, a Butterworth filter is selected. From (4.33) the required n is as follows:

$$n = \left\lceil \frac{\log_{10}[(10^{0.30102} - 1)/(10^{1.5} - 1)]}{2 \log_{10}\{[\tan(0.5\pi/2)]/\tan(0.75\pi/2)\}} \right\rceil = \lceil 1.9412 \rceil = 2$$

The ω_p is determined from (4.35) as

$$\omega_p = 2 \tan^{-1}[(10^{0.30102} - 1)^{-1/4} \tan(0.5\pi/2)] = 0.5\pi$$

The required low-pass filter is determined by performing a digital-to-digital transformation on the $H_{B2}(z)$ of Table 4.2. This low pass-to-low pass digital-to-digital transformation from Table 4.1 is

$$z^{-1} \to \frac{z^{-1} - \alpha}{1 - \alpha z^{-1}}$$

where α is determined using $\theta_p = 1$ and $\omega_p = 0.5\pi$ as

$$\alpha = \sin[(1 - 0.5\pi)/2]/\sin[(1 + 0.5\pi)/2] = -0.293401993$$

Therefore the required filter is

$$H(z) = H_{B2}(z)\big|_{z^{-1} \to (z^{-1} + 0.293401992)/(1 + 0.293401992\, z^{-1})}$$

Upon performing the transformation, $H(z)$ is found to be

$$H(z) = \frac{(1 + z^{-1})^2}{3.4142 + 0.5858\, z^{-2}}$$

which agrees with the transfer function determined in Example 4.1 by the bilinear transformation method.

4.4 IMPULSE INVARIANT DESIGN

Early in the development of digital filter design procedures a method based on preserving the response to an impulse was used. If $h_a(t)$ represents the response of an analog filter to a unit impulse $\delta(t)$, then the unit sample response of a discrete-time filter used in an A/D–$H(z)$–D/A structure is selected to be the sampled version of $h(t)$. Therefore the discrete-time filter is characterized by the system function $H(z)$ given by

$$H(z) = \mathcal{Z}[h(n)] = \mathcal{Z}[h_a(t)|_{t=nT}]$$

If we are given an analog filter with system function $H_a(s)$ the corresponding impulse invariant design filter has an $H(z)$ seen from above as

$$H(z) = \mathcal{Z}(\{\mathcal{L}^{-1}[H_a(s)]\}|_{t=nT}) \tag{4.36}$$

In Example 4.6 a simple single-pole analog filter is used for $H_a(s)$ to illustrate the use of (4.36).

EXAMPLE 4.6

Find the $H(z)$ corresponding to the impulse invariant design using a sample rate of $1/T$ samples/sec for an analog filter $H_a(s)$ specified as follows:

$$H_a(s) = A/(s + \alpha)$$

Solution. The analog system's impulse response is obtained by taking the inverse Laplace transform of $H_a(s)$ to give $h_a(t)$ as

$$h_a(t) = Ae^{-\alpha t}u(t)$$

The corresponding $h(n)$ is then given by

$$h(n) = Ae^{-\alpha nT}u(nT) = A(e^{-\alpha T})^n u(n)$$

and therefore the discrete-time filter has the following $\mathcal{Z}$ transform:

$$H(z) = \mathcal{Z}[h(n)] = \mathcal{Z}[A(e^{-\alpha T})^n u(n)]$$

$$= \frac{Az}{z - e^{-\alpha T}}$$

In many cases the transfer function $H_a(s)$ is given by a sum of N terms with unique α_k as follows:

$$H_a(s) = \sum_{k=1}^{N} \frac{A_k}{s + \alpha_k}$$

For this case the impulse invariant design $H(z)$ is given by

$$H(z) = \sum_{k=1}^{N} \frac{A_k z}{z - e^{-\alpha_k T}}$$

The above result is easily shown by using the linearity of the $\mathbb{Z}$ transform and the results of Example 4.6. If the $H_a(s)$ has repeated poles, then $H(z)$ must be found by using (4.36) directly.

The logical question at this point is: How does the equivalent frequency response of the A/D–$H(z)$–D/A structure using this $H(z)$ compare to the frequency response of the original system specified by $H_a(s)$? Using Example 4.6 for discussion purposes we have that the frequency response and the magnitude of the frequency response of the given analog filter are as follows:

$$H_a(j\Omega) = \frac{A}{j\Omega + \alpha}, \qquad |H_a(j\Omega)| = \frac{A}{(\Omega^2 + \alpha^2)^{1/2}}$$

To obtain the equivalent frequency response of the A/D–$H(z)$–D/A structure one must first find the frequency response of the discrete-time filter specified by $H(z)$. This can be obtained by replacing the z in $H(z)$ by $e^{j\omega}$ to give

$$H(e^{j\omega}) = \frac{A e^{j\omega}}{e^{j\omega} - e^{-\alpha T}}$$

The analog frequency response of the equivalent analog filter is then determined by replacing ω by ΩT to give

$$H_{eq}(j\Omega) = H(e^{j\omega})\big|_{\omega \to \Omega T} = \frac{A e^{j\Omega T}}{e^{j\Omega T} - e^{-\alpha T}}, \qquad \Omega < \pi/T$$

Multiplying numerator and denominator by $e^{-j\Omega T}$, writing $e^{-j\Omega T}$ as the sum of $\cos \Omega T$ and $-j \sin \Omega T$, and rearranging $H_{eq}(j\Omega)$ and $|H_{eq}(j\Omega)|$ are determined to be

$$H_{eq}(j\Omega) = \frac{A}{(1 - e^{-\alpha T} \cos \Omega T) + j e^{-\alpha T} \sin \Omega T}, \qquad \Omega < \pi/T$$

$$|H_{eq}(j\Omega)| = \frac{A}{(1 + e^{-\alpha 2 T} - 2 e^{-\alpha T} \cos \Omega T)^{1/2}}, \qquad \Omega < \pi/T$$

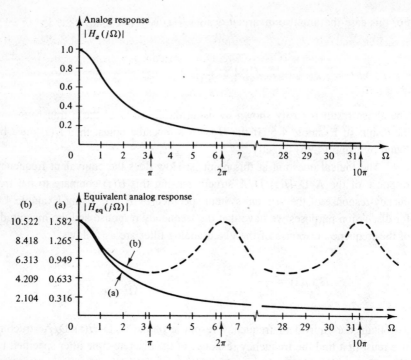

Figure 4.10 Plots of $H_a(j\Omega)$ and $H_{eq}(j\Omega)$ for impulse invariant design. Curve (a): $\alpha = 1$, $T = 0.1$, curve (b): $\alpha = 1$, $T = 1$.

The plots of $|H_{eq}(j\Omega)|$ and $|H_a(j\Omega)|$ are shown in Fig. 4.10 for two different cases. It is seen in Figure 4.10 curve (a) that the magnitude of the two frequency responses $|H_{eq}(j\Omega)|$ and $H_a(j\Omega)|$ are very close, while in Fig. 4.10 curve (b) that the magnitude plots are dramatically different. Therefore, good results using the impulse invariant design are obtained provided the time between samples is selected small enough. What is small enough may be difficult to assess when the $H_a(s)$ has several poles or repeated poles, and when it is found it may be so small that implementation may be costly. In general, the transformational methods described earlier allow designs with sample rates that are less than those required by the impulse invariant method and also allow flexibility with respect to selection of the sample rate size.

4.5 MINIMIZATION OF MEAN SQUARED ERROR IIR FILTER DESIGN

In the previous sections design procedures for digital filters were based on transformations of analog filter prototypes and digital transformations of digital filter prototypes. There are many digital filter design procedures that do not rely on analog methods or transformations. Since a digital filter is specified by its unit sample response, difference equation, or frequency response, it is possible to just adjust their parameters in some methodical fashion to satisfy some measure of closeness to some desired properties. An example of one of these methods is the minimization of mean squared error procedure for the design of an IIR filter as given in Oppenheim and Schafer [8].

Suppose we specify a set of digital frequencies $\{\omega_i : i = 1, 2, \ldots, M\}$ and a desired frequency response magnitude at those frequencies of $|H_d(e^{j\omega_i})|$ as

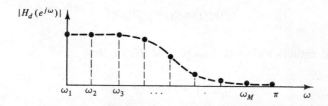

shown below. Assume we want to get as close as we can to matching the magnitudes at the ω_i's with a filter of the following form:

$$H(z) = A \cdot \prod_{k=1}^{K} \frac{1 + a_k z^{-1} + b_k z^{-2}}{1 + c_k z^{-1} + d_k z^{-2}} = AG(z) \tag{4.37}$$

Closeness is measured by the mean squared error E defined by

$$E = \sum_{i=1}^{M} [|H(e^{j\omega_i})| - |H_d(e^{j\omega_i})|]^2 \tag{4.38}$$

The problem is now one of selecting $A, a_1, a_2, \ldots, a_k, b_1, b_2, \ldots, b_k, c_1, c_2, \ldots, c_k$ and $d_1, d_2, \ldots, d_k$ such that E is minimized subject to constraints that all the poles of $H(z)$ are inside the unit circle to give a stable filter. By taking the partials of E with respect to the $4k + 1$ parameters, a set of nonlinear equations results which can be solved using the Fletcher–Powell method ignoring the constraints. If the resulting constants result in poles outside the unit circle, reciprocal poles are used, preserving the magnitude response.

4.6 FIR FILTER DESIGN

In the previous sections, digital filters were designed to give a desired frequency response magnitude without regard to the phase response. In many cases a linear phase characteristic is required throughout the passband of the filter to preserve the shape of a given signal within the passband. For example, assume a low-pass filter with frequency response $H(e^{j\omega})$ given by

$$H(e^{j\omega}) = \begin{cases} e^{-j\omega\alpha}, & |\omega| < \omega_0 \\ 0, & \omega_0 < |\omega| < \pi \\ \text{periodic} & \text{for all other } \omega \end{cases} \qquad (4.39)$$

If $X(e^{j\omega})$ represents the Fourier transform of an input sequence $x(n)$, then the transform $Y(e^{j\omega})$ of the output sequence $y(n)$, as seen in Eq. (1.27), is given by

$$Y(e^{j\omega}) = X(e^{j\omega}) \cdot H(e^{j\omega})$$

If $X(e^{j\omega})$ is entirely within the bandpass of $H(e^{j\omega})$, then

$$Y(e^{j\omega}) = X(e^{j\omega}) \cdot e^{-j\omega\alpha}$$

Therefore the output signal $y(n)$ can be obtained by taking the inverse Fourier transform, resulting in

$$y(n) = x(n - \alpha)$$

The linear phase filter did not alter the shape of the original signal, simply translated it by an amount α. If the phase response had not been linear, the output signal would have been a distorted version of $x(n)$.

In Fig. 4.11 the responses of two different filters to the same input (a sum of two sinusoidal signals) is presented. The filters have the same magnitude frequency responses but differ in their phases as one has linear and the other a quadratic phase. For the filter with linear phase, the sinusoidal components each go through a steady state phase change, but in such a way that the output signal is just a delayed version of the input while the quadratic phase filter causes phase shifts in the two sinusoidal signals resulting in an output that is a distorted version of the input signal.

It can be shown that a causal IIR filter cannot produce a linear phase characteristic and that only special forms of causal FIR filters can give linear phase. This result is clarified in the following theorem.

Theorem 8. If $h(n)$ represents the impulse response of a discrete-time system, a necessary and sufficient condition for linear phase is that $h(n)$ have finite duration N, and that it be symmetric about its midpoint.

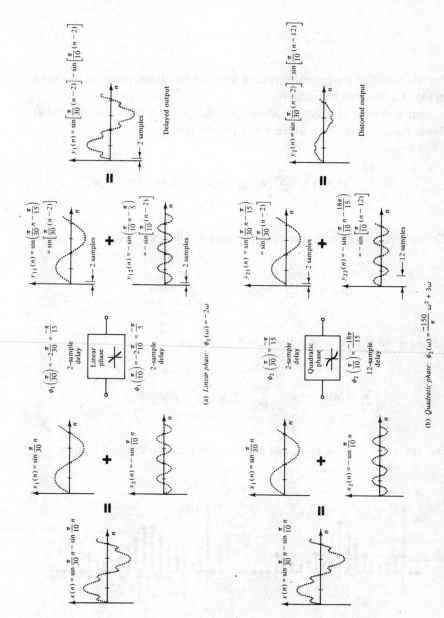

Figure 4.11 Illustrations of the effect of (a) linear phase and (b) nonlinear phase characteristics on steady state outputs with identical magnitude frequency response curves.

For a causal FIR filter whose impulse response begins at zero and ends at $N - 1$, $h(n)$ must satisfy the following:

$$h(n) = h(N - 1 - n) \qquad \text{for } n = 0, 1, \ldots, N - 1 \qquad (4.40)$$

For this condition the general shapes of $h(n)$ that give linear phase are as shown in Fig. 4.12 for odd and even N.

If $h(n)$ is as given in the above theorem, we now show that $H(e^{j\omega})$ has linear phase. For the case where N is even, we have

$$H(e^{j\omega}) = \sum_{n=-\infty}^{\infty} h(n)e^{-j\omega n} = \sum_{n=0}^{N-1} h(n)e^{-j\omega n}$$

The summation can be broken into two parts as follows:

$$H(e^{j\omega}) = \sum_{n=0}^{N/2-1} h(n)e^{-j\omega n} + \sum_{n=N/2}^{N-1} h(n)e^{-j\omega n}$$

Letting $m = N - 1 - n$ in the second sum gives

$$\sum_{n=N/2}^{N-1} h(n)e^{-j\omega n} = \sum_{m=N/2-1}^{0} h(N - 1 - m)e^{-j\omega(N-1-m)}$$

But $h(N - 1 - m) = h(m)$, and the summation can be reversed to give

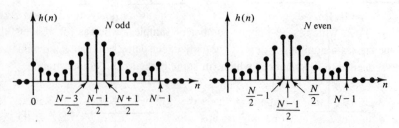

Figure 4.12 General shapes of impulse response $h(n)$ that give linear phase for odd and even N.

$$H(e^{j\omega}) = \sum_{n=0}^{N/2-1} h(n)e^{-j\omega n} + \sum_{m=0}^{N/2-1} h(m)e^{-j\omega(N-1-m)}$$

Combining yields

$$H(e^{j\omega}) = \sum_{n=0}^{N/2-1} h(n)\exp\left[-j\frac{\omega(N-1)}{2}\right]\left\{\exp\left[-j\left(\omega n - \frac{\omega(N-1)}{2}\right)\right]\right.$$
$$\left. + \exp\left[-j\left(\omega(N-1-n) - \frac{\omega(N-1)}{2}\right)\right]\right\}$$

$$= \sum_{n=0}^{N/2-1} 2h(n)\exp\{-j[\omega(N-1)/2]\}\cos\{\omega[n-(N-1)/2]\}$$

By factoring we are able to separate $H(e^{j\omega})$ into two parts as follows:

$$H(e^{j\omega}) = \underbrace{e^{-j\omega(N-1)/2}}_{\text{Linear phase}} \underbrace{\sum_{n=0}^{N/2-1} 2h(n)\cos\{\omega[n-(N-1)/2]\}}_{\text{Magnitude}}, \qquad N \text{ even} \quad (4.41)$$

Therefore, if the sum remains positive, $H(e^{j\omega})$ has a linear phase with slope $-(N-1)/2$. For N an odd number, a similar derivation leads to

$$H(e^{j\omega}) = e^{-j\omega(N-1)/2}\left\{h\left(\frac{N-1}{2}\right) + \sum_{n=0}^{(N-3)/2} 2h(n)\cos\left[\omega\left(n-\frac{N-1}{2}\right)\right]\right\}, N \text{ odd}$$
$$(4.42)$$

For N odd, the slope of $-(N-1)/2$ causes a delay in the output of $(N-1)/2$, which is an integer number of samples, whereas for N even, the slope causes a noninteger delay. The noninteger delay will cause the values of the sequence to be changed, which, in some cases, may be undesirable.

We see from the theorem that linear phase is easily obtainable if we constrain our impulse response to have the symmetry indicated. Procedures for the design of discrete-time filters under this constraint are described in the following sections. One approach is to define catalogs of so-called "windows" that can be selected to satisfy a certain set of design specifications.

4.6.1 Design of FIR filters Using Windows

The easiest way to obtain an FIR filter is to simply truncate the impulse response of an IIR filter. If $h_d(n)$ represents the impulse response of a desired IIR filter, then an FIR filter with impulse response $h(n)$ can be obtained as follows:

$$h(n) = \begin{cases} h_d(n), & N_1 \le n \le N_2 \\ 0, & \text{Otherwise} \end{cases} \qquad (4.43)$$

In general, $h(n)$ can be thought of as being formed by the product of $h_d(n)$ and a "window function," $w(n)$, as follows:

$$h(n) = h_d(n) \cdot w(n) \qquad (4.44)$$

For the $h(n)$ of (4.43), $w(n)$ is said to be a rectangular window and is given by

$$w(n) = \begin{cases} 1, & N_1 \le n \le N_2 \\ 0, & \text{otherwise} \end{cases}$$

If we let $H(e^{j\omega})$, $H_d(e^{j\omega})$, and $W(e^{j\omega})$ represent the Fourier transforms of $h(n)$, $h_d(n)$, and $w(n)$, respectively, the frequency response $H(e^{j\omega})$ of the resulting filter is the convolution of $H_d(e^{j\omega})$ and $W(e^{j\omega})$ given by

$$H(e^{j\omega}) = \frac{1}{2\pi} \int_{-\pi}^{\pi} H_d(e^{j\theta}) W(e^{j(\omega - \theta)}) \, d\theta = H_d(e^{j\omega}) * W(e^{j\omega}) \qquad (4.45)$$

For example, if $H_d(e^{j\omega})$ represents an ideal low-pass filter with cutoff frequency ω_0 and $w(n)$ is a rectangular window positioned about the origin, the $H(e^{j\omega})$ is as shown in Fig. 4.13.

Therefore, it is seen that the convolution produces a smeared version of the ideal low-pass frequency response $H_d(e^{j\omega})$. In general, the wider the main

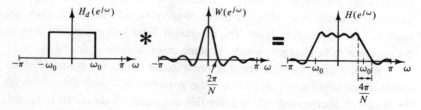

Figure 4.13 Frequency response obtained by rectangularly windowing ideal low-pass impulse response.

lobe of $W(e^{j\omega})$, the more spreading, whereas the narrower the main lobe (larger N), the closer $|H(e^{j\omega})|$ comes to $|H_d(e^{j\omega})|$.

In general, we are left with a trade-off of making N large enough so that smearing is minimized, yet small enough to allow reasonable implementation. Much work has been done on adjusting $w(n)$ to satisfy certain main lobe and side lobe requirements. Some of the most commonly used windows are the rectangular, Bartlett, Hanning, Hamming, Blackman, and Kaiser windows. These are defined mathematically as follows:

Rectangular:
$$w_R(n) = \begin{cases} 1, & 0 \le n \le N - 1 \\ 0, & \text{elsewhere} \end{cases} \tag{4.46}$$

Bartlett:
$$w_B(n) = \begin{cases} 2n/(N - 1), & 0 \le n \le (N - 1)/2 \\ 2 - 2n/(N - 1), & (N - 1)/2 \le n \le N - 1 \\ 0, & \text{elsewhere} \end{cases} \tag{4.47}$$

Hanning:
$$w_{\text{Han}}(n) = \begin{cases} \{1 - \cos[2\pi n/(N - 1)]\}/2, & 0 \le n \le N - 1 \\ 0, & \text{elsewhere} \end{cases} \tag{4.48}$$

Hamming:
$$w_{\text{Ham}}(n) = \begin{cases} 0.54 - 0.46\cos[2\pi n/(N - 1)], & 0 \le n \le N - 1 \\ 0, & \text{elsewhere} \end{cases} \tag{4.49}$$

Blackman:
$$w_{Bl}(n) = \begin{cases} 0.42 - 0.5\cos[2\pi n/(N - 1)] \\ \quad + 0.08\cos[4\pi n/(N - 1)], & 0 \le n \le N - 1 \\ 0, & \text{elsewhere} \end{cases} \tag{4.50}$$

Kaiser:
$$w_K(n) = \begin{cases} \dfrac{I_0\left\{\omega_a\left[\left(\dfrac{N-1}{2}\right)^2 - \left(n - \dfrac{N-1}{2}\right)^2\right]^{1/2}\right\}}{I_0\left[\omega_a\left(\dfrac{N-1}{2}\right)\right]} & 0 \le n \le N - 1 \\ \\ 0, & \text{elsewhere} \end{cases} \tag{4.51}$$

where $I_0(x)$ is the modified zero order Bessel function of the first kind given by $I_0(x) = \int_0^{2\pi} \exp(x\cos\theta)\, d\theta/(2\pi)$.

Except for the rectangular window, the windows exhibit a symmetrical tapering away from the center. Plots of the windows and their Fourier transform magnitudes (in decibels) are shown in Fig. 4.14 for an $N = 51$. An excellent

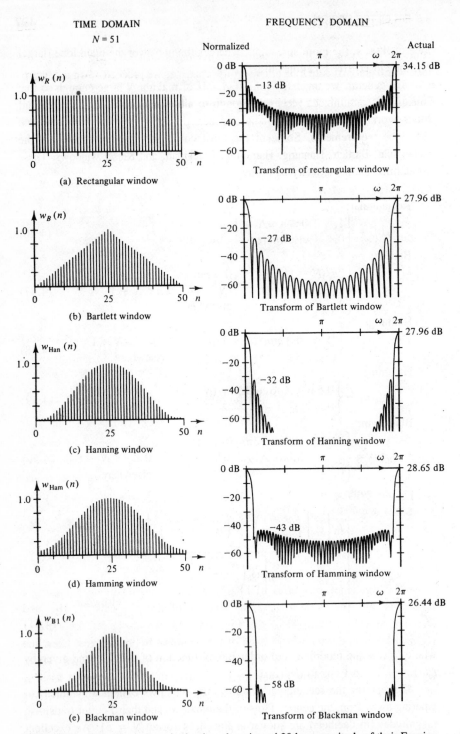

Figure 4.14 Plots of windows in the time domain and 20 log magnitude of their Fourier tranforms in the frequency domain. Reproduced with modifications from Harris, Frederic J., "On the Use of Windows for Harmonic Analysis with the Discrete Fourier Transform," Proceedings of the IEEE, Vol. 66, No. 1, Jan 1978.

study on windows and their properties is given by Harris [7]. Note that the main lobe width and first side lobe attenuation increase as we proceed down the figure.

Design Procedure An ideal low-pass filter with liner phase of slope $-\alpha$ and cutoff ω_c can be characterized in the frequency domain by

$$H_d(e^{j\omega}) = \begin{cases} e^{-j\omega\alpha}, & |\omega| \leq \omega_c \\ 0, & \omega_c < |\omega| < \pi \end{cases}$$

The corresponding impulse response $h_d(n)$ can be obtained by taking the inverse Fourier transform of $H_d(e^{j\omega})$ and easily shown to be

$$h_d(n) = \frac{\sin[\omega_c(n - \alpha)]}{\pi(n - \alpha)}$$

A causal FIR filter with impulse response $h(n)$ can be obtained by multiplying $h_d(n)$ by a window beginning at the origin and ending at $N - 1$ as follows:

$$h(n) = \frac{\sin[\omega_c(n - \alpha)]}{\pi(n - \alpha)} w(n)$$

For $h(n)$ to be a linear phase filter, α must be selected so that the resulting $h(n)$ is symmetric. As $\sin[\omega_c(n - \alpha)]/\pi(n - \alpha)$ is symmetric about $n = \alpha$ and the window symmetric about $n = (N - 1)/2$, a linear phase filter results if the product is symmetric. This requires that

$$\alpha = (N - 1)/2$$

The frequency responses of low-pass filters with cutoff ω_c equals $\pi/2$ formed in this way using a Hamming window of size N equals 51, 101, and 201, are shown on the right side of Fig. 4.15. Shown on the left are the normalized frequency responses of each of the windows by themselves. Note that the stopband attenuation is approximately 55 dB for $N = 101$, slightly lower for $N = 51$, and slightly higher for $N = 201$. It is also seen that as N is increased, the transition widths of the designed filters are decreased and have approximate widths of $8\pi/N$. The 3-dB cutoff frequency is seen to be slightly less than the ω_c of $\pi/2$ and the attenuation is approximately 6 dB at ω equals

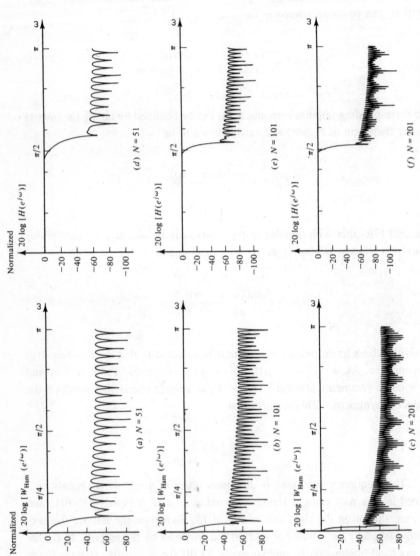

Figure 4.15 Dependence of the Fourier transform of a Hamming window upon window length (a) $N = 51$; (b) $N = 101$; (c) $N = 201$. Corresponding low-pass filter frequency response of a low-pass filter with cutoff frequency $\omega_c = \pi/2$ using the windows (d) $N = 51$; (e) $N = 101$; (f) $N = 201$. (Reproduced by permission from Oppenheim, A. V., and R. W. Schafer, *Digital Signal Processing*, Prentice-Hall, Englewood Cliffs, NJ, 1975.)

(a) $N = 51$

(b) $N = 101$

(c) $N = 201$

(d) $N = 51$

(e) $N = 101$

(f) $N = 201$

Normalized
20 log $[W_{\text{Ham}}(e^{j\omega})]$

Normalized
20 log $[H(e^{j\omega})]$

20 log $[W_{\text{Ham}}(e^{j\omega})]$

20 log $[H(e^{j\omega})]$

20 log $[W_{\text{Ham}}(e^{j\omega})]$

20 log $[H(e^{j\omega})]$

$\pi/2$. Similar trends are observed for the other types of windows allowing us to state the following general properties:

1. The stopband gain for the low-pass filter designed is relatively insensitive to the size of the window and the selection of ω_c depending mainly on the type of window.
2. The transition width of the designed low-pass filter is approximately equal to the main lobe of the window used.

These results are summarized quantitatively for the various types of windows in Table 4.3. This table although a crude approximation may be used to design an FIR low-pass filter from a given set of digital frequency requirements. If we let K_1, ω_1 and K_2, ω_2 represent the cutoff and stopband requirements for the digital filter, an iterative procedure is as follows:

Step 1. Select the window type from the table to be the one highest up the list such that the stop band gain exceeds K_2.

Step 2. Select the number of points in the window to satisfy the transition width for the type of window used. If ω_c is the transition width, we must have

$$\omega_t = \omega_2 - \omega_1 \geq k \cdot 2\pi/N$$

where k depends upon the type of window used. Rearranging the above equation, N is seen to satisfy

$$N \geq k \cdot 2\pi/(\omega_2 - \omega_1) \tag{4.52}$$

Step 3. Select the ω_c and α for the impulse response as

$$\omega_c = \omega_1 \tag{4.53a}$$
$$\alpha = (N - 1)/2 \tag{4.53b}$$

TABLE 4.3   DESIGN TABLE FOR FIR LOW-PASS FILTER DESIGN

	Transition width	Minimum stopband attenuation
Rectangular	$4\pi/N$	-21 dB
Bartlett	$8\pi/N$	-25 dB
Hanning	$8\pi/N$	-44 dB
Hamming	$8\pi/N$	-53 dB
Blackman	$12\pi/N$	-74 dB
Kaiser	Variable	

Thus, a trial filter impulse response is given by

$$h(n) = \frac{\sin\left[\omega_c(n - (N - 1)/2)\right]}{\pi[n - (N - 1)/2]} \cdot w(n) \tag{4.54}$$

Step 4. Plot the frequency response $H(e^{j\omega})$ which, for odd N, is as follows:

$$H(e^{j\omega}) = e^{-j\omega(N-1)/2}\left\{ h((N - 1)/2) + \sum_{n=0}^{(N-3)/2} 2h(n)\cos\left[\omega(n - (N - 1)/2)\right] \right\} \tag{4.55}$$

and check to see if the given specifications are satisfied. For N even use Eq. (4.41).

Step 5. If the attenuation requirement at ω_1 is not satisfied, adjust ω_c accordingly, normally larger on the first iteration, and return to Step 4 using Eq. (4.54) with the new ω_c; otherwise continue.

Step 6. If the frequency response requirements are satisfied, check to see if a further reduction in N might be possible. If a further reduction in N is not possible, the $h(n)$ found is the desired design. If a reduction in N appears possible, reduce N in Eq. 4.54 and return to Step 4 for verification.

The overall procedure may be terminated any time a filter is found that satisfies the given digital specifications. This procedure is a trial and error method and performs satisfactorily; however, no claim of optimality can be made. Normally a filter found using this procedure will have a higher N than what could be designed by various computer techniques; however, the method presented is both simple and instructive and provides useful designs.

If the filter is to be used in an A/D–$H(z)$–D/A structure, the set of equivalent analog specifications must first be converted to digital specifications before the procedure described above is begun. For analog critical frequencies Ω_1 and Ω_2, the corresponding digital specifications using a sampling rate of $1/T$ samples/sec are given by

$$\omega_i = \Omega_i T$$

A low-pass FIR filter design using the above described technique is presented in the following example.

EXAMPLE 4.7

Design a low-pass digital filter to be used in an A/D–$H(z)$–D/A structure that will have a -3-dB cutoff of 30π rad/sec and an attenuation of

50 dB at 45π rad/sec. The filter is required to have linear phase and the system will use a sampling rate of 100 samples/sec.

Solution. The desired equivalent analog frequency response is shown in Fig. 4.16 and the digital specifications obtained are as follows:

$$\omega_c = \Omega_c \cdot T = 30\pi(0.01) = 0.3\pi \, \text{rad}, \qquad K_c \geq -3 \, \text{dB}$$
$$\omega_r = \Omega_r \cdot T = 45\pi(0.01) = 0.45\pi \, \text{rad}, \qquad K_r \leq -50 \, \text{dB}$$

1. To obtain a stopband attenuation of -50 dB or more, a Hamming, Blackman, or Kaiser window could be used. The Hamming window is chosen since it has the smallest transition band thus giving the smallest N.
2. The approximate number of points needed to satisfy the transition band requirement can be found for $\omega_1 = 0.30\pi$ and $\omega_2 = 0.45\pi$ using the Hamming window ($k = 4$) to be

$$N \geq k \cdot 2\pi/(\omega_2 - \omega_1) = 4 \cdot 2\pi/(0.45\pi - 0.3\pi) = 53.3$$

To obtain an integer delay the next odd number ($N = 55$) is selected.
3. From Eq. (4.53), ω_c and α are selected as follows:

$$\omega_c = \omega_1 = 0.3\pi$$
$$\alpha = (N - 1)/2 = 27$$

thus giving a trial impulse response $h(n)$ for a Hamming window as

$$h(n) = \frac{\sin [0.3\pi(n - 27)]}{\pi(n - 27)} [0.54 - 0.46 \cos (2\pi n/54)], \qquad 0 \leq n \leq 54$$

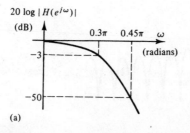

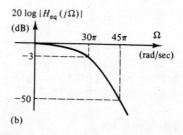

Figure 4.16 Desired frequency response for Example 4.7; (a) digital frequency response; (b) equivalent analog frequency response.

4. Using this $h(n)$ in (4.54), the magnitude of the frequency response is obtained as shown in Fig. 4.17(a).

5. The attenuation is seen to be too much at ω_1, so if we increase ω_c slightly we can hope to satisfy our cutoff frequency requirement.

6. It can also be shown by trial and error that N can be reduced to $N = 29$ and still satisfy the transition region and cutoff requirement with

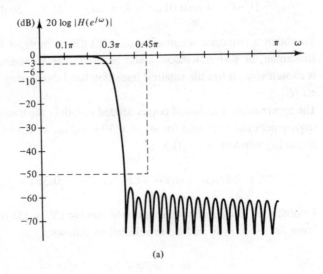

(a)

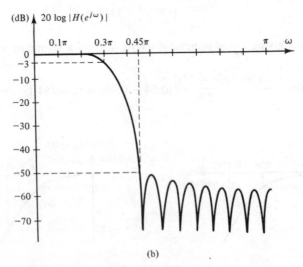

(b)

Figure 4.17 Magnitude of the frequency response for the FIR filter designed for Example 4.7; (a) initial design, (b) improved design.

frequency response shown in Fig. 4.17(b) and impulse response given
by

$$h(n) = \frac{\sin[0.33\pi(n-14)]}{\pi(n-14)}[0.54 - 0.46\cos(2\pi n/28)], \quad 0 \le n \le 28$$

Once the $h(n)$ has been determined, a corresponding difference equation can be obtained for realizing the filter. The output of the filter can be written as a convolution sum as follows:

$$y(n) = \sum_{k=-\infty}^{\infty} h(k)x(n-k)$$

Since the filter impulse response is finite and is zero outside the interval from 0 to $N-1$, $y(n)$ can be written as

$$y(n) = \sum_{k=0}^{N-1} h(k)x(n-k)$$
$$= h(0)x(n) + h(1)x(n-1) + \cdots + h(N-1)x(n-(N-1))$$

To realize the filter in this form requires N multiplications and $N-1$ two-input additions, however, using the symmetry of the impulse response, the output can be written for N odd as

$$y(n) = h(0)[x(n) + x(n-(N-1))] + h(1)[x(n-1) + x(n-(N-2))]$$
$$+ \cdots + h((N-1)/2)[x((N-1)/2) + x((N+1)/2)]$$

thus halving the number of multiplications while keeping the number of two-input additions the same. Implementation and realization of FIR and IIR filters are discussed in detail in Chapter 5.

4.6.2 FIR Filter Design Using Computer Techniques

In the window procedure described previously, the desired impulse response was obtained by an iterative process using changes in ω_c and N for a fixed window type. The impulse responses were restricted to be of the class formed by using classical windows together with ideal low-pass responses. Various computer techniques by many authors have been used to iterate directly on the impulse response subject to the constraint of linear phase, to obtain filters that are not any of the classical window forms yet perform excellently [4], [8], [9].

Many computer software packages are available for the purpose of FIR filter design, one of which is available through IEEE [11].

One of the best approaches to the computer generation of FIR filters is the one based on the use of the Remez exchange algorithm described by Gold and Rabiner [9]. This approach starts with an initial guess for N and $h(n)$ and is based on the fact that the frequency response for an FIR system can be expressed as a trigonometric polynomial. The problem can then be changed to one of Chebyshev approximation where the best approximation to a given function is known to have an alternation property. For more details consult Gold and Rabiner [9] or Oppenheim and Schafer [8].

4.7 SUMMARY

A number of methods have been presented for the design of IIR and FIR filters. It was not the purpose of the chapter to exhaustively present existing methods but to present a small "sampling" of existing methods. In particular, emphasis was placed on the bilinear transformation approach for the design of IIR filters and the classical window approach for the design of FIR filters. Various computer methods were briefly discussed for both FIR and IIR filters, and many more exist. The use of existing "canned" filter design software in many cases requires some basic knowledge of filter design procedure, but hardly the details presented in this chapter. It is hoped, however, that the background presented will allow the student to use these programs more effectively.

The question of whether to use an IIR or an FIR filter is a very important one. The number of multiplications and memory required for an FIR filter is usually much greater than that for an IIR filter, however, the FIR filter can give us linear phase where the IIR filter cannot. Also, the FIR filter is not as sensitive to changes in coefficients and is unconditionally stable, whereas the IIR filter may exhibit sensitivity to coefficients and rounding and have problems with stability. Normally, the decision between the two will be based on whether linear phase is absolutely necessary and whether real time processing is needed.

REFERENCES

1. Antoniou, Andreas. *Digital Filters: Analysis and Design*. McGraw-Hill, New York, 1979.
2. Bogner, R. E., and A. G. Constantinides. *Introduction to Digital Filtering*. Wiley, New York, 1975.

3. Cadzow, James A. *Discrete-Time Systems.* Prentice-Hall, Englewood Cliffs, NJ, 1973.

4. Chen, Chi-Tsong. *One-Dimensional Digital Signal Processing.* Marcel Dekker, New York, 1979.

5. Childers, D. G., and A. E. Durling. *Digital Filtering and Signal Processing.* West Publishing Co., St. Paul, MN, 1975.

6. Hamming, R. W. *Digital Filters.* Prentice-Hall, Englewood Cliffs, NJ, 1975.

7. Harris, Fredric J. "On the Use of Windows for Harmonic Analysis with the Discrete Fourier Transform," *Proceedings of the IEEE,* vol. 66, No. 1 (January 1978), 51–83.

8. Oppenheim, A. V., and R. W. Schafer. *Digital Signal Processing.* Prentice-Hall, Englewood Cliffs, NJ, 1975.

9. Rabiner, L., and B. Gold. *Theory and Application of Digital Signal Processing.* Prentice-Hall, Englewood Cliffs, NJ, 1975.

10. Stanley, W. D. *Digital Signal Processing.* Reston Publishing Co., Reston, VA, 1975.

11. Digital Signal Processing Committee, IEEE Acoustics, Speech and Signal Processing Society, Eds. *Programs for Digital Signal Processing.* IEEE Press, New York, 1979.

PROBLEMS

4.1 Suppose we are given the following differential equation:

$$\sum_{k=0}^{N} c_k \frac{d^k}{dt^k} y_a(t) = \sum_{k=0}^{M} d_k \frac{d^k}{dt^k} x_a(t)$$

The first forward difference $\Delta^{(1)} x(n)$ is defined by $\Delta^{(1)} x(n) = [x(n + 1) - x(n)]/T$ and the nth forward difference is obtained by successive first forward differences.

(a) If the derivatives are approximated by forward differences, find the mapping from the s-plane to the z-plane necessary to obtain the digital transfer function directly from the analog transfer function.

(b) Using this transformation find the z-plane image of $s = j\Omega$ as Ω goes from $-\infty$ to $+\infty$.

(c) Also with this same transformation, find the z-plane image of $s = re^{j3\pi/4}$ as r goes from $r = 0$ to $r = \infty$. Use this result to tell how a stable analog system can give rise to an unstable digital filter.

4.2 Given $[y(n) - y(n - 1)] = \dfrac{T}{2} \left\{ \dfrac{-a_0}{a_1} [y(n) + y(n - 1)] + \dfrac{b_0}{a_1} [x(n) + x(n - 1)] \right\}$

Take the $\mathcal{Z}$ transform ignoring initial conditions and solve for $H(z)$ to verify Eq. (4.24).

4.3 An analog filter with system function $H_a(s)$ has the magnitude plot shown below.

A digital filter is formed by letting $s = 2(1 - z^{-1})/(1 + z^{-1})$ as follows:

$$H(z) = H_a(s)|_{s=2(1-z^{-1})/(1+z^{-1})}$$

Plot the magnitude of the frequency response for the resulting digital filter.

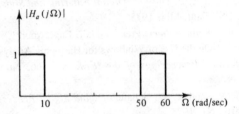

4.4 An analog filter has a system function $H_a(s) = 1/(s + 1)$. A digital filter is found by letting $s = (1 - z^{-1})/(1 + z^{-1})$, that is,

$$H(z) = H_a(s)|_{s=(1-z^{-1})/(1+z^{-1})}$$

At what digital frequency, ω_0, will $20 \log |H(e^{j\omega_0})|$ equal -3.01 dB?

4.5 An analog filter specified by an $H(s)$ has the frequency response shown below. A digital filter $H(z)$ is formed by replacing s by $4(1 - z^{-1})/(1 + z^{-1})$. Roughly draw the 20 log magnitude of the frequency response of the digital filter $H(z)$.

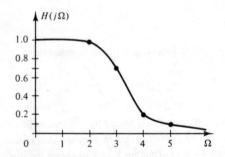

4.6 $H(s)$ is a system function representing a 1-rad third-order analog Chebyshev filter with $\epsilon = 0.5088471$.

(a) Give $H(s)$ in quadratic factored form.

(b) What is the gain in decibels at $\Omega = 0$?

(c) If a digital filter is formed by replacing s by $8(1 - z^{-1})/(1 + z^{-1})$, what is the -1-dB cutoff for the resulting digital filter?

(d) Will the digital filter formed in (c) be BIBO stable? Why or why not?

4.7 Using the bilinear transformation, $s = (1 - z^{-1})/(1 + z^{-1})$, what is the image of $s = e^{j\pi/2}$ in the z plane?

4.8 Using the bilinear transformation $z = (2 + s)/(2 - s)$,

(a) Find the image of $s = -1 + j$ in the z plane.

(b) Plot the image of $s = -1 + j\Omega$ as Ω varies from $-\infty$ to ∞.

(c) If $H(s)$ is a stable analog filter, would this type of transformation lead to a stable digital filter? Explain.

4.9 Develop a transformation for the solution of a first-order linear constant coefficient difference equation similar to that developed in Section 4.1.1 by using Simpson's rule for the integral approximation rather than the trapezoidal approximation.

4.10 Design a digital filter $H(z)$, using the bilinear transformation method, to be used in an A/D–$H(z)$–D/A structure to satisfy the following equivalent analog specifications (use a sampling rate of 1000 samples/sec).
(S1) Maximally flat low-pass frequency response with a -2-dB cutoff at 5 rad/sec.
(S2) At least 23 dB of attenuation at all radian frequencies greater than 10 rad/sec.

4.11 Repeat Problem 4.10 for a -3-dB cutoff at $\Omega_c = 20\pi$ rad/sec and a stopband attenuation of at least 10 dB at all Ω greater than 28.1932π rad/sec using a sampling rate of 40 samples/sec.

4.12 Design a digital filter $H(z)$, using the bilinear transformation method to be used in an A/D–$H(z)$–D/A structure to satisfy the following analog specifications (use Chebyshev prototype):
(S1) Sampling rate of 60,000 samples/sec
(S2) Low-pass filter with -2-dB cutoff at 15,000 Hz
(S3) Stopband attenuation of 10 dB at 30,000 Hz and greater
[Give $H(z)$ in transformational form. DO NOT SIMPLIFY!]

4.13 Design a digital filter $H(z)$ that when used in a prefilter $-$ A/D–$H(z)$–D/A structure will satisfy the following equivalent analog specifications:
(S1) Low-pass filter with a -2-dB cutoff of 10π rad/sec
(S2) Stopband attenuation of 20 dB or greater at and past 60π rad/sec
(S3) Sampling rate of 300 samples/sec for A/D and D/A
(S4) Monotonic passband.

4.14 Roughly draw the equivalent analog frequency response of the A/D–$H(z)$–D/A structure using requirements for $H(z)$ of Problem 4.13. What is the largest cutoff frequency that the prefilter can have so that the prefilter $-$A/D–$H(z)$–D/A structure gives the desired equivalent analog specifications?

4.15 Design an IIR digital filter specified by $H(z)$ that, when used in a prefilter $-$A/D–$H(z)$–D/A structure, will satisfy the following equivalent analog specifications:
(S1) -2-dB cutoff at 75π rad/sec
(S2) 40 dB attenuation at and past 500π rad/sec
(S3) Monotonic stopband and passband
(S4) Sampling rate of 1500 samples/sec
[Give $H(z)$ in transformational form. DO NOT SIMPLIFY!]

4.16 Design a digital high-pass filter $H(z)$ to be used in an A/D–$H(z)$–D/A structure to satisfy the specifications given in the figure below. The sampling rate is fixed at 1000 samples/sec.

(a) Find $H(z)$ and plot $|H(e^{j\omega})|$ and arg $H(e^{j\omega})$.

(b) Give the equivalent analog frequency response showing critical frequencies.

(c) What limitations must be placed on the input signals so that the A/D–$H(z)$–D/A structure truly acts like a high-pass filter to them?

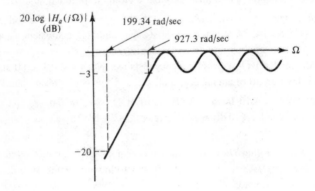

4.17 (a) Design a low-pass digital filter that will operate on sampled analog data such that the analog cutoff frequency is 200 Hz (1 dB acceptable ripple) and at 400 Hz the attenuation is at least 20 dB with monotonic shape past 400 Hz. The sample rate is 2000 samples/sec.

(b) For the designed filter in part (a), plot 20 log $|H(e^{j\omega})|$ and arg $H(e^{j\omega})$.

(c) Plot the equivalent frequency response in decibels for the A/D–$H(z)$–D/A structure.

(d) Give a difference equation realization of your filter.

(e) How many real multiplications and additions are required for implementation in this form?

4.18 Rework Problem 4.17 with a sampling rate of 1000 samples/sec and discuss the effect that the sampling rate plays on the design.

4.19 Design an IIR digital filter $H(z)$ that when used in a prefilter –A/D–$H(z)$–D/A structure will satisfy the following equivalent analog specifications (leave answer in transformational form):

(S1) Low-pass filter with -1-dB cutoff at 100π rad/sec

(S2) Stopband attenuation of 35 dB or greater at 1000π rad/sec

(S3) Monotonic stopband and passband

(S4) Sampling rate of 2000 samples/sec

4.20 (a) Design an IIR digital filter $H(z)$ that when used in a prefilter –A/D–$H(z)$–D/A structure will satisfy the following equivalent analog specifications (use Chebyshev prototype):

(S1) Low-pass filter with -2-dB cutoff at 100 Hz

(S2) Stopband attenuation of 20 dB or greater at 500 Hz

(S3) Sampling rate 4000 samples/sec

(b) Obtain a difference equation realization.

(c) How many real multiplications are required for the implementation?

(d) If a Butterworth prototype were used instead of the Chebyshev filter, the number of multiplications is .... (use $>$, $<$, $\geq$, $\leq$) in general.

(e) Determine roughly how many multiplications are required to satisfy the same specifications given above with the additional requirement that the filter be linear phase.

4.21 Design an IIR digital bandpass filter with -3-dB lower cutoff of 0.4π rad and upper -3-dB cutoff of 0.5π rad. The transition band for both upper and lower frequencies is 0.1π rad with minimum stopband attenuation of at least 40 dB. See Problem 4.37 for FIR design of same filter.

4.22 Given a signal $x(t) = 5 \sin 2\pi t + 2 \sin 10\pi t$,

(a) Plot $x(t)$ for $t = nT$, where $T = 1/50$ sec and $n = 0, 1, \ldots, 150$.

(b) Design a digital filter that will pass the 1-Hz signal with attenuation less than 1 dB and suppress the 5-Hz signal to at least 40 dB down from the magnitude of the 1-Hz signal.

(c) From the $H(z)$ of part (b), write a difference equation realization for the output $y(n)$ and give the number of multiplications and additions required to determine the output at each value of n.

(d) Plot the poles and zeros of the $H(z)$ determined in (b) and comment on results.

(e) Implement the filter on the sampled $x(t)$ for $t = nT$, $n = 0, 1, \ldots, 150$ (implementation by computer program) and plot $y(nT)$ the filtered version of $x(nT)$ for $n = 0, 1, \ldots, 150$ and compare with the 1-Hz part of $x(nT)$ of part (a). Discuss the results, especially with respect to transients and the success of your filtering.

(f) Compute the predicted amplitude and phase change of the $5 \sin 2\pi t$ part of $x(nT)$ by using the $H(z)$ of part (b) and compare with that determined from the plots of part (e).

4.23 A digital low-pass filter with -3-dB unit bandwidth has the transfer function $H(z)$ given by

$$H(z) = (1 + z^{-1})/(2.83305 - 0.83045z^{-1})$$

(a) Find the low-pass digital filter $H_d(z)$ with -3-dB cutoff frequency of $\pi/2$ rad by using a digital-to-digital transformation.

(b) Evaluate $|H(e^{j3\pi/4})|$.

(c) At what value of ω for the low-pass filter in (a) is the $|H_d(e^{j\omega})|$ equal to the $|H(e^{j3\pi/4})|$?

4.24 Suppose you are given a digital filter, $H_1(z)$, with magnitude frequency response shown on the top in the figure at the top of the next page.

(a) Tell how you can obtain a digital filter $H_2(z)$ with magnitude response shown on the bottom in the figure at the top of the next page.

(b) Also find the ω_r.

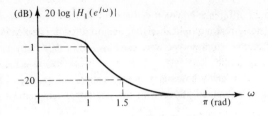

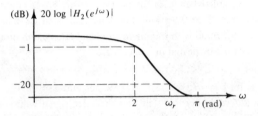

4.25 We have available a digital low-pass filter with the following characteristics: $H(z)$, attenuation of 1 dB at 0.1π, and attenuation of 40 dB at 0.2π. A high-pass digital filter with -1-dB cutoff of 0.5π can be obtained by a digital to digital transformation.

(a) Give the required transformation and show how you obtain the required high-pass transfer function $G(z)$ as a function of z^{-1}.

(b) At what digital frequency will the high-pass filter be down 40 dB?

4.26 Suppose you are given a digital filter $H(z)$ with magnitude frequency response shown on the left of the figure. Suppose an LP $\rightarrow$ HP digital transformation is applied to $H(z)$ to give an $H_d(z)$ with magnitude frequency response shown on the right.

(a) Give the form of the transformation needed.

(b) Solve for the ω_r of the filter $H(z)$.

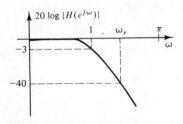

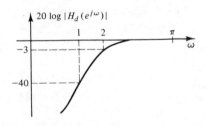

4.27 Let $w_r(n)$ be a 50 point rectangular window given by

$$w_r(n) = \begin{cases} 1, & 0 \le n \le 49 \\ 0, & \text{elsewhere} \end{cases}$$

(a) Give the simplest formula (showing phase and magnitude) for $W_r(e^{j\omega})$, the frequency response of $w_r(n)$.

(b) Find $20 \log |W_r(e^{j\omega})|$ for $\omega = 0$ rad.

4.28 Compute $20 \log |W(e^{j\omega})|$ at $\omega = 0$ for the following windows:

(a) Bartlett window for odd N. (Hint: use summation formula from Appendix B).

(b) Rectangular window.

(c) Hanning window.

(d) Hamming window.

(e) Blackman window.

4.29 For a rectangular window $w_R(n)$,

(a) Plot $20 \log |W_R(e^{j\omega})|$ vs. ω, for ω from 0 to π and $N = 11$.

(b) Plot arg $W_R(e^{j\omega})$ vs. ω, for ω from 0 to π and $N = 11$.

(c) Let $h(n)$ be the unit impulse response of an ideal low-pass filter with cutoff frequency $\omega_c = \pi/2$. Find and plot the frequency response (magnitude and phase) of $w_R(n)$ times the shifted version of $h(n)$ using $N = 11$. The shift is to give linear phase.

(d) Repeat (c) using the Hamming window.

(e) Compare the magnitude plots of parts (c) and (d) with each other and the ideal low-pass characteristics, and discuss results in terms of transition bandwidth and minimum stopband attenuation.

4.30 A filter is specified by its unit sample response $h(n)$, given by

$$h(n) = -\tfrac{1}{3}\delta(n+1) + \tfrac{1}{2}\delta(n) - \tfrac{1}{3}\delta(n-1)$$

(a) Is it a linear phase filter?

(b) Is it a low-pass or high-pass filter?

(c) Is it a causal filter?

4.31 Given the ideal frequency response shown,

(a) Find $h_d(n)$ that gives this frequency response.

(b) Suppose $h_d(n)$ is multiplied by the Hamming window defined by Eq. (4.49). What is the size of this window to give linear phase for a filter with impulse response $h(n) = h_d(n) \cdot w_{HAM}(n)$?

(c) Roughly sketch the frequency response for $h(n)$. Be sure to give values for cutoff frequency, stopband attenuation, and transition bandwidth.

(d) Repeat (b) for a rectangular window and comment on your result.

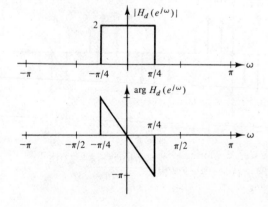

4.32 We are given that

$$h(n) \begin{cases} \dfrac{1}{2}[1 - \cos(2\pi n/100)]\dfrac{\sin[0.2\pi(n - 50)]}{\pi(n - 50)}, & 0 \leq n \leq 100 \\ \\ 0, & \text{elsewhere} \end{cases}$$

(a) Roughly draw the $|H(e^{j\omega})|$ in decibels and label the critical values of cutoff, stopband attenuation, and transition width.

(b) If $x(n) = \sin(\pi n/4)$ is the input to this filter, how many samples is the steady output displaced from $x(n)$?

(c) Give the difference equation representation of the filter specified by $h(n)$. [Use $h(n)$ to represent the values of the impulse response but do not calculate the values.]

4.33 (a) What is the log magnitude in decibels for the Kaiser window alone ($N = 256$) at $\omega = 0.2\pi$ rad?

(b) Roughly how many times more points will the Kaiser window require to have the same transition width as a rectangular window?

(c) Repeat (b) for the Hamming window vs. the rectangular window.

(d) What is the approximate slope (db/rad) of a Hamming LPF in the stop band for $N = 21$? If the sampling rate is 1000 samples/sec, what is the equivalent analog attenuation slope in dB/(rad/sec)?

4.34 Given

$$h(n) = \begin{cases} [0.54 - 0.46\cos(2\pi n/74)]\sin\dfrac{0.35\pi(n - 37)}{\pi(n - 37)}, & 0 \leq n \leq 74 \\ \\ 0, & \text{elsewhere} \end{cases}$$

(a) Roughly draw, labeling critical points, the $|H(e^{j\omega})|$ in decibels

• (b) If the $x(n)$ shown is an input to the system above, is it in the stopband or passband of the filter given by $h(n)$? Explain why.

(c) Calculate exactly $20 \log |H(e^{j\omega})|$ for $\omega = 0$.

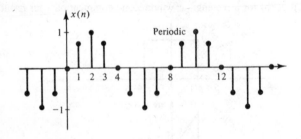

4.35 Given

$$h(n) = \frac{1}{2}[1 - \cos(2\pi n/46)]\frac{\sin[0.25\pi(n - 23)]}{\pi \cdot (n - 23)}, \qquad 0 \leq n \leq 46$$

(a) Roughly draw the plot of 20 log $|H(e^{j\omega})|$ labeling critical frequencies and attenuation values.

(b) Compute exactly 20 log $|H(e^{j\omega})|$ at $\omega = 0$. (Be careful.)

(c) Approximately what is 20 log $|H(e^{j\omega})|$ at $\omega = 0.25\pi$?

(d) Give a difference equation realization of the digital filter with the above $h(n)$. Use $h_i = h(i)$ in your answer (do not compute the values).

(e) To implement this filter requires how many multiplications? Additions?

(f) Give a formula for the calculation of the frequency response of the filter.

(g) Plot the equivalent analog frequency response if this $h(n)$ is used in an A/D–$H(z)$–D/A structure with a sampling rate of 100 samples/sec.

4.36 An analog signal with frequency band 0–10 kHz is sampled at 50K samples/sec. We desire to pass the signal by the use of an FIR digital filter requiring that the transition band be no more than 5 kHz with a stopband attenuation of at least 40 dB. Furthermore we want the phase to be linear throughout the passband. Design such an FIR filter and check your design by plotting its frequency response.

4.37 A digital bandpass filter with -3-dB lower cutoff of 0.4π rad and -3-dB upper cutoff of 0.5π rad is required. The transition band for both upper and lower frequencies is 0.1π rad with a stopband attenuation of at least 40 dB.

(a) Find the impulse response $h(n)$ for an FIR filter that satisfies these requirements by using a Hamming window.

(b) What is the size of the window? Number of multiplications required?

(c) What is the approximate attenuation in the stopband?

(d) What is the main tradeoff made in the selection of the type of window in FIR filter design?

(e) Compare your results to that determined in Problem 4.21 for an IIR filter.

4.38 What is the length of an impulse response $h(n)$ for an FIR filter that will satisfy the requirements given in Problem 4.13, but with the additional requirement of linear phase?

4.39 Suppose that we add to the specifications of Problem 4.19 the fact that the filter should be linear phase.

(a) Do we have to use an FIR filter? Explain.

(b) Find an FIR filter to satisfy the specifications of Problem 4.19 with the additional assumption of linear phase.

4.40 The ideal discrete-time high-pass filter has the frequency response shown in the figure at the top of the next page.

(a) Find the unit sample response of such a system in general and plot $h(n)$ for $\omega_{c_0} = \pi/2$.

(b) Repeat (a) using the linear phase shown and the same ideal magnitude.

(c) With results of (a) for $\omega_{c_0} = \pi/2$, define a truncated version of $h(n)$ as

$$h_T(n) = h(n)[u(n + 4) - u(n - 4)]$$

(d) Plot $|H_T(e^{j\omega})|$ and compare with $|H(e^{j\omega})|$ of part (a).

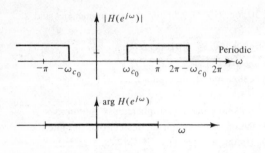

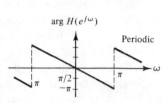

4.41 Answer the following collection of questions.

(a) Why choose an IIR filter instead of an FIR filter?

(b) Give a sufficient condition for an IIR filter to have linear phase.

(c) Give a sufficient condition for an FIR filter to have linear phase.

(d) What does linear phase do to the response of an input signal within the passband of the filter?

(e) Which of the Chebyshev or Butterworth filter implementations give the lowest-order filter assuming equal cutoff attenuation and stopband attenuation?

(f) To get the same filter cutoff and stopband attenuation, does the IIR or FIR filter give a lower-order difference equation?

(g) To get the same transition bandwidth for a low-pass linear phase digital filter, the Hamming window design would require roughly how many times the number of terms of the Rectangular design? What is the advantage of the Hamming window design over the rectangular design?

(h) The most straightforward approach to FIR digital filter design is to truncate the impulse response of an ideal IIR. Why is this usually an undesirable approach?

Chapter 5

Realizations of Digital Filters

In Chapter 4, a procedure for the design of digital filters to satisfy a given set of frequency requirements was presented. Once a system function $H(z)$ or an impulse response $h(n)$ was specified, the design was completed. The digital filter was used in an A/D–$H(z)$–D/A structure and could be implemented or synthesized by a difference equation obtained directly from $H(z)$ or $h(n)$. The difference equation could be implemented by computer program, special purpose digital circuitry, or special programmable integrated circuit. It was indicated that the direct evaluation of the difference equation was not the only possible realization of the digital filter. The purpose of this chapter is to provide alternative realizations of the digital filter by breaking up the direct realization in some fashion. Particular attention will be given to various direct, cascade and parallel forms, and state variable realizations.

5.1 DIRECT FORM REALIZATIONS OF IIR FILTERS

An important class of linear shift invariant systems can be characterized by the following rational system function with $M \leq N$:

$$H(z) = \frac{\displaystyle\sum_{k=0}^{M} b_k z^{-k}}{1 + \displaystyle\sum_{k=1}^{N} a_k z^{-k}} \tag{5.1}$$

217

Using this system function and the properties of the Z transform, it was shown previously that the system with input $x(n)$ and output $y(n)$ could be realized by the following linear constant coefficient difference equation:

$$y(n) = -\sum_{k=1}^{N} a_k y(n-k) + \sum_{k=0}^{M} b_k x(n-k) \qquad (5.2)$$

A realization of the filter using Eq. (5.2) as shown will be called the direct form I realization. The output $y(n)$ is seen to be a weighted sum of the input $x(n)$ at the present time n, past inputs $x(n-k)$ for $k = 1, 2, \ldots, M$ and past outputs $y(n-k)$ for $k = 1, 2, \ldots, N$. The direct form I realization is shown in block diagram form in Fig. 5.1. The delay blocks represent a form of storage and delay, the $\otimes$ a multiplying operation, and the Σ a summing operation. The number of delay blocks is easily seen to be $M + N$ for this particular case.

Another realization of Eq. (5.2) can be obtained by breaking $H(z)$ into a product of two transfer functions $H_1(z)$ and $H_2(z)$, where $H_1(z)$ contains only the denominator or poles and $H_2(z)$ contains only the numerator or zeros, as follows:

$$H(z) = H_1(z) \cdot H_2(z) = Y(z)/X(z) \qquad (5.3a)$$

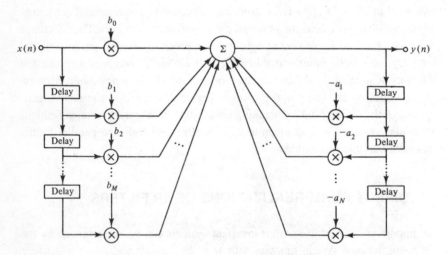

Figure 5.1 Direct form I realization of discrete-time system given by difference equation

$$y(n) = \sum_{k=0}^{M} b_k x(n-k) - \sum_{k=1}^{N} a_k y(n-k)$$

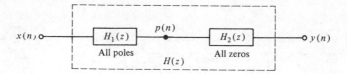

Figure 5.2   Decomposition for direct form II realization.

where
$$H_1(z) = 1 / \left(1 + \sum_{k=1}^{N} a_k z^{-k} \right) \tag{5.3b}$$

$$H_2(z) = \sum_{k=0}^{M} b_k z^{-k} \tag{5.3c}$$

The output of the filter is obtained by calculating an intermediate result $p(n)$, obtained from operating on the input with filter $H_1(z)$, and then operating on $p(n)$ with filter $H_2(z)$ as shown in Fig. 5.2. The transforms of $p(n)$ and $y(n)$ are as follows:

$$P(z) = H_1(z) \cdot X(z) \tag{5.4}$$
$$Y(z) = H_2(z) \cdot P(z) \tag{5.5}$$

Substituting (5.3b) and (5.3c) into (5.4) and (5.5) gives the following:

$$P(z) = \left[1 / \left(1 + \sum_{k=1}^{N} a_k z^{-k} \right) \right] X(z) \tag{5.6}$$

$$Y(z) = \left(\sum_{k=0}^{M} b_k z^{-k} \right) P(z) \tag{5.7}$$

Taking the inverse transforms of (5.6) and (5.7) gives the following pair of difference equations for the realization which is shown in Fig. 5.3:

$$p(n) = x(n) - \sum_{k=1}^{N} a_k p(n - k) \tag{5.8}$$

$$y(n) = \sum_{k=0}^{M} b_k p(n - k) \tag{5.9}$$

Upon closer examination of Fig. 5.3, it is seen that the two branches of delay elements can be combined as they both refer to delayed versions of $p(n)$, and,

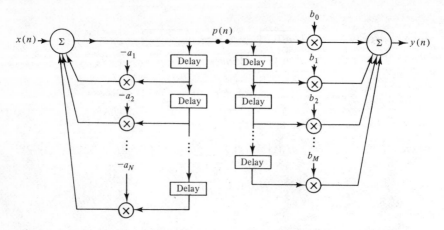

Figure 5.3   Realization using intermediate result.

upon simplification, the direct form II canonical block diagram shown in Fig. 5.4(a) is obtained. In this form, the number of delay blocks is seen to be N, the order of the difference equation. It can be shown that N is the minimum number and that this particular realization is just one of many that contains the minimum number of delays as well as multipliers and adders. This does not mean that it is the best realization, however, for immunity to round off and quantization errors are very important considerations.

Substituting (5.9) into (5.8), the output $y(n)$ can be written entirely in terms of the $p(n - k)$ and $x(n)$ as follows:

$$y(n) = \sum_{k=1}^{N} (b_k - b_0 a_k)p(n - k) + b_0 x(n) \tag{5.10}$$

In the above expression, $b_k = 0$ for $M \leq k \leq N$. The realization in block diagram form is presented in Fig. 5.4(b).

An important special case that is used as a building block occurs when $N = M = 2$. Thus, $H(z)$ is a ratio of two quadratics in z^{-1}, called a biquadratic section, and is given by

$$H(z) = \frac{b_0 + b_1 z^{-1} + b_2 z^{-2}}{1 + a_1 z^{-1} + a_2 z^{-2}} = \frac{b_0(1 + b_1' z^{-1} + b_2' z^{-2})}{1 + a_1 z^{-1} + a_2 z^{-2}} \tag{5.11}$$

The direct form II and alternative realization of the biquadratic $H(z)$ are shown in Fig. 5.5. The alternative realization has been shown to be useful for

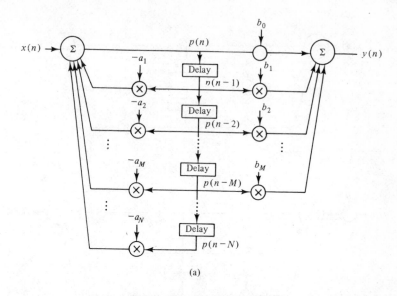

(a)

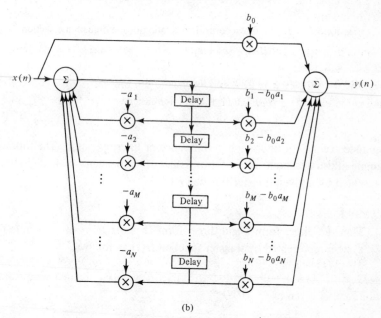

(b)

Figure 5.4 (a) Direct form II canonic realization. (b) Alternative realization obtained from the direct form II realization.

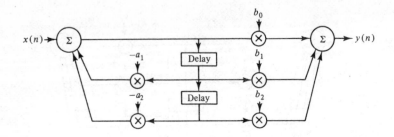

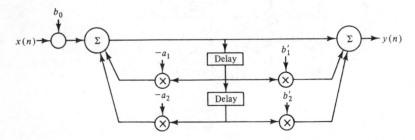

Figure 5.5   (a) The direct form II realization of the biquadratic section

$$H(z) = (b_0 + b_1 z^{-1} + b_2 z^{-2})/(1 + a_1 z^{-1} + a_2^{-2})$$

(b) An alternative realization of the biquadratic section

$$H(z) = b_0(1 + b_1' z^{-1} + b_2' z^{-2})/(1 + a_1 z^{-1} + a_2 z^{-2}).$$

amplitude scaling for improving performance of filter operation. The following example illustrates both direct forms I and II realizations of a digital filter represented by a given transfer function $H(z)$.

EXAMPLE 5.1

Find the direct form I and direct form II realizations of a discrete-time system represented by transfer function $H(z)$ as follows:

$$H(z) = \frac{8z^3 - 4z^2 + 11z - 2}{(z - \frac{1}{4})(z^2 - z + \frac{1}{2})}$$

Solution.   To put in direct form I and II, $H(z)$ must first be changed to the rational polynomial in z^{-1} form given in Eq. (5.1). Multiplying out the denominator of $H(z)$ and dividing both numerator and denominator by z^3, we find

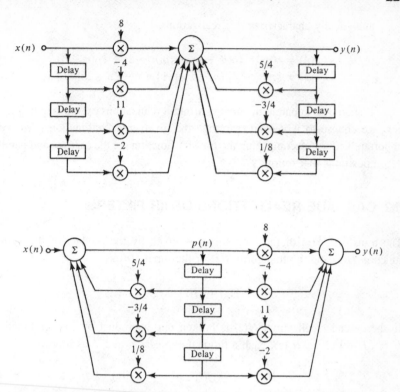

Figure 5.6 (a) Direct form I realization for Example 5.1. (b) Direct form II realization for Example 5.1.

$$H(z) = \frac{\overset{b_0}{8} \overset{b_1}{- 4} z^{-1} \overset{b_2}{+ 11} z^{-2} \overset{b_3}{- 2} z^{-3}}{1 - \underset{a_1}{\tfrac{5}{4}} z^{-1} + \underset{a_2}{\tfrac{3}{4}} z^{-2} - \underset{a_3}{\tfrac{1}{8}} z^{-3}}$$

Recognizing the a_i and b_i's, the direct form I block diagram is easily drawn from Fig. 5.1 as shown in Fig. 5.6(a) and realized by the following formula:

$$y(n) = 8x(n) - 4x(n - 1) + 11x(n - 2) - 2x(n - 3)$$
$$+ \tfrac{5}{4}y(n - 1) - \tfrac{3}{4}y(n - 2) + \tfrac{1}{8}y(n - 3)$$

Using Fig. 5.4(a), Eq. (5.8) and (5.9), the direct form II block diagram is as shown in Fig. 5.6(b), and the following two equations analytically characterize this realization:

$$p(n) = x(n) + \tfrac{5}{4}p(n - 1) - \tfrac{3}{4}p(n - 2) + \tfrac{1}{8}p(n - 3)$$
$$y(n) = 8p(n) - 4p(n - 1) + 11p(n - 2) - 2p(n - 3)$$

It should be apparent that many different realizations exist depending upon how we choose to write and rearrange the given transfer function. Two very important ways of decomposing the transfer function are the cascade and parallel decompositions that follow.

5.2 CASCADE REALIZATIONS OF IIR FILTERS

The transform $Y(z)$ of the output $y(n)$ is given by the product of the system function $H(z)$ and the transform $X(z)$ of the input $x(n)$ as

$$Y(z) = H(z) \cdot X(z)$$

In the cascade realization, $H(z)$ is broken into a product of transfer functions $H_1(z), H_2(z), \ldots, H_k(z)$, each a rational expression in z^{-1} as follows:

$$H(z) = H_K(z)H_{K-1}(z) \cdots H_1(z) \qquad (5.12)$$

Therefore, $Y(z)$ can be written as

$$Y(z) = H_K(z)H_{K-1}(z) \cdots H_1(z)X(z) \qquad (5.13)$$

In this way $y(n)$ is thought of as the output of a cascade of K successive filters with $x(n)$ as the input to the first as shown in Fig. 5.7. The outputs of each stage give intermediate results which are defined as $y_1(n), y_2(n), \ldots, y_{K-1}(n)$ for convenience. $H(z)$ could be broken up in many different ways; however, the most common cascade realization is to require each of the $H_k(z)$'s to be a biquadratic section. In many cases the design procedure yields a product of

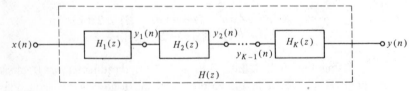

Figure 5.7 Cascade representation of $H(z)$.

biquadratic expressions so no further work is necessary to put $H(z)$ in the required form. The $H_k(z)$ take the following form:

$$H_k(z) = \frac{b_{0k} + b_{1k}z^{-1} + b_{2k}z^{-2}}{1 + a_{1k}z^{-1} + a_{2k}z^{-2}}, \quad k = 1, 2, \ldots, K \quad (5.14)$$

By letting b_{1k} and b_{2k} equal zero, $H_k(z)$ contains just poles; letting a_{1k} and a_{2k} equal zero, $H_k(z)$ contains only zeros; and letting either a_{2k} or b_{2k} or both equal zero, $H_k(z)$ contains a single pole or zero (linear expressions). Each of the $H_k(z)$ could then be realized using either the direct form I or the direct form II. The general cascade realization using direct form II is shown in Fig. 5.8, where K equals $N/2$, if N is even, and $(N + 1)/2$ if N is odd. For N odd, one of the $H_k(z)$ will be bilinear ($a_{2k} = b_{2k} = 0$). For the cascade arrangement, the intermediate results of $y_1(n), y_2(n), \ldots, y_K(n)$ can be calculated consecutively according to the following equations:

$$y_1(n) = -a_{11}y_1(n - 1) - a_{21}y_1(n - 2) + b_{01}x(n)$$
$$+ b_{11}x(n - 1) + b_{21}x(n - 2)$$
$$y_2(n) = -a_{12}y_2(n - 1) - a_{22}y_2(n - 2) + b_{02}y_1(n)$$
$$+ b_{12}y_1(n - 1) + b_{22}y_1(n - 2)$$

$$\vdots$$

$$y(n) = y_K(n) = -a_{1K}y_K(n - 1) - a_{2K}y_K(n - 2) + b_{0K}y_{K-1}(n)$$
$$+ b_{1K}y_{K-1}(n - 1) + b_{2K}y_{K-1}(n - 2) \quad (5.15)$$

Each of the above equations could then be written, if desired, using the direct form II realization as follows:

$$\left. \begin{array}{l} p_1(n) = x(n) - a_{11}p_1(n - 1) - a_{21}p_1(n - 2) \\ y_1(n) = b_{01}p_1(n) + b_{11}p_1(n - 1) + b_{21}p_1(n - 2) \end{array} \right\} \text{Section 1}$$

$$\left. \begin{array}{l} p_2(n) = y_1(n) - a_{12}p_2(n - 1) - a_{22}p_2(n - 2) \\ y_2(n) = b_{02}p_2(n) + b_{12}p_2(n - 1) + b_{22}p_2(n - 2) \end{array} \right\} \text{Section 2}$$

$$\vdots$$

$$\left. \begin{array}{l} p_K(n) = y_{K-1}(n) - a_{1K}p_K(n - 1) - a_{2K}p_K(n - 2) \\ y(n) = y_K(n) = b_{0K}p_K(n) + b_{1K}p_K(n - 1) + b_{2K}p_K(n - 2) \end{array} \right\} \text{Section } K$$

$$(5.16)$$

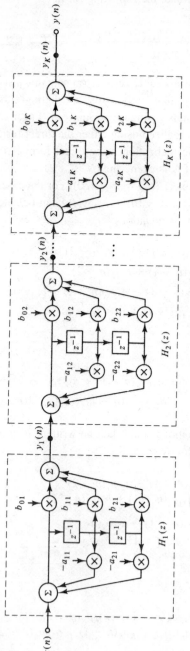

Figure 5.8 Cascade realization using direct form II for each section.

Substituting the $y_k(n)$ of Section k, for every section k, into the $p_{k+1}(n)$ equation of Section $k + 1$, we have

$$p_1(n) = x(n) - a_{11}p_1(n - 1) - a_{21}p_1(n - 2)$$
$$p_2(n) = b_{01}p_1(n) + b_{11}p_1(n - 1) + b_{21}p_1(n - 2)$$
$$\quad - a_{12}p_2(n - 1) - a_{22}p_2(n - 2)$$

$$\cdot$$
$$\cdot$$
$$\cdot$$

$$p_K(n) = b_{0K-1}p_{K-1}(n) + b_{1K-1}p_{K-1}(n - 1) + b_{2K-1}p_{K-1}(n - 2)$$
$$\quad - a_{1K}p_K(n - 1) - a_{2K}p_K(n - 2)$$
$$y(n) = b_{0K}p_K(n) + b_{1K}p_K(n - 1) + b_{2K}p_K(n - 2) \qquad (5.17)$$

This cascade formulation is illustrated in Example 5.2.

EXAMPLE 5.2

Find a cascade realization in terms of quadratic sections for the $H(z)$ given below. Show the block diagram and give the equations for realization in that form:

$$H(z) = \frac{8z^3 - 4z^2 + 11z - 2}{(z - 0.25)(z^2 - z + 0.5)}$$

Solution. Desiring only real coefficients, $H(z)$ can be decomposed into two sections, one bilinear and one biquadratic, by first factoring $H(z)$ as follows:

$$H(z) = \frac{8(z - 0.18995438)(z^2 - 0.31004562z + 1.31610547)}{(z - 0.25)(z^2 - z + 0.5)}$$

Dividing through both numerator and denominator by z^3 and factoring the 8 as $2 \cdot 4$, one possible rearrangement for $H(z)$ is as follows:

$$H(z) = \frac{\overset{b_0}{\overbrace{(2}} - \overset{b_1}{\overbrace{0.37990876z^{-1})}}}{1 - \underbrace{0.25z^{-1}}_{a_1}} \cdot \frac{\overset{b_0}{\overbrace{(4}} - \overset{b_1}{\overbrace{1.24018248z^{-1}}} + \overset{b_2}{\overbrace{5.26442188z^{-2})}}}{1 - \underbrace{1z^{-1}}_{a_1} + \underbrace{0.5z^{-2}}_{a_2}}$$

Therefore, using the direct form II realization for each section, $H(z)$ can be realized by the block diagram shown in Fig. 5.9. For this example the separation of the 8 into 2 for the bilinear section and 4 for the biquadratic section is arbitrary. In general, however, the selection is made to reduce

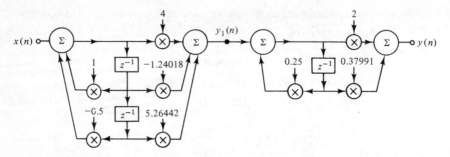

Figure 5.9 A cascade realization of system given in Example 5.2.

errors due to roundoff and provide immunity to noise. For further discussion on amplitude scaling and roundoff errors, consult Antoniou [1] and Oppenheim and Schafer [6].

The equations for realization using direct form I for each stage of the cascade combination are easily seen to be

$$1: \quad y_1(n) = y_1(n-1) - 0.5y_1(n-2) + 4x(n) - 1.2401824x(n-1)$$
$$+ 5.26442188x(n-2)$$
$$2: \quad y(n) = 0.25y(n-1) + 2y_1(n) - 0.37990876y_1(n-1)$$

The equations for $y_1(n)$ and $y(n)$ can, if desired, be written using intermediate results for the direct form II equations as follows:

$$p_1(n) = x(n) + p_1(n-1) - 0.5p_1(n-2)$$
$$y_1(n) = 4p_1(n) - 1.24018248p_1(n-1) + 5.26442188p_1(n-2)$$
$$p_2(n) = y_1(n) + 0.25p_2(n-1)$$
$$y(n) = y_2(n) = 2p_2(n) - 0.37990876p_2(n-1)$$

which can be reduced to

$$p_1(n) = x(n) + p_1(n-1) - 0.5p_1(n-2)$$
$$p_2(n) = 4p_1(n) - 1.24018248p_1(n-1) + 5.26442188p_1(n-1)$$
$$+ 0.25p_2(n-1)$$
$$y(n) = 2p_2(n) - 0.37990876p_2(n-1)$$

5.3 PARALLEL REALIZATIONS OF IIR FILTERS

The transfer function $H(z)$ could be written as a sum of transfer functions $H_1(z)$, $H_2(z)$, . . ., $H_K(z)$ as follows:

$$H(z) = H_1(z) + H_2(z) + \cdots + H_K(z) \qquad (5.18)$$

Therefore, the output transform $Y(z)$ can be written in terms of $H_1(z)$, $H_2(z), \ldots, H_K(z)$, and $X(z)$ by

$$Y(z) = H_1(z)X(z) + H_2(z)X(z) + \cdots + H_K(z)X(z) \cdot \qquad (5.19)$$

In this way, the system is realized by passing the input sequence $x(n)$ through K discrete-time filters and summing the results to obtain the output $y(n)$. This is called a parallel configuration. One parallel form results when the $H_k(z)$ are all selected to be of the following form:

$$H_k(z) = \frac{b_{0k} + b_{1k}z^{-1}}{1 + a_{1k}z^{-1} + a_{2k}z^{-2}} \qquad (5.20)$$

Note that this form for $H_k(z)$ includes the following as special cases:

(a) $H_k(z) = c$ if $b_{0k} = c$ and $b_{1k} = a_{1k} = a_{2k} = 0$ (5.21)

(b) $H_k(z) = \dfrac{b_{0k}}{1 + a_{1k}z^{-1}}$ if $b_{1k} = a_{2k} = 0$ (5.22)

(c) $H_k(z) = b_{1k}z^{-1}$ if $b_{0k} = a_{1k} = a_{2k} = 0$ (5.23)

This particular form of $H(z)$ can be obtained by performing a partial fraction expansion of $H(z)$, where the quadratic terms are used for each pair of complex poles so that the coefficients of $H_i(z)$ turn out to be real. Each $H_i(z)$ could then be realized using either direct form I or II. The general block diagram for this realization is shown in Fig. 5.10. Again selecting $y_1(n)$, $y_2(n)$, . . ., $y_K(n)$ as partial results the filter can be realized by evaluating as follows

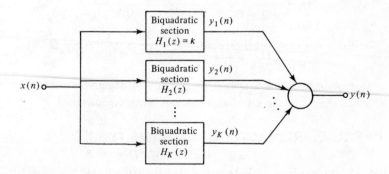

Figure 5.10 Parallel realization in terms of biquadratic sections.

$$y_1(n) = -a_{11}y_1(n-1) - a_{21}y_1(n-2) + b_{01}x(n) + b_{11}x(n-1)$$
$$y_2(n) = -a_{12}y_2(n-1) - a_{22}y_2(n-2) + b_{02}x(n) + b_{12}x(n-1) \qquad (5.24)$$

.
.
.

$$y_K(n) = -a_{1K}y_K(n-1) - a_{2K}y_K(n-2) + b_{0K}x(n) + b_{1K}x(n-1)$$

$$y(n) = y_1(n) + y_2(n) + \cdots + y_K(n) \qquad (5.25)$$

Notice that the outputs $y_1(n)$, $y_2(n)$, . . ., $y_K(n)$ are no longer coupled and, if desired, may be calculated in any order or all at the same time. Each of the equations above for $y_k(n)$ can then be broken into intermediate results for a direct form II realization by equations, if desired. Example 5.3 illustrates the decomposition process and shows one possible parallel realization for a given $H(z)$.

EXAMPLE 5.3

Obtain a parallel realization for the following $H(z)$ using a partial fraction expansion in terms of $H_i(z)$ of the form given in Eq. (5.20):

$$H(z) = \frac{8z^3 - 4z^2 + 11z - 2}{(z - \frac{1}{4})(z^2 - z + \frac{1}{2})}$$

Show the block diagram realization using direct form II and give the equations for realization using direct form I for each section.

Solution. To obtain a parallel realization, $H(z)$ is first put in negative powers of z^{-1} and then expanded in terms of expressions like (5.21) to (5.23).

$$H(z) = \frac{8 - 4z^{-1} + 11z^{-2} - 2z^{-3}}{(1 - 0.25z^{-1})(1 - z^{-1} + 0.5z^{-2})}$$

$$= \frac{A}{(1 - 0.25z^{-1})} + \frac{Bz^{-1} + C}{(1 - z^{-1} + 0.5z^{-2})} + D$$

Multiplying out and equating coefficients of negative powers of z, we obtain the following set of simultaneous linear equations:

$$A + C + D = 8$$
$$-A + B - 0.25C - 1.25D = -4$$
$$0.5A - 0.25B + 0.75D = 11$$
$$-0.125D = -2$$

Upon solving we find $A = 8$, $B = 20$, $C = -16$, and $D = 16$. Therefore $H(z)$ can be written as

$$H(z) = 16 + \frac{8}{1 - 0.25z^{-1}} + \frac{-16 + 20z^{-1}}{1 - z^{-1} + 0.5z^{-2}}$$

Using Fig. 5.5(a) and the direct form II realization of each section specified above, the total realization for $H(z)$ is easily found to be as shown below.

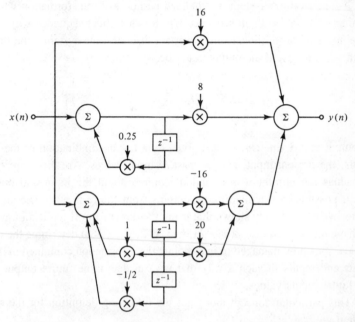

A direct form I realization of each section might be used instead of the direct form II realizations. The direct form I governing equations are easily found from (5.24) and (5.25) to be

$$y_1(n) = 16x(n)$$
$$y_2(n) = 0.25y_2(n - 1) + 8x(n)$$
$$y_3(n) = y_3(n - 1) - 0.5y_3(n - 2) - 16x(n) + 20x(n - 1)$$

$$y(n) = y_1(n) + y_2(n) + y_3(n)$$

5.4 STATE VARIABLE REALIZATIONS

In the first part of this chapter, various forms of realizations were investigated that involved breaking down the transfer function into parts or sections. Many other realizations are possible and most of them can be presented within the general time domain representation of the system under consideration. In continuous time it was convenient for analysis purposes to represent a linear constant coefficient differential equation by a system of first-order differential equations. It is, therefore, not surprising that it is useful to represent a linear constant coefficient difference equation by a system of first-order linear constant coefficient difference equations. Such a formulation results in a special form of representation of the difference equation called a state variable representation.

A discrete-time system was defined earlier as a transformation which assigned an acceptable output sequence $y(n)$ for each input sequence $x(n)$. For the special class of causal linear time invariant discrete-time systems, the transformation was linear and the output could be written as

$$y(n) = -\sum_{k=1}^{N} a_k y(n-k) + \sum_{k=0}^{M} b_k x(n-k) \tag{5.26}$$

The output at any time, n, is thus given by a linear combination of the past N outputs, the present input, and the past M input values. Are these present and past values the minimum information required about the input and output to make it possible to calculate future outputs from future inputs? The answer is NO, for we have already shown that using the direct form II realization we need only have the N values for $p(n-k)$, $k = 1, 2, \ldots, N$ to determine the output. We have somehow managed to incorporate the information contained in the past outputs and inputs in such a way that we can determine future output values from future inputs values.

This particular concept motivates the following definition for the state of a system.

Definition. The state of a system is the minimal information required that along with the input allows the determination of the output.

The state of a system at time n will be represented by an N vector $\mathbf{v}(n)$ with components $v_1(n)$, $v_2(n)$, $\ldots$, $v_N(n)$ called the state variables as follows:

$$\mathbf{v}(n) = \begin{bmatrix} v_1(n) \\ v_2(n) \\ \vdots \\ v_N(n) \end{bmatrix} \tag{5.27}$$

The use of $\mathbf{v}(n)$ for the state variable rather than the standard control system notation of $\mathbf{x}(n)$ allows us to use $x(n)$ and $y(n)$ for input and output, respectively as done previously in this text. For a general system, the output $y(n)$ may be any function of the state vector and the input $x(n)$; however, we concentrate on the important special case where the output is a linear function of the present state and present input expressible by

$$y(n) = [c_1 \quad c_2 \quad \cdots \quad c_N] \begin{bmatrix} v_1(n) \\ v_2(n) \\ \cdot \\ \cdot \\ \cdot \\ v_N(n) \end{bmatrix} + dx(n) \qquad (5.28)$$

By letting $\mathbf{C}^T = [c_1 \; c_2 \; \cdots \; c_N]$ and using the state vector $\mathbf{v}(n)$, the above equation can be written more compactly as

$$y(n) = \mathbf{C}^T \mathbf{v}(n) + dx(n) \qquad (5.29)$$

To calculate the output at time $n + 1$ we require the state at time $n + 1$; therefore, we must have some way of generating the new state from the old state. In a general system the new state could be any function of the previous state and inputs; however, we consider the special case of linear time invariance where the so-called state equation takes the following form:

$$\begin{bmatrix} v_1(n + 1) \\ v_2(n + 1) \\ \cdot \\ \cdot \\ v_N(n + 1) \end{bmatrix} = \begin{bmatrix} a_{11} & a_{12} & \cdots & a_{1N} \\ a_{21} & a_{22} & \cdots & a_{2N} \\ \cdot & \cdot & & \cdot \\ \cdot & \cdot & & \cdot \\ a_{N1} & a_{N2} & \cdots & a_{NN} \end{bmatrix} \begin{bmatrix} v_1(n) \\ v_2(n) \\ \cdot \\ \cdot \\ v_N(n) \end{bmatrix} + \begin{bmatrix} b_1 \\ b_2 \\ \cdot \\ \cdot \\ b_N \end{bmatrix} x(n) \qquad (5.30)$$

Using matrix notation, (5.30) can be written with the obvious assignments for A and B as

$$\mathbf{v}(n + 1) = \mathbf{A}\mathbf{v}(n) + \mathbf{B}x(n) \qquad (5.31)$$

In (5.31): $\mathbf{A}$, called the state transition matrix, is an $N \times N$ matrix; $\mathbf{v}(n)$, the state vector, is an N vector; $\mathbf{B}$, defined as the excitation transition matrix, is an N vector; and $x(n)$, the input, is a scalar representing the input at time n.

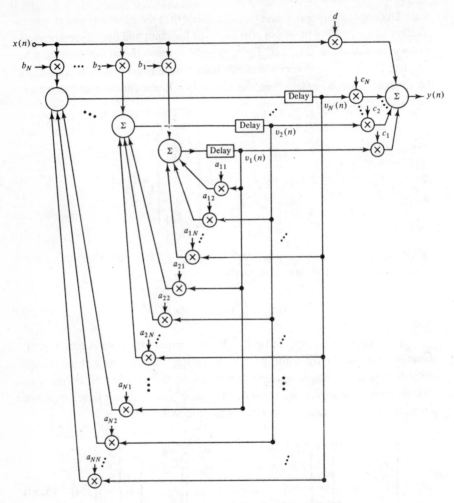

Figure 5.11   State space realization for an IIR filter.

The block diagram for realization of the system in state variable form appears in Fig. 5.11. Basically, it consists of two parts: First, the calculation of the output using the present state and the input, and second, the calculation of the next state. There are many established methods for determining the state variable representation for a linear shift invariant causal system represented by the rational system function given by (5.1). Before presenting some of these methods, a special case of (5.26) will be investigated where the state variable assignment is intuitive and simple.

Consider the special case of (5.26) where $b_i = 0$ for $i = 1, 2, \ldots, M$. Therefore the output $y(n)$ can be written as

$$y(n) = -\sum_{k=1}^{N} a_k y(n - k) + b_0 x(n) \qquad (5.32)$$

We wish to determine the state variable representation of such a system. An obvious candidate for the state vector consists of delayed versions of the output since together with the present input, the output can be determined. Such a selection for the state vector is as follows:

$$\mathbf{v}(n) = \begin{bmatrix} y(n - N) \\ y(n - (N - 1)) \\ \cdot \\ \cdot \\ \cdot \\ y(n - 1) \end{bmatrix} \triangleq \begin{bmatrix} v_1(n) \\ v_2(n) \\ \cdot \\ \cdot \\ \cdot \\ v_N(n) \end{bmatrix} \qquad (5.33)$$

With this state vector assignment, the output from $y(n)$ can be written as

$$y(n) = [-a_N - a_{N-1} \cdots - a_1] \begin{bmatrix} v_1(n) \\ v_2(n) \\ \cdot \\ \cdot \\ \cdot \\ v_N(n) \end{bmatrix} + b_0 x(n) \qquad (5.34)$$

To establish that the $\mathbf{v}(n)$ given in Eq. (5.33) is a state vector, we need to find an equation of the form of (5.30), the state equation, thus showing that each next state can be determined from the input and the previous state. Evaluating $\mathbf{v}(n)$ of (5.33) at $n + 1$ and recognizing the various states gives

$$\mathbf{v}(n + 1) = \begin{bmatrix} y(n - (N - 1)) \\ y(n - (N - 2)) \\ \cdot \\ \cdot \\ \cdot \\ y(n - 1) \\ y(n) \end{bmatrix} = \begin{bmatrix} v_2(n) \\ v_3(n) \\ \cdot \\ \cdot \\ \cdot \\ v_N(n) \\ y(n) \end{bmatrix} \qquad (5.35)$$

Substituting (5.32) into (5.35), $\mathbf{v}(n + 1)$ becomes

$$\mathbf{v}(n + 1) = \begin{bmatrix} v_2(n) \\ v_3(n) \\ \cdot \\ \cdot \\ \cdot \\ -a_N v_1(n) - a_{N-1} v_2(n) - \cdots - a_1 v_N(n) + b_0 x(n) \end{bmatrix} \tag{5.36}$$

The above equation can be expressed in the standard state transition form as follows:

$$\mathbf{v}(n + 1) = \begin{bmatrix} 0 & 1 & 0 & \cdots & 0 & 0 \\ 0 & 0 & 1 & \cdots & 0 & 0 \\ \cdot & \cdot & \cdot & & \cdot & \cdot \\ \cdot & \cdot & \cdot & & \cdot & \cdot \\ \cdot & \cdot & \cdot & & \cdot & \cdot \\ 0 & 0 & 0 & \cdots & 0 & 1 \\ -a_N & -a_{N-1} & -a_{N-2} & \cdots & -a_2 & -a_1 \end{bmatrix} \begin{bmatrix} v_1(n) \\ v_2(n) \\ \cdot \\ \cdot \\ v_{N-1}(n) \\ v_N(n) \end{bmatrix} + b_0 x(n) \tag{5.37}$$

We have thus found the state space formulation, (5.34) and (5.37), of the system represented by (5.32). For this simple case the selection of the state variables was easy and direct; however, for the case where the difference equation has delayed versions of the input as well as delayed versions of the output, it is not quite as obvious what the state vector should be.

Two of many possible state variable assignments for representation of a causal linear time invariant discrete-time system represented by the general form (5.26) are now presented.

5.4.1 Direct Programming Realization

For convenience in formulation, it will be assumed that $M = N$; however, if $M < N$, then the expressions developed are still valid but $b_{M+1}, b_{M+2}, \cdots,$ b_N are all zero. When $H(z)$ is written in rational form, (5.1), an assignment for the state variables becomes apparent. It was previously shown that the output $y(n)$, given by (5.9), could be written as a weighted sum of an intermediate result $p(n)$, given by (5.8). So if we select the state variables $v_k(n)$ as

$$v_k(n) = p(n - k), \qquad k = 1, 2, \ldots, N \tag{5.38}$$

we have immediately from (5.10) the following:

$$
\begin{bmatrix} v_1(n+1) \\ v_2(n+1) \\ v_3(n+1) \\ \cdot \\ \cdot \\ \cdot \\ v_N(n+1) \end{bmatrix} = \begin{bmatrix} p(n) \\ p(n-1) \\ p(n-2) \\ \cdot \\ \cdot \\ \cdot \\ p(n-(N-1)) \end{bmatrix} = \begin{bmatrix} -a_1 & -a_2 & -a_3 & \cdots & -a_N \\ 1 & 0 & 0 & \cdots & 0 \\ 0 & 1 & 0 & \cdots & 0 \\ \cdot & \cdot & \cdot & \cdot & \cdot \\ \cdot & \cdot & \cdot & \cdot & \cdot \\ \cdot & \cdot & \cdot & \cdot & \cdot \\ 0 & 0 & 0 & \cdots & 0 \end{bmatrix}
$$

$$
\times \begin{bmatrix} p(n-1) \\ p(n-2) \\ p(n-3) \\ \cdot \\ \cdot \\ \cdot \\ p(n-N) \end{bmatrix} + \begin{bmatrix} 1 \\ 0 \\ 0 \\ \cdot \\ \cdot \\ \cdot \\ 0 \end{bmatrix} x(n)
$$

$$(5.39)$$

The output equation is obtained from (5.10) as

$$
y(n) = [b_1 - b_0 a_1 \; \vdots \; b_2 - b_0 a_2 \; \vdots \; \cdots \; \vdots \; b_N - b_0 a_N] \begin{bmatrix} p(n-1) \\ p(n-2) \\ \cdot \\ \cdot \\ \cdot \\ p(n-N) \end{bmatrix} + b_0 x(n)
$$

$$(5.40)$$

Therefore, we have sufficient information to determine the next state, (5.39), and calculate the output (5.40).

The selection of the state variables in this fashion has been called the direct programming formulation in classical control theory, and (5.39) and (5.40) representing the system can be put in the standard form

$$
\mathbf{v}(n+1) = \mathbf{A}\mathbf{v}(n) + \mathbf{B}x(n)
$$
$$
y(n) = \mathbf{C}^T \mathbf{v}(n) + dx(n)
$$

where $\mathbf{A}$, $\mathbf{B}$, $\mathbf{C}^T$, and d are as follows:

$$
\mathbf{A} = \begin{bmatrix}
-a_1 & -a_2 & \cdots & -a_{N-1} & -a_N \\
1 & 0 & \cdots & 0 & 0 \\
0 & 1 & \cdots & 0 & 0 \\
\cdot & \cdot & \cdot & \cdot & \cdot \\
\cdot & \cdot & \cdot & \cdot & \cdot \\
\cdot & \cdot & \cdot & \cdot & \cdot \\
0 & 0 & \cdots & 1 & 0
\end{bmatrix}, \qquad
\mathbf{B} = \begin{bmatrix} 1 \\ 0 \\ \cdot \\ \cdot \\ \cdot \\ 0 \end{bmatrix} \qquad (5.41)
$$

$$
\mathbf{C}^T = [b_1 - b_0 a_1 \vdots b_2 - b_0 a_2 \vdots \cdots \vdots b_N - b_0 a_N], \qquad d = b_0 \qquad (5.42)
$$

This particular state variable assignment leads to a realization equivalent to the alternative direct form II realization described earlier and is shown in Fig. 5.12. Another classical assignment of the state variable results from the method of nested programming described as follows.

5.4.2 Nested Programming Realization

The method of nested programming assumes that system can be defined by (5.1).

If the first state variable, $v_1(n)$, is selected as follows:

$$
v_1(n) \triangleq \sum_{k=1}^{N} [-a_k y(n-k) + b_k x(n-k)] \qquad (5.43)
$$

then the output equation (5.3) for the system becomes

$$
y(n) = v_1(n) + b_0 x(n) \qquad (5.44)
$$

To characterize the system it is necessary to assign $\{v_k(n){:}k = 2, 3, \ldots, N\}$ the other $N - 1$ state variables in a consistent manner. By writing the next state equation, a hint at a proper selection becomes available. Evaluating (5.43) at $n + 1$ gives

$$
v_1(n+1) = \sum_{k=1}^{N} [-a_k y(n+1-k) + b_k x(n+1-k)] \qquad (5.45)
$$

Separating out the $k = 1$ term from the summation gives us the next state equation as follows:

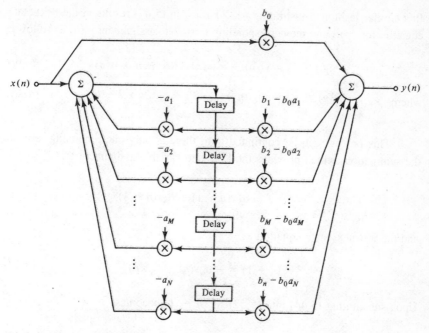

Figure 5.12 Direct programming realization of a discrete-time system specified by difference equation (5.26).

$$v_1(n + 1) = \sum_{k=2}^{N} [-a_k y(n + 1 - k) + b_k x(n + 1 - k)] + b_1 x(n) - a_1 y(n)$$

$$(5.46)$$

The $y(n)$ term was already expressed by (5.44) in terms of the first state variable. Also, $b_1 x(n)$, the next to last term, is in terms of the input $x(n)$. Therefore, a reasonable choice for $v_2(n)$ is all the remaining terms or

$$v_2(n) \triangleq \sum_{k=2}^{N} [-a_k y(n + 1 - k) + b_k x(n + 1 - k)] \qquad (5.47)$$

Substituting (5.47) and (5.44) into (5.46), the next state equation for $v_1(n)$ is as follows:

$$\begin{aligned} v_1(n + 1) &= v_2(n) + b_1 x(n) - a_1[v_1(n) + b_0 x(n)] \\ &= v_2(n) - a_1 v_1(n) + (b_1 - a_1 b_0) x(n) \end{aligned} \qquad (5.48)$$

In a similar fashion, substituting $n + 1$ for n in (5.47) results in the next state equation for $v_2(n)$ and the corresponding next state assignment $v_3(n)$ as follows:

$$v_2(n + 1) = v_3(n) - a_2 v_1(n) + (b_2 - a_2 b_0)x(n) \qquad (5.49)$$

where
$$v_3(n) \triangleq \sum_{k=3}^{N} [-a_k y(n + 2 - k) + b_k x(n + 2 - k)] \qquad (5.50)$$

This process can be continued until there is only one remaining term in the summation, which provides the last state variable assignment as

$$v_N(n) \triangleq -a_N y(n - 1) + b_N x(n - 1) \qquad (5.51)$$

and the last next state equation as

$$v_N(n + 1) = -a_N y(n) + b_N x(n) \qquad (5.52)$$

Upon substituting (5.44) into (5.52), $v_N(n + 1)$ becomes

$$v_N(n + 1) = -a_N v_1(n) + (b_N - a_N b_0)x(n) \qquad (5.53)$$

Collecting (5.43), (5.47), (5.50), and (5.51), the components of this state vector assignment are seen to be

$$v_1(n) = \sum_{k=1}^{N} [-a_k y(n - k) + b_k x(n - k)]$$

$$v_2(n) = \sum_{k=2}^{N} [-a_k y(n + 1 - k) + b_k x(n + 1 - k)]$$

$$\cdot$$
$$\cdot$$
$$\cdot$$

$$v_m(n) = \sum_{k=m}^{N} [-a_k y(n + m - k) + b_k x(n + m - k)]$$

$$\cdot$$
$$\cdot$$
$$\cdot$$

$$v_{N-1}(n) = -a_{N-1} y(n - 1) - a_N y(n - 2) + b_{N-1} x(n - 1) + b_N x(n - 2)$$
$$v_N(n) = -a_N y(n - 1) + b_N x(n - 1) \qquad (5.54)$$

The next state equations, (5.48), (5.49), and (5.53) can be put in the following state transition form:

$$
\begin{bmatrix} v_1(n+1) \\ v_2(n+1) \\ \cdot \\ \cdot \\ \cdot \\ v_{N-1}(n+1) \\ v_N(n+1) \end{bmatrix} = \begin{bmatrix} -a_1 & 1 & 0 & 0 & \cdots & 0 & 0 \\ -a_2 & 0 & 1 & 0 & \cdots & 0 & 0 \\ \cdot & & & \cdot & & \cdot & \cdot \\ \cdot & & \cdot & & \cdot & & \cdot \\ \cdot & & \cdot & & \cdot & & \cdot \\ -a_{N-1} & 0 & 0 & 0 & \cdots & 0 & 1 \\ -a_N & 0 & 0 & 0 & \cdots & 0 & 0 \end{bmatrix} \begin{bmatrix} v_1(n) \\ v_2(n) \\ \cdot \\ \cdot \\ \cdot \\ v_{N-1}(n) \\ v_N(n) \end{bmatrix}
$$

$$
+ \begin{bmatrix} b_1 - a_1 b_0 \\ b_2 - a_2 b_0 \\ \cdot \\ \cdot \\ \cdot \\ b_{N-1} - a_{N-1} b_0 \\ b_N - a_N b_0 \end{bmatrix} x(n)
$$

or

$$\mathbf{v}(n+1) = \mathbf{A}\mathbf{v}(n) + \mathbf{B}x(n) \tag{5.55}$$

and the output equation written as

$$
y(n) = \begin{bmatrix} 1 & 0 & \cdots & 0 \end{bmatrix} \begin{bmatrix} v_1(n) \\ v_2(n) \\ \cdot \\ \cdot \\ \cdot \\ v_N(n) \end{bmatrix} + b_0 x(n) = \mathbf{C}^T \mathbf{v}(n) + dx(n) \tag{5.56}
$$

The block diagram for the system realization determined by the nested programming approach is shown in Fig. 5.13, and the following example illustrates the method.

EXAMPLE 5.4

A system is represented by its system function $H(z)$ given by

$$H(z) = \frac{8z^3 - 4z^2 + 11z - 2}{(z - \frac{1}{4})(z^2 - z + \frac{1}{2})}$$

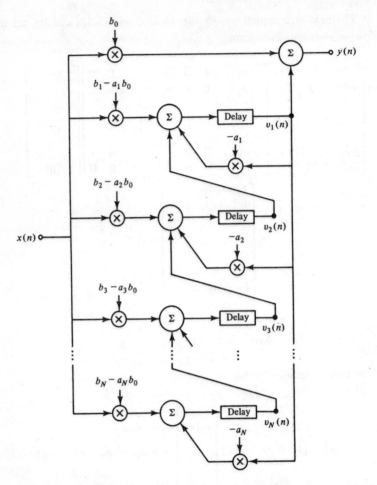

Figure 5.13 State space realization from nested programming method.

Find state variable realizations for the system using both the method of direct programing and the method of nested programming and present the block diagram for each realization.

Solution. First, put the $H(z)$ in the standard form given in (5.1) by multiplying out the factors and expressing the result in negative powers of z as

$$H(z) = \frac{\overset{b_0}{8} \overset{b_1}{-4}z^{-1} \overset{b_2}{+11}z^{-2} \overset{b_3}{-2}z^{-3}}{1 \underset{a_1}{-\tfrac{5}{4}z^{-1}} \underset{a_2}{+\tfrac{3}{4}z^{-1}} \underset{a_3}{-\tfrac{1}{8}z^{-3}}}$$

Recognizing b_k and a_k, the direct programming state variable form, including both state transition equation and output equation, can be written from (5.39) and (5.40) as

$$\begin{bmatrix} v_1(n+1) \\ v_2(n+1) \\ v_3(n+1) \end{bmatrix} = \begin{bmatrix} \frac{5}{4} & -\frac{3}{4} & \frac{1}{8} \\ 1 & 0 & 0 \\ 0 & 1 & 0 \end{bmatrix} \begin{bmatrix} v_1(n) \\ v_2(n) \\ v_3(n) \end{bmatrix} + \begin{bmatrix} 1 \\ 0 \\ 0 \end{bmatrix} x(n)$$

$$y(n) = \begin{bmatrix} 6 & 5 & -1 \end{bmatrix} \begin{bmatrix} v_1(n) \\ v_2(n) \\ v_3(n) \end{bmatrix} + 8x(n)$$

The state variable block diagram appears in Fig. 5.14(a). Using (5.55) and (5.56), the state variable realization by the nested programming approach is as follows:

$$\begin{bmatrix} v_1(n+1) \\ v_2(n+1) \\ v_3(n+1) \end{bmatrix} = \begin{bmatrix} \frac{5}{4} & 1 & 0 \\ -\frac{3}{4} & 0 & 1 \\ \frac{1}{8} & 0 & 0 \end{bmatrix} \begin{bmatrix} v_1(n) \\ v_2(n) \\ v_3(n) \end{bmatrix} + \begin{bmatrix} 6 \\ 5 \\ -1 \end{bmatrix} x(n)$$

$$y(n) = \begin{bmatrix} 1 & 0 & 0 \end{bmatrix} \begin{bmatrix} v_1(n) \\ v_2(n) \\ v_3(n) \end{bmatrix} + 8x(n)$$

and the state variable realization appears in Fig. 5.14(b).

5.4.3 Transformed State Vector Realizations

There are many other state variable representations of linear discrete-time invariant systems. Other methods of special interest include a partial fraction expansion in terms of linear factors and an iterative programming method which uses cascade sections. These forms are easily obtained once the numerator and denominator of the transfer function have been put in factored form. An excellent presentation of these forms appears in Cadzow [2] if more information is desired.

We see that the state variables have been assigned as outputs of linear systems resulting from a particular type of decomposition of the transfer function.

It will now be shown that an infinite number of state variable representations can be obtained by performing a special type of linear transformation on an existing state variable representation. Consider the state model and output equation given by

$$\mathbf{v}(n+1) = \mathbf{A}\mathbf{v}(n) + \mathbf{B}x(n) \tag{5.57}$$
$$y(n) = \mathbf{C}^T\mathbf{v}(n) + dx(n) \tag{5.58}$$

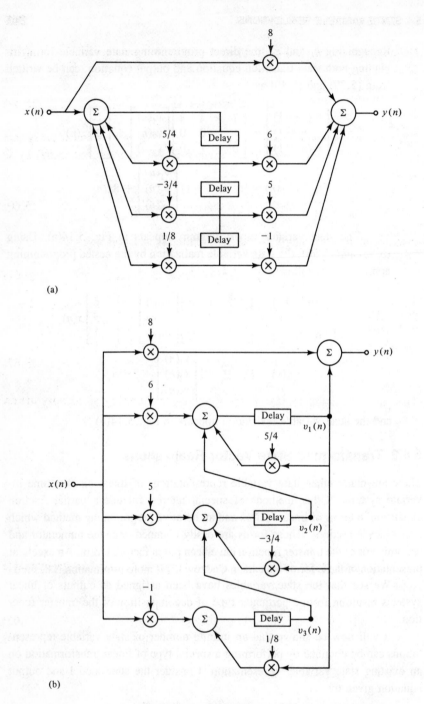

Figure 5.14 (a) Direct programming state variable realization for Example 5.4; (b) Nested programming state variable realization for Example 5.4.

Define a new vector $\mathbf{v}'(n)$ as the product of $\mathbf{v}(n)$ by a nonsingular $N \times N$ matrix $\mathbf{Q}$ as follows:

$$\mathbf{v}'(n) = \mathbf{Q}\mathbf{v}(n) \tag{5.59}$$

We wish to show that we have another state variable representation of the system with $\mathbf{v}'(n)$ as the new state vector. Premultiplying both sides of (5.57) by $\mathbf{Q}$ gives

$$\mathbf{Q}\mathbf{v}(n + 1) = \mathbf{Q}\mathbf{A}\mathbf{v}(n) + \mathbf{Q}\mathbf{B}x(n) \tag{5.60}$$

Inserting an identity matrix $\mathbf{Q}^{-1}\mathbf{Q}$ between the $\mathbf{A}$ and $\mathbf{v}(n)$ gives

$$\mathbf{Q}\mathbf{v}(n + 1) = \mathbf{Q}\mathbf{A}\mathbf{Q}^{-1}\mathbf{Q}\mathbf{v}(n) + \mathbf{Q}\mathbf{B}x(n) \tag{5.61}$$

When we recognize $\mathbf{Q}\mathbf{v}(n + 1)$ from (5.59) as $\mathbf{v}'(n + 1)$, and $\mathbf{Q}\mathbf{v}(n)$ as $\mathbf{v}'(n)$, (5.61) becomes

$$\mathbf{v}'(n + 1) = \mathbf{Q}\mathbf{A}\mathbf{Q}^{-1}\mathbf{v}'(n) + \mathbf{Q}\mathbf{B}x(n) \tag{5.62}$$

The output equation (5.58) can be rewritten by inserting an identity matrix between $\mathbf{C}^T$ and $\mathbf{v}(n)$ as

$$y(n) = \mathbf{C}^T\mathbf{Q}^{-1}\mathbf{Q}\mathbf{v}(n) + dx(n) \tag{5.63}$$

Using (5.59) again, (5.63) is seen to be

$$y(n) = \mathbf{C}^T\mathbf{Q}^{-1}\mathbf{v}'(n) + dx(n) \tag{5.64}$$

Equations (5.62) and (5.64) can now be put in the standard state variable form as

$$\mathbf{v}'(n + 1) = \mathbf{A}'\mathbf{v}'(n) + \mathbf{B}'x(n) \tag{5.65}$$
$$y(n) = \mathbf{C}'\mathbf{v}'(n) + d'x(n) \tag{5.66}$$

where $\mathbf{A}'$, $\mathbf{B}'$, $\mathbf{C}'$, and d' are given by

$$\mathbf{A}' = \mathbf{Q}\mathbf{A}\mathbf{Q}^{-1}, \qquad \mathbf{B}' = \mathbf{Q}\mathbf{B}$$
$$\mathbf{C}' = \mathbf{C}^T\mathbf{Q}^{-1}, \qquad d' = d$$

Therefore, we have shown that $\mathbf{Q}\mathbf{v}(n)$ for any nonsingular matrix $\mathbf{Q}$ is a legitimate choice for a state vector, thus providing an infinite number of possible state variable realizations.

5.5 REALIZATIONS OF FIR FILTERS

In the first part of this chapter, techniques have been presented for realizing IIR filters. We have also seen previously that many problems require an FIR filter or linear phase FIR filter for reasons of stability and steady state requirements. Certain realizations of FIR filters can be developed as special cases of the IIR structures previously presented, while others like the linear phase FIR filter have even simpler structures because of the symmetry of their impulse response. Several realizations for FIR filters are now discussed.

5.5.1 Realization of General FIR Filters

A causal FIR filter was characterized by its transfer function $H(z)$ given by

$$H(z) = \sum_{k=0}^{M} b_k z^{-k}$$

or a difference equation of the following form:

$$y(n) = \sum_{k=0}^{M} b_k x(n - k) \tag{5.67}$$

We see that the output is simply a weighted sum of present and past input values and does not depend upon previous output values. The block diagram for a direct realization of (5.67) is shown in Fig. 5.15. It is noticed that this is just a special case of the direct form I realization for an IIR filter where all the a_i are set equal to zero. This form corresponds to a tapped delay line structure.

An alternative realization can be obtained by writing the $H(z)$ in the following special factored form:

$$H(z) = b_0 + z^{-1}[b_1 + z^{-1}\{b_2 + z^{-1}\langle \cdots + z^{-1}(b_{M-1} + b_M z^{-1}) \cdots \rangle\}]$$

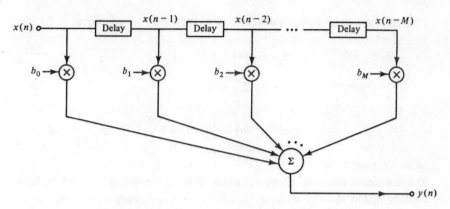

Figure 5.15 Direct form I realization of FIR filter.

$Y(z)$, which is $H(z)$ multiplied by $X(z)$, can be separated into two terms as follows:

$$Y(z) = b_0 X(z) + z^{-1} Y_1(z) \qquad (5.68a)$$

where $Y_1(z)$ is the transform of an intermediate result given by

$$Y_1(z) = b_1 X(z) + z^{-1} \{ b_2 + z^{-1} \langle \cdots + z^{-1} (b_{M-1} + b_M z^{-1}) \cdots \rangle \} X(z)$$

Taking the inverse transform of both sides of (5.68a), the output $y(n)$ is determined as the sum of $b_0 x(n)$ and a delayed version of the inverse of $Y_1(z)$ as follows:

$$y(n) = b_0 x(n) + y_1(n-1)$$

In a similar fashion, $Y_1(z)$ can be broken down into two terms as

$$Y_1(z) = b_1 X(z) + z^{-1} Y_2(z) \qquad (5.68b)$$

where $Y_2(z)$ is given by

$$Y_2(z) = b_2 X(z) + z^{-1} \langle \cdots + z^{-1} (b_{M-1} + b_M z^{-1}) \cdots \rangle X(z)$$

The inverse transform of both sides of (5.68b) gives

$$y_1(n) = b_1 x(n) + y_2(n-1)$$

This process can be continued, resulting in the following equations:

$$y_2(n) = b_2x(n) + y_3(n - 1)$$

$$\cdot$$
$$\cdot$$
$$\cdot$$

$$y_{M-1}(n) = b_{M-1}x(n) + y_M(n - 1)$$
$$y_M(n) = b_Mx(n)$$

The realization using the above equations showing the output $y(n)$ and the intermediate results $y_k(n)$ is given in Fig. 5.16. This realization corresponds to the IIR realization obtained by the nested programming method. In this alternative realization, it is observed that it is not the past values of the input that are stored but equivalent information via the weighted versions of previous outputs.

Various cascade realizations for the FIR filter also exist. The simplest form occurs when the system function is factored in terms of quadratic expressions in z^{-1} as follows:

$$H(z) = \prod_{k=1}^{K} (b_{0k} + b_{1k}z^{-1} + b_{2k}z^{-2}) \qquad (5.69)$$

Selecting the quadratic terms to correspond to the complex zeros of $H(z)$ allows a realization in terms of real coefficients b_{0k}, b_{1k}, and b_{2k}. Each quadratic could then be realized using the direct or alternative structures for the FIR filter given in Fig. 5.15 and 5.16. Using the direct form I for the realization of each quadratic section gives the realization shown in Fig. 5.17.

There also exist many parallel realizations for FIR filters based on Lagrange, Newton, and Hermite interpolation formulas. In general, these realizations require more multiplications, additions, and delays than the previously discussed realizations. If further information is desired, Oppenheim [6] should prove as a useful starting point.

5.5.2 Realization of Linear Phase FIR Filters

An important special subset of FIR filters has the linear phase characteristic. In Chapter 4, it was shown that if the impulse response was symmetric about its center, linear phase resulted. For a causal filter whose impulse response begins at zero and ends at $N - 1$, this symmetry was expressed for all $n = 0, 1, \ldots,$ $N - 1$ as follows:

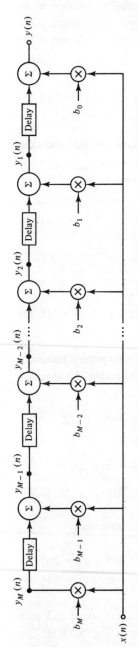

Figure 5.16 Alternative realization for FIR filter corresponding to nested programming form.

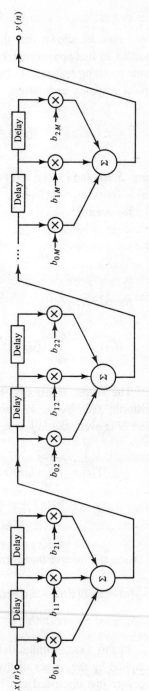

Figure 5.17 Alternative realization for an FIR system using cascaded quadratic sections.

$$h(n) = h(N - 1 - n) \cdot \tag{5.70}$$

It will now be shown that this symmetry allows the transfer function to be rewritten so that approximately half the number of multiplications are required for the resulting realization. The transfer function $H(z)$ is the $\mathcal{Z}$ transform of the impulse response as follows:

$$H(z) = \mathcal{Z}[h(n)] = \sum_{n=0}^{N-1} h(n)z^{-n} \tag{5.71}$$

Using (5.70) and (5.71), $H(z)$ can be easily written as follows for odd and even N.

for even N:

$$H(z) = \sum_{n=0}^{N/2-1} h(n)[z^{-n} + z^{-(N-1-n)}] \tag{5.72}$$

for odd N:

$$H(z) = \sum_{n=0}^{(N-3)/2} h(n)[z^{-n} + z^{-(N-1-n)}] + h((N-1)/2)z^{-(N-1)/2} \tag{5.73}$$

The output transform, $Y(z)$, can be obtained by multiplying the input transform, $X(z)$, by the system function given in (5.72) to obtain, for the case where N is even, the following:

$$Y(z) = \sum_{n=0}^{N/2-1} h(n)[z^{-n} + z^{-(N-1-n)}]X(z)$$

$$= h(0)[1 + z^{-(N-1)}]X(z) + h(1)[z^{-1} + z^{-(N-2)}]X(z)$$

$$+ \cdots + h(N/2 - 1)[z^{-(N/2-1)} + z^{-N/2}]X(z) \tag{5.74}$$

By taking the inverse transform of (5.74), the output $y(n)$ becomes

$$y(n) = h(0)[x(n) + x(n - (N - 1))] + h(1)[x(n - 1) + x(n - (N - 2))]$$

$$+ \cdots + h(N/2 - 1)[x(n - (N/2 - 1)) + x(n - N/2)] \tag{5.75}$$

In this case, various delayed versions of $x(n)$ are first added and then multiplied by the values of the impulse response as seen in Fig. 5.18(a). The procedure just discussed can be applied to (5.73), resulting in the realization

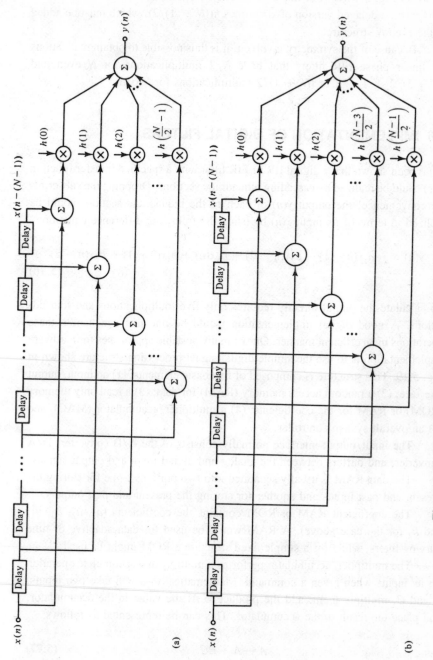

Figure 5.18 Realization of FIR linear phase filter; (a) N even, (b) N odd.

shown in Fig. 5.18(b). For this case of odd N, the last term of (5.73) gives an extra term, a delayed version of $x(n)$ times $h((N - 1)/2)$, which must be added to the even N structure.

Because of the symmetry involved, it is thus possible to obtain realizations for linear phase FIR filters that have $N/2$ multiplications for N even and $(N - 3)/2 + 1 + 1$ or $(N + 1)/2$ multiplications for N odd.

5.6 IMPLEMENTATION OF DIGITAL FILTERS

It has been shown that a digital IIR or FIR filter with a specified transfer function $H(z)$ could be realized by cascading biquadratic sections. Dropping the subscripts for convenience, the output $y(n)$ of each of the biquadratic sections could be realized in terms of its input $x(n)$ by using the following difference equation:

$$y(n) = -a_1 y(n - 1) - a_2 y(n - 2) + b_0 x(n) + b_1 x(n - 1) + b_2 x(n - 2) \tag{5.76}$$

To calculate the output directly requires only five multiplications and four additions. A good digital implementation would be one that can handle these operations in an efficient manner. One of many possible approaches for hardware implementation is to use the simple digital signal processing structure shown in Fig. 5.19. This structure is composed of five basic elements: (1) an input/output interface; (2) a random access memory (RAM) for data; (3) a read only memory (ROM) or RAM for the coefficients; (4) a multiplier/accumulator (MAC); and (5) an overall system controller.

The input/output interface normally consists of the A/D converter, D/A converter, and buffers between the analog and digital input and output signals.

The data RAM is usually separated into two portions—one for storing the present and past inputs and another for storing the present and past outputs.

The coefficient RAM or ROM contains the coefficients (a_1, a_2, b_0, b_1, and b_2 for the case above). A RAM would be used for data adaptive or time varying filters, while for a simple fixed structure a ROM might be used.

The multiplier/accumulator performs a multiply and accumulate operation on its inputs when given a command. Mathematically, it will take two inputs, B and C, multiply them, add the product to A, the value in the accumulator, and place the result in the accumulator. This can be represented as follows:

$$A \leftarrow A + BC \tag{5.77}$$

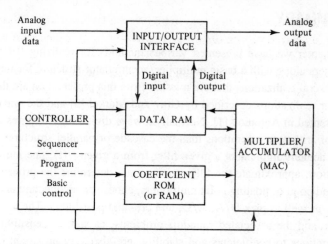

Figure 5.19   Simple digital signal processing structure.

From (5.76) we see that each biquadratic section of a filter can be implemented by a sequence of five such operations after the accumulator has been cleared. Figure 5.20 gives the operations and the resulting accumulator values at each step. Notice that B and C change at each multiplication, so we must have a way of providing these different inputs to the MAC in the proper sequence.

The controller serves this purpose and can be thought of as an element which is composed of a "program" for the filter, sequencer for the program, and control unit to provide all the necessary signals for implementation.

To implement the cascade form of K sections we require that the output of each proceding seciton be determined before the calculation of the next section output. If, however, the filter were realized in a parallel structure, a MAC could be used for each section and all section outputs calculated at the same time. Thus, a reduction in time is obtained for the calculations at the expense of having K-duplicated MACs.

Number	Operation	B	C	A
0	Clear	−	−	0
1	MA	$-a_1$	$y(n-1)$	$-a_1y(n-1)$
2	MA	$-a_2$	$y(n-2)$	$-a_1y(n-1)-a_2y(n-2)$
3	MA	b_0	$x(n)$	$-a_1y(n-1)-a_2y(n-2)+b_0x(n)$
4	MA	b_1	$x(n-1)$	$-a_1y(n-1)-a_2y(n-2)+b_0x(n)+b_1x(n-1)$
5	MA	b_2	$x(n-2)$	$-a_1y(n-1)-a_2y(n-2)+b_0x(n)+b_1x(n-1)+b_2x(n-2)$

Figure 5.20   Operations for implementation of a biquadratic section using a multiply and accumulate ($A \leftarrow A + BC$) function.

5.7 SUMMARY

In this chapter we have presented various methods for realizing IIR and FIR digital filters along with a basic multiplier/accumulator structure for implementation. Several realizations that are missing from this presentation are the lattice and ladder realizations described by Gray and Markel [3] and the wave digital filter presented in Antoniou [1]. Normally a wave digital filter requires a greater number of arithmetic operations than the cascade or parallel structures.

To actually implement a given filter from a given $H(z)$, one must choose a realization, approximate the filter coefficients, determine the proper scaling, select fixed point or floating point multipliers, choose an architecture and control structure, as well as pick D/A, A/D, and pre- and post-filters. The use of finite arithmetic and the associated roundoff problems, as well as sensitivity of the filter operations to coefficients and stability, are also very important considerations. It was not the purpose of this text to investigate these problems, and in many cases they can only be analyzed using the theory of random processes and probability, which are beyond the scope of this book. The interested reader will find beginning and at times still incomplete treatments of these problems in Refs. Oppenheim and Schafer [6] and Antoniou [1].

REFERENCES

1. Antoniou, Andreas. *Digital Filters: Analysis and Design.* McGraw-Hill, New York, 1979.
2. Cadzow, James A., and H. R. Martens. *Discrete-Time Computer Control Systems.* Prentice-Hall, Englewood Cliffs, NJ, 1970.
3. Gray, A. H., and J. D. Markel, "Digital Lattice and Ladder Filter Synthesis," *IEEE Transactions on Audio and Electroacoustics* Au-21 (December 1973), 491–500.
4. Gupta, S. C. *Transform and State Variable Methods in Linear Systems.* Wiley, New York, 1966.
5. Markhoul, J. "A Class of all Zero Lattice Digital Filters: Properties and Applications," *IEEE Transactions on Acoustics, Speech, and Signal Processing* ASSP-26 (August 78), 304–314.
6. Oppenheim, A. V., and R. W. Schafer. *Digital Signal Processing*, Prentice-Hall, Englewood Cliffs, NJ, 1975.
7. Peled, A., and B. Liu. *Digital Signal Processing.* Wiley, New York, 1976.
8. Shirm, L., IV, "Multiplier/Accumulator Application Notes," TRW LSI Products, La Jolla, CA, January 1980.
9. Stinaff, K. D., "A Microprocessor-Based Implementation for Programmble Digital Filters," NAECON'76 Record, pp. 463–470.

10. "Terminology in Digital Signal Processing," *IEEE Transactions on Audio and Electroacoustics* Au-20 (December 1972), 322–337.

PROBLEMS

5.1 A system is specified by its transfer function $H(z)$ given by

$$H(z) = \frac{(z - 1)(z - 2)(z + 1)z}{[z - (\frac{1}{2} + j\frac{1}{2})][z - (\frac{1}{2} - j\frac{1}{2})](z - j1/4)(z + j\frac{1}{4})}$$

Realize the system in the following forms:
- **(a)** Direct form I
- **(b)** Direct form II
- **(c)** Cascade of two biquadratic sections
- **(d)** A parallel realization in constant, linear, and biquadratic sections.

5.2 Realize an FIR filter with impulse response $h(n)$ given by

$$h(n) = (\tfrac{1}{2})^n [u(n) - u(n - 5)]$$

using (a) direct form I, (b) form corresponding to that obtained by nested programming, (c) cascade of quadratic sections.

5.3 Realize linear phase filters with the following impulse responses using the structures shown in Fig. 5.18:
- **(a)** $h(n) = \delta(n) + \frac{1}{2}\delta(n - 1) - \frac{1}{4}\delta(n - 2) + \frac{1}{2}\delta(n - 3) + \delta(n - 4)$
- **(b)** $h(n) = \delta(n) - \frac{1}{2}\delta(n - 1) + \frac{1}{4}\delta(n - 2) + \frac{1}{4}\delta(n - 3) - \frac{1}{2}\delta(n - 4) + \delta(n - 5)$

5.4 Find the system function $H(z)$ for each impulse response in Problem 5.3 and plot its poles and zeros. Realize each filter as a cascade of linear and quadratic sections.

5.5 A system is specified by the following $H(z)$:

$$H(z) = \frac{z + 1}{(z - \frac{5}{4})(z - \frac{1}{4})}$$

- **(a)** Obtain the state space representation for the system in the form of Eqs. (5.28) and (5.30) and draw the corresponding state space realization (i) by the method of direct programming, (ii) by the method of nested programming, and (iii) as a parallel combination of simple poles.
- **(b)** Find a matrix **Q** that will change the parallel state variable form to the form obtained by the direct programming method.

5.6 A system is represented by its transfer function $H(z)$ given by

$$H(z) = 3 + \frac{4z}{z - \frac{1}{2}} - \frac{2}{z - \frac{1}{4}}$$

- **(a)** Does this $H(z)$ represent a FIR or an IIR filter? Why?
- **(b)** Give a difference equation realization of this system using direct form 1.
- **(c)** Draw the block diagram for the direct form II canonic realization, and give the governing equations for implementation.

(d) Find the state space realization using the method of nested programming and illustrate.

(e) Form a new state vector that is $\mathbf{Q}$ times the state vector from the realization in (d), where $\mathbf{Q}$ is

$$\mathbf{Q} = \begin{bmatrix} 1 & 2 \\ 0 & 4 \end{bmatrix}$$

Find the resulting state space realization and illustrate.

(f) Using the $H(z)$ above, realize and illustrate the parallel form which is shown in Figure 5.10 using constants and single pole terms.

Chapter 6

The Discrete Fourier Transform

6.0 INTRODUCTION

In solving many engineering problems it is convenient to decompose a given signal into the weighted sum of basis functions. If such a signal is the input to a linear time invariant system then superposition can be applied and the output can be obtained by adding up the weighted responses to each of the basic functions. Since we have procedures for calculating the output of linear systems to complex exponentials and sinusoidal signals a reasonable choice for the basic functions for analyzing linear systems would be complex exponentials and sinusoidal signals.

For periodic signals this choice gives the Fourier series. In this chapter the Fourier series expansion for continuous-time periodic signals will be reviewed followed by a discussion of the Fourier series expansion for discrete-time signals. Next, the definition and properties of the discrete Fourier transform along with efficient methods for computing the discrete Fourier transform are developed. These efficient methods, called fast Fourier transform algorithms, can be used to advantage in many digital signal processing applications. In particular the functions of linear filtering and spectral analysis are addressed by interpreting DFT results in terms of the frequency content of discrete-time signals and by using the DFT to perform linear convolution, which is the basic linear filtering operation.

6.1 CONTINUOUS-TIME FOURIER SERIES

It is well known that a *periodic* continuous time signal $f(t)$ with period T can be expressed as a weighted sum of a *countable* number of complex exponential *continuous* time functions in the following way

$$f(t) = \sum_{k=-\infty}^{\infty} F_k e^{jk\Omega_0 t} \qquad \text{for all } t \tag{6.1}$$

where F_k, the so-called coefficients of the expansion, and Ω_0, the fundamental frequency, are determined by

$$F_k = \frac{1}{T} \int_0^T f(t) e^{-jk\Omega_0 t} dt \qquad \text{for all } k$$
$$\Omega_0 = 2\pi/T \tag{6.2}$$

The expansion given by Eq. (6.1) is called the complex exponential Fourier series representation of a continuous-time periodic signal $f(t)$. In many cases it is convenient to use the trigonometric form of the Fourier series which expresses a periodic signal in terms of a weighted sum of sines and cosines as

$$f(t) = a_0 + \sum_{k=1}^{\infty} (a_k \cos k\Omega_0 t + b_k \sin k\Omega_0 t) \tag{6.3}$$

where the constants a_0, a_k, and b_k can be determined as

$$a_0 = \frac{1}{T} \int_0^T f(t) \, dt \tag{6.4a}$$

$$a_k = \frac{2}{T} \int_0^T f(t) \cos k\Omega_0 t \, dt, \qquad k = 1, 2, \ldots \tag{6.4b}$$

$$b_k = \frac{2}{T} \int_0^T f(t) \sin k\Omega_0 t \, dt, \qquad k = 1, 2, \ldots \tag{6.4c}$$

The constants a_k, b_k, of the trigonometric form, and F_k, of the complex exponential form, are easily shown to be related by

$$a_0 = F_0 \tag{6.5a}$$
$$a_k = F_k + F_{-k}, \qquad k = 1, 2, \ldots \tag{6.5b}$$
$$b_k = (F_{-k} - F_k)/j, \qquad k = 1, 2, \ldots \tag{6.5c}$$

6.2 DISCRETE-TIME FOURIER SERIES

In a similar fashion a real periodic discrete-time signal $x(n)$ of period N can be expressed as a weighted sum of complex exponential *sequences*. Because of the fact that sinusoidal sequences are unique only for digital frequencies from 0 to 2π, the expansion contains only a *finite* number of complex exponentials as follows:

$$x(n) = \frac{1}{N} \sum_{k=0}^{N-1} X(k)e^{jk\omega_0 n} \qquad \text{for all } n \qquad (6.6)$$

where the coefficients of the expansion $X(k)$ and the fundamental digital frequency ω_0 are given by

$$X(k) = \sum_{n=0}^{N-1} x(n)e^{-jk\omega_0 n} \qquad \text{for all } k$$
$$\omega_0 = 2\pi/N \qquad (6.7)$$

Equations (6.6) and (6.7) are called the discrete Fourier series (DFS) pair. The expression given in Eq. (6.6) is referred to as the exponential form of the Fourier series for a periodic discrete-time signal. Equation (6.7) is sometimes written in the following equivalent form:

$$X(k) = \sum_{n=0}^{N-1} x(n) W_N^{kN} \qquad (6.7a)$$

where W_N is defined by

$$W_N = e^{-j2\pi/N} \qquad (6.7b)$$

An alternative form of the discrete Fourier series, comparable to Eq. (6.3) for the continuous-time periodic signal, can be easily found. However, it has different expressions for odd and even N. We have for odd and even N the following:

Even N:

$$x(n) = A(0) + \sum_{k=1}^{N/2-1} A(k)\cos\left(k\frac{2\pi}{N}n\right) + \sum_{k=1}^{N/2-1} B(k)\sin\left(k\frac{2\pi}{N}n\right)$$
$$+ A\left(\frac{N}{2}\right)\cos\pi n \qquad (6.8a)$$

Odd N:

$$x(n) = A(0) + \sum_{k=1}^{(N-1)/2} A(k) \cos\left(k\frac{2\pi}{N}n\right) + \sum_{k=1}^{(N-1)/2} B(k) \sin\left(k\frac{2\pi}{N}n\right) \quad (6.8b)$$

The expressions given in (6.8a) and (6.8b) are called the trigonometric form of the Fourier series for a periodic discrete-time signal. In Eq. (6.8a) the last term contains $\cos \pi n$, which is just $(-1)^n$, the highest-frequency sequence possible. The constants $A(0)$, $A(k)$, and $B(k)$ can be shown, for N even, to be

$$A(0) = \frac{1}{N}\sum_{n=0}^{N-1} x(n) \quad (6.9a)$$

$$A(k) = \frac{2}{N}\sum_{n=0}^{N-1} x(n)\cos\left(k\frac{2\pi}{N}n\right), \qquad k = 1, 2, \ldots, \frac{N}{2} - 1 \quad (6.9b)$$

$$B(k) = \frac{2}{N}\sum_{n=0}^{N-1} x(n)\sin\left(k\frac{2\pi}{N}n\right), \qquad k = 1, 2, \ldots, \frac{N}{2} - 1 \quad (6.9c)$$

$$A(N/2) = \frac{1}{N}\sum_{n=0}^{N-1} x(n)\cos \pi n \quad (6.9d)$$

If N is odd, $A(0)$ remains the same as Eq. (6.9a); however, the $A(k)$ and $B(k)$ of Eqs. (6.9b) and (6.9c) are good for $k = 1, 2, \ldots, (N - 1)/2$, and there will be no $A(N/2)$ coefficent.

The relationships between the $A(k)$, $B(k)$, of the trigonometric form, and the $X(k)$ of the exponential form for a real $x(n)$ with an even N are given by

$$A(0) = X(0)/N \quad (6.10a)$$

$$A(k) = [X(k) + X(N - k)]/N, \qquad k = 1, 2, \ldots, \frac{N}{2} - 1 \quad (6.10b)$$

$$B(k) = j[X(k) - X(N - k)]/N, \qquad k = 1, 2, \ldots, \frac{N}{2} - 1 \quad (6.10c)$$

$$A(N/2) = X(N/2)/N \quad (6.10d)$$

Again, if N is odd, (6.10a) remains the same, (6.10b) and (6.10c) will be good for $k = 1, 2, \ldots, (N - 1)/2$, and there will be no $A(N/2)$. Example 6.1 shows both the exponential and trigonometric forms of the discrete Fourier series for a particular periodic signal.

EXAMPLE 6.1

Find both the exponential and trigonometric forms of the discrete Fourier series representation of the $x(n)$ shown.

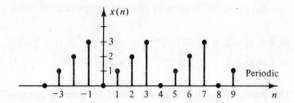

Solution. The exponential form of the Fourier series is specified once the $X(k)$ are calculated from either Eq. (6.7) or Eq. (6.7a). Before carrying out these calculations, using Eq. (6.7a) the powers of W_N must be determined for $N = 4$. W_4 is given by

$$W_4 = e^{-j(2\pi/4)} = \cos\frac{\pi}{2} - j\sin\frac{\pi}{2} = -j$$

The other powers of W_4 are easily seen to be

$$W_4^2 = W_4^1 \cdot W_4^1 = (-j)(-j) = -1$$
$$W_4^3 = W_4^2 \cdot W_4^1 = (-1)(-j) = j$$
$$W_4^4 = W_4^3 \cdot W_4^1 = (j)(-j) = 1$$

Using these powers of W_4, the $X(k)$ for $k = 0, 1, 2, 3$ are calculated as follows:

$$X(0) = \sum_{n=0}^{4-1} x(n)W_4^{0 \cdot n} = x(0) + x(1) + x(2) + x(3) = 0 + 1 + 2 + 3$$
$$= 6$$

$$X(1) = \sum_{n=0}^{4-1} x(n)W_4^{1 \cdot n} = x(0)W_4^0 + x(1)W_4^1 + x(2)W_4^2 + x(3)W_4^3$$
$$= 0(1) + 1(-j) + 2(-1) + 3(j) = -2 + 2j$$

$$X(2) = \sum_{n=0}^{4-1} x(n)W_4^{2 \cdot n} = x(0)W_4^0 + x(1)W_4^2 + x(2)W_4^4 + x(3)W_4^6$$
$$= 0(1) + 1(-1) + 2(1) + 3(-1) = -2$$

$$X(3) = \sum_{n=0}^{4-1} x(n)W_4^{3 \cdot n} = x(0)W_4^0 + x(1)W_4^3 + x(2)W_4^6 + x(3)W_4^9$$
$$= x(0)W_4^0 + x(1)W_4^3 + x(2)W_4^2 + x(3)W_4^1$$
$$= 0(1) + 1(j) + 2(-1) + 3(-j) = -2 - 2j$$

Therefore $x(n)$ can be written in complex exponential form as

$$x(n) = \tfrac{3}{2} + [(-2 + 2j)/4]W_4^{-n} + (-\tfrac{1}{2})W_4^{-2n} + [(-2 - 2j)/4]W_4^{-3n}$$
$$= \tfrac{3}{2} + [(-1 + j)/2]e^{j(\pi/2)n} - (-\tfrac{1}{2})e^{j\pi n} - [(1 + j)/2]e^{j(3\pi)n}$$

The complex multiplications could be carried out and simplified to give the corresponding trigonometric form; however, as that is the procedure that developed Eq. (6.10), we might as well use it directly. From Eq. (6.10), the $A(0)$, $A(1)$, $B(1)$, and $A(2)$ can be found as

$$A(0) = X(0)/N = \tfrac{6}{4} = \tfrac{3}{2}$$
$$A(1) = [X(1) + X(4 - 1)]/4 = [(-2 + 2j) + (-2 - 2j)]/4 = -1$$
$$B(1) = j[X(1) - X(4 - 1)]/4 = j[(-2 + 2j) - (-2 - 2j)]/4 = -1$$
$$A(2) = X(\tfrac{4}{2})/4 = X(2)/4 = -\tfrac{1}{2}$$

Therefore the trigonometric form of the discrete Fourier series of Example 6.1 is found from Eq. (6.8) to be

$$x(n) = \frac{3}{2} - \cos\left(\frac{\pi}{2}n\right) - \sin\left(\frac{\pi}{2}n\right) - \frac{1}{2}\cos \pi n$$

From the calculations in Example 6.1, it is apparent that the $X(k)$'s can be written in matrix form as follows:

$$\begin{bmatrix} X(0) \\ X(1) \\ X(2) \\ X(3) \end{bmatrix} = \begin{bmatrix} W_4^0 & W_4^0 & W_4^0 & W_4^0 \\ W_4^0 & W_4^1 & W_4^2 & W_4^3 \\ W_4^0 & W_4^2 & W_4^0 & W_4^2 \\ W_4^0 & W_4^3 & W_4^2 & W_4^1 \end{bmatrix} \begin{bmatrix} x(0) \\ x(1) \\ x(2) \\ x(3) \end{bmatrix}$$

The entries in the coefficient matrix have been reduced in powers of W_4 by using the fact that $W_4^4 = 1$.

Fast ways of making the calculations above can be shown to be equivalent to decomposing the square matrix into products of relatively sparse symmetric matrices [1]. This area will be explored further in the sections on the fast Fourier transform.

6.3 THE DISCRETE FOURIER TRANSFORM

Many times it is convenient to expand a finite duration sequence in a discrete Fourier series. The expansion that is obtained periodically extends the given finite duration sequence and so is correct for only the finite duration. Such a representation is referred to as the discrete Fourier transform (DFT) of the sequence. In this way, if $x(n)$ is a sequence defined only over the interval from 0 to $N - 1$, the DFT, $X(k)$, of $x(n)$ is defined only over the same interval from 0 to $N - 1$ by

$$X(k) = \sum_{n=0}^{N-1} x(n)e^{-jk\omega_0 n} \qquad 0 \le k \le N - 1 \qquad (6.11)$$

where $\omega_0 = 2\pi/N$. The corresponding inverse discrete Fourier transform (IDFT) of the sequence $X(k)$ gives a sequence $x(n)$ defined only on the interval from 0 to $N - 1$ as follows:

$$x(n) = \frac{1}{N} \sum_{k=0}^{N-1} X(k)e^{jk\omega_0 n}, \qquad 0 \le n \le N - 1 \qquad (6.12)$$

Equations (6.11) and (6.12) constitute the discrete Fourier transform pair and are sometimes rewritten in terms of W_N as follows:

$$X(k) = \sum_{n=0}^{N-1} x(n)W_N^{kn}, \qquad 0 \le k \le N - 1 \qquad (6.13)$$

$$x(n) = \frac{1}{N} \sum_{k=0}^{N-1} X(k)W_N^{-kn}, \qquad 0 \le n \le N - 1 \qquad (6.14)$$

where
$$W_N = e^{-j2\pi/N}$$

For notational purposes the discrete Fourier transform and inverse discrete Fourier transform given in Eqs. (6.13) and (6.14) are sometimes represented by

$$X(k) = \text{DFT}[x(n)]$$

and
$$x(n) = \text{IDFT}[X(k)]$$

A pictorial summary of the equations for the nonperiodic continuous-time Fourier transform, periodic continuous-time Fourier series, nonperiodic discrete-time Fourier transform, periodic discrete-time Fourier series and the fixed length discrete-time discrete Fourier transform appears in Fig. 6.1.

A number of basic properties for the discrete Fourier transform are now presented. These properties, including linearity, multiplication by exponentials, and the two new operations of circular translation and convolution, are summarized in Table 6.1. Also included in the table are two sinusoidal forms that will prove useful in spectral analysis.

NONPERIODIC CONTINUOUS-TIME

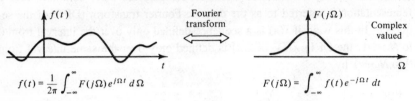

$$f(t) = \frac{1}{2\pi} \int_{-\infty}^{\infty} F(j\Omega) e^{j\Omega t} \, d\Omega \qquad\qquad F(j\Omega) = \int_{-\infty}^{\infty} f(t) e^{-j\Omega t} \, dt$$

PERIODIC CONTINUOUS-TIME

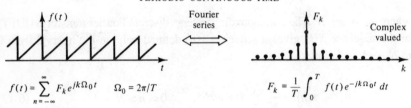

$$f(t) = \sum_{n=-\infty}^{\infty} F_k e^{jk\Omega_0 t} \quad \Omega_0 = 2\pi/T \qquad\qquad F_k = \frac{1}{T} \int_{0}^{T} f(t) e^{-jk\Omega_0 t} \, dt$$

NONPERIODIC DISCRETE-TIME

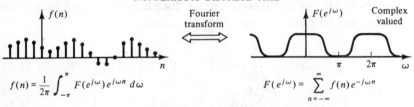

$$f(n) = \frac{1}{2\pi} \int_{-\pi}^{\pi} F(e^{j\omega}) e^{j\omega n} \, d\omega \qquad\qquad F(e^{j\omega}) = \sum_{n=-\infty}^{\infty} f(n) e^{-j\omega n}$$

PERIODIC DISCRETE-TIME

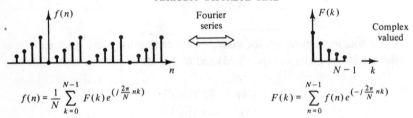

$$f(n) = \frac{1}{N} \sum_{k=0}^{N-1} F(k) e^{\left(j \frac{2\pi}{N} nk \right)} \qquad\qquad F(k) = \sum_{n=0}^{N-1} f(n) e^{\left(-j \frac{2\pi}{N} nk \right)}$$

FIXED LENGTH DISCRETE-TIME

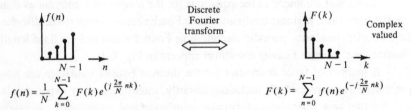

$$f(n) = \frac{1}{N} \sum_{k=0}^{N-1} F(k) e^{\left(j \frac{2\pi}{N} nk \right)} \qquad\qquad F(k) = \sum_{n=0}^{N-1} f(n) e^{\left(-j \frac{2\pi}{N} nk \right)}$$

Figure 6.1 A pictorial summary of equations for the Fourier representation of continuous-time and discrete-time signals.

TABLE 6.1 USEFUL PROPERTIES OF THE DFT

1. Linearity

$$DFT[a_1x_1(n) + a_2x_2(n)] = a_1DFT[x_1(n)] + a_2DFT[x_2(n)]$$

2. Circular shift

$$DFT[x((n - m) \bmod N)] = W_N^{km}X(k)$$

3. Multiplication by complex exponential sequence

$$DFT[W_N^{-ln}x(n)] = X[(k - l) \bmod N]$$

4. Circular convolution (time domain)

$$DFT \sum_{m=0}^{N-1} x(m)y[(n - m) \bmod N] = X(k)Y(k)$$

5. Product (time domain)

$$DFT[x(n)y(n)] = \frac{1}{N} \sum_{l=0}^{N-1} X(l \bmod N) \, Y((k - l) \bmod N)$$

6. Conjugation

$$DFT[x^*(n)] = X^*(-k \bmod N)$$
$$DFT[x^*(-n \bmod N)] = X^*(k)$$

The following properties are true for *real x(n)* only:

7. $X(k) = X^*(N - k)$ for $k = 1, 2, \ldots, N - 1$, which implies the following:
   (a) Re $[X(k)]$ = Re $[X(N - k)]$
   (b) Im $[X(k)]$ = $-$ Im $[X(N - k)]$
   (c) arg $[X(k)]$ = $-$ arg $[X(N - k)]$

8. *Even N*:

$$x(n) = \frac{X(0)}{N} + \frac{1}{N} \sum_{k=1}^{N/2-1} 2|X(k)| \cos\left\{ \frac{2\pi k}{N}n + \tan^{-1}\left[\frac{X_I(k)}{X_R(k)} \right] \right\} + \frac{X(N/2)}{N} \cos \pi n$$

Odd N:

$$x(n) = \frac{X(0)}{N} + \frac{1}{N} \sum_{k=1}^{(N-1)/2} 2|X(k)| \cos\left\{ \frac{2\pi k}{N}n + \tan^{-1}\left[\frac{X_I(k)}{X_R(k)} \right] \right\}$$

6.3.1 Linearity

Because the DFT is given by a sum, it is easy to see for two sequences, $x_1(n)$ and $x_2(n)$, defined on interval $[0, N - 1]$ that

$$DFT[a_1x_1(n) + a_2x_2(n)] = a_1DFT[x_1(n)] + a_2DFT[x_2(n)]$$

Therefore the DFT is linear.

6.3.2 Circular Translation

Given a sequence $x(n)$ defined for all n, the shifted version of $x(n)$ is written $x(n - n_0)$, where n_0 represents the number of indices that the sequence $x(n)$ is shifted to the right. When working with the DFT the sequences are only defined for time index values from 0 to $N - 1$. Therefore, if a regular shift is attempted, parts of the sequence would fall outside the region of interest and the first part of the sequence would be undefined. When using the DFT, it was noted earlier that although not defined outside the interval from 0 to $N - 1$, the mathematical framework was periodic. Because of this, the fundamental form of sequence translation useful in a mathematical sense is a circular translation.

The circular translation or shift of $x(n)$, defined on 0 to $N - 1$, by an amount n_0 to the right is denoted by $x((n - n_0) \bmod N)$. The operation can be thought of as wrapping the part that falls outside the region of interest around to the front of the sequence, or equivalently just a straight translation of the periodic extension outside of 0 to $N - 1$ of the given sequence. Circularly shifted versions of $x(n)$ are shown below for $n_0 = 1, 2, \ldots, N$:

$$x(n) \sim [x(0), x(1), \ldots, x(N - 3), x(N - 2), x(N - 1)]$$

$$x((n - 1) \bmod N) \sim [x(N - 1), x(0), x(1), \ldots, x(N - 3), x(N - 2)]$$
$$x((n - 2) \bmod N) \sim [x(N - 2), x(N - 1), x(0), x(1), \ldots, x(N - 3)]$$

$$\cdot$$
$$\cdot$$
$$\cdot$$

$$x((n - n_0) \bmod N) \sim [x(N - n_0), x(N - n_0 + 1), \ldots, x(N - n_0 - 1)]$$

$$\cdot$$
$$\cdot$$
$$\cdot$$

$$x((n - N) \bmod N) \sim [x(0), x(1), \ldots, x(N - 3), x(N - 2), x(N - 1)]$$

It is seen that a shift of $n_0 = N$ results in the original signal $x(n)$.

Another way of visualizing circular translation is to consider the sequence values to be placed on the rim of a wheel with fixed reference as shown in Fig. 6.2. A rotation of the wheel gives a new assignment to the reference, thus giving the circularly translated signal.

6.3.3 Multiplication by Exponentials

If a sequence $x(n)$ defined on $[0, N - 1]$ is circularly translated by an amount m, its DFT can be determined from the DFT of $x(n)$ as follows:

$$\text{DFT}[x((n - m) \bmod N)] = W_N^{km} X(k)$$

Correspondingly, if the $X(k)$ is circularly shifted, the resulting inverse transform will be the multiplication of the inverse of $X(k)$ by a complex exponential as seen in the following property.

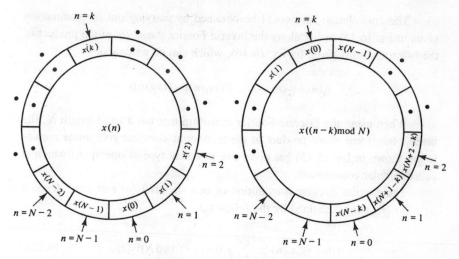

Figure 6.2 Circular translation.

If the DFT$[x(n)] = X(k)$, then

$$\text{DFT}\,[W_N^{-ln}x(n)] = X((k-l)\,\text{mod}\,N)$$

It is important to note from the above property that a shift in the frequency domain values $X(k)$ generally results in a complex inverse sequence $x(n)$ even though the original sequence in the time domain could have been real.

6.3.4 Circular Convolution

In working with discrete-time signals that could exist for index times possibly infinite in extent, it was shown previously that the Fourier transform of the convolution sum of the signals was the product of the transforms of each of the signals.

The convolution sum for two signals $x_1(n)$ and $x_2(n)$, sometimes referred to as linear convolution, is given by

$$x_1(n) * x_2(n) = \sum_{k=-\infty}^{\infty} x_1(n-k)x_2(k) \quad \text{or} \quad \sum_{k=-\infty}^{\infty} x_1(k)x_2(n-k) \quad (6.15)$$

The corresponding Fourier transform of the new sequence $x_1(n) * x_2(n)$ was shown to be the product of $X_1(e^{j\omega})$ and $X_2(e^{j\omega})$, the Fourier transform of $x_1(n)$ and $x_2(n)$, respectively:

$$\mathcal{F}[x_1(n) * x_2(n)] = X_1(e^{j\omega})X_2(e^{j\omega}) \quad (6.16)$$

The convolution sum could be obtained by carrying out the summation given in Eq. (6.15) or by taking the inverse Fourier transform of the product of the two transforms given by Eq. (6.16), which can be written as

$$x_1(n) * x_2(n) = \mathcal{F}^{-1}\{\mathcal{F}[x_1(n)] \cdot \mathcal{F}[x_2(n)]\} \qquad (6.17)$$

When using the discrete Fourier transform that has a fixed length N, the inverse transform of the product of the transforms does not give linear convolution shown in Eq. (6.15) but gives rise to a new type of operation, which is called circular convolution.

The N-point circular convolution of two signals $x_1(n)$ and $x_2(n)$ denoted by $x_1(n) \circledast_N x_2(n)$ is defined by the following:

$$x_1(n)\circledast_N x_2(n) \triangleq \sum_{k=0}^{N-1} x_1((n-k)\bmod N)x_2(k) \qquad (6.18a)$$

$$= \sum_{k=0}^{N-1} x_1(k)x_2((n-k)\bmod N) \qquad (6.18b)$$

where $x_1((n-k)\bmod N)$ is the reflected and circularly translated version of $x_1(n)$. The mechanics of circular convolution are shown in Example 6.2.

EXAMPLE 6.2

Let $x_1(n)$ and $x_2(n)$ be the two sequences shown below. Find $x_1(n)\circledast_4 x_2(n)$, the four-point circular convolution of $x_1(n)$ and $x_2(n)$, by using the circular convolution formula shown in Eq. (6.18b).

Solution.   Let $y(n)$ be the circular convolution as follows:

$$y(n) = x_1(n)\circledast_4 x_2(n)$$

The values of $y(n)$ for each of $n = 0$, 1, 2, and 3 can be determined by sequence term-by-term multiplication, followed by an addition as shown in Fig. 6.3. For $n = 0$, $x_2(n)$ is circularly reflected to give the sequence (0, 3, 2, 1). The term-by-term multiplication with (1, 2, 2, 0) is easily obtained and the sum of its terms gives 10, which is the value of $y(0)$, the circular convolution evaluated at $n = 0$. For $n = 1$, the sequence (0, 3, 2, 1) is circularly translated by 1 to give (1, 0, 3, 2), and corresponding multiplication and summing gives a $y(1)$ of 7. Values for $y(2)$ and $y(3)$ are similarly calculated and the final sequence is given as (10, 7, 4, 9) as plotted in Fig. 6.3.

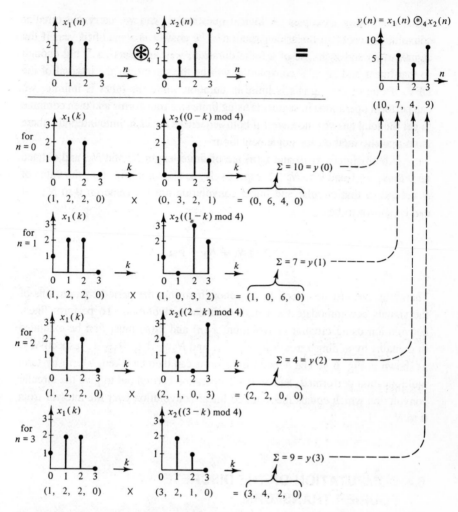

Figure 6.3 Four-point circular convolution of discrete-time signals $x_1(n)$ and $x_2(n)$ for Example 6.2.

It can be shown that if $X_1(k)$ and $X_2(k)$ represent the N-point DFTs of $x_1(n)$ and $x_2(n)$, respectively, circular convolution can be obtained by taking the IDFT of the product of the DFTs of $x_1(n)$ and $x_2(n)$.

$$x_1(n) \circledast_N x_2(n) = \text{IDFT}_N\{\text{DFT}_N[x_1(n)] \cdot \text{DFT}_N[x_2(n)]\}$$

This means that circular convolution is the natural operation when working with DFTs. However, linear convolution is many times necessary in digital signal processing examples. A logical question is, can we carry out circular convolution to obtain linear convolution? The answer is a qualified yes. If the signals $x_1(n)$ and $x_2(n)$ are of a finite duration, we can select an N big enough so that linear and circular convolution will be the same over the interval of the DFT. If one of the signals is finite in duration while the other is infinite, we can perform operations in segments using finite size transforms and then combine to get the total answer, however, if both sequences are of infinite intervals, there is no realistic method for implementation.

If both signals $x_1(n)$ and $x_2(n)$ are of finite length N_1 and N_2, and defined on 0 to $N_1 - 1$ and 0 to N_{2-1}, respectively, as shown in Fig. 6.4, the value of N needed so that circular and linear convolution are the same on 0 to $N - 1$ can be shown to be

$$N \geqq N_1 + N_2 - 1$$

Therefore, N must be selected large enough so that the periodic extensions of the signals accommodate the length of linear convolution. To perform linear convolution using circular convolution, $x_1(n)$ and $x_2(n)$ must first be extended in domains by adding zeros from $N_1 - 1$ and $N_2 - 1$ to $N - 1$, respectively, as shown in Fig. 6.4. The N-point transforms can then be taken, a term-by-term multiplication performed, and an IDFT operation carried out to give the circular convolution which equals the required linear convolution over the interval from 0 to $N - 1$.

6.4 COMPUTATION OF THE DISCRETE FOURIER TRANSFORM

To calculate the DFT we need to evaluate the following:

$$X(k) = \sum_{n=0}^{N-1} x(n) W_N^{kn}, \qquad k = 0, 1, \ldots, N - 1 \qquad (6.19)$$

where
$$W_N = e^{-j(2\pi/N)}$$

If $x(n)$ is considered to have complex values, Eq. (6.19) can be rewritten in terms of real and imaginary parts as

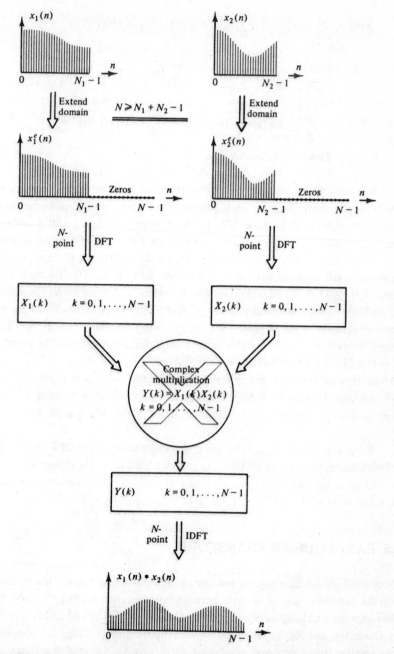

Figure 6.4 Computation of linear convolution by using circular convolution computed with the discrete Fourier transform.

$$X(k) = \sum_{n=0}^{N-1} \{\text{Re}\,[x(n)] + j\,\text{Im}\,[x(n)]\}\{\text{Re}\,[W_N^{kn}] + j\,\text{Im}\,[W_N^{kn}]\}$$

$$= \left\{ \sum_{n=0}^{N-1} \overbrace{\text{Re}\,[x(n)] \cdot \text{Re}\,[W_N^{kn}]}^{\text{mult.}} - \sum_{n=0}^{N-1} \overbrace{\text{Im}\,[x(n)] \cdot \text{Im}\,[W_N^{kn}]}^{\text{mult.}} \right\}$$

$$+ j\left\{ \sum_{n=0}^{N-1} \overbrace{\text{Re}\,[x(n)] \cdot \text{Im}\,[W_N^{kn}]}^{\text{mult.}} + \sum_{n=0}^{N-1} \overbrace{\text{Im}\,[x(n)] \cdot \text{Re}\,[W_N^{kn}]}^{\text{mult.}} \right\}$$

$$\text{for } k = 0, 1, \ldots, N-1 \qquad\qquad (6.20)$$

It is easy to see that the direct evaluation of $X(k)$ for each k requires $4N$ *real* multiplications, four for each pair of terms above and N pairs per summation. Therefore, to evaluate $X(k)$ for all k from zero to $N-1$ requires N times $4N$ or $4N^2$ real multiplications. Each of the four sums of N terms requires $N-1$ real two-input additions, and to combine the sum to get the real part and imaginary parts requires two more. Therefore, to evaluate $X(k)$ for each k requires $4(N-1) + 2$ real additions, which—when multiplied by N, since there are N $X(k)$s—gives the total number of real additions as $N(4N-2)$ for the direct evaluation of the N-point DFT. In obtaining the above results, we have treated the W_N^{kn} as always complex, even though for some values of kn it equals $1, -1, j,$ or $-j$, which would give a trivial complex multiplication. Therefore, the numbers of complex multiplications and additions presented are only approximations. However, as N gets larger, the approximation is quite good as there is a small percentage of values of kn such that W_N^{kn} equals $1, -1, j,$ or $-j$.

From Eq. (6.19) we see that the direct calculation of the DFT requires N^2 complex multiplications and $N(N-1)$ complex additions. These numbers serve as references for examining the efficiency of various so-called fast techniques for calculation of the DFT.

6.5 FAST FOURIER TRANSFORM

Many methods for reducing the number of multiplications, usually more critical than the number of additions, have been investigated over the last 50 years. The most important technique, popularized by Cooley and Tukey and earlier reported by Danielson and Runga, is based on decomposing or breaking the transform into smaller transforms and combining them to give the total transform. As Cooley and Tukey have shown, this decimation process can be done in both the time and frequency domains.

6.5.1 Decimation-in-Time FFT

In the following presentation, the number of points is assumed as a power of 2, that is, $N = 2^\nu$. The decimation-in-time approach is one of breaking the N-point transform into two $(N/2)$-point transforms, then breaking each $(N/2)$-point transform into two $(N/4)$-point transforms, and continuing this process until two-point transforms are obtained. The first step of this process is described below. Let $x(n)$ represent a sequence of N values, where N is a power of 2. Decimate or break this sequence into two sequences of length $N/2$, one composed of the even-indexed values of $x(n)$ and the other of odd-indexed values of $x(n)$:

Given
$$\text{sequence: } x(0)\,x(1)\,x(2)\cdots x\left(\frac{N}{2}-1\right)\cdots\cdots\cdots\cdots x(N-1)$$

Even
$$\text{indexed: } x(0)\,x(2)\,x(4)\cdots x(N-2)$$

Odd
$$\text{indexed: } x(1)\,x(3)\,x(5)\cdots x(N-1)$$

We know that the transform $X(k)$ of the N-point sequence $x(n)$ is given by

$$X(k) = \sum_{n=0}^{N-1} x(n)W_N^{nk}, \qquad k = 0, 1, \ldots, N-1$$

Breaking the sum into two parts, one for the even and one for the odd-indexed values, gives

$$X(k) = \sum_{\substack{n=0 \\ n \text{ even}}}^{N-2} x(n)W_N^{kn} + \sum_{\substack{n=1 \\ n \text{ odd}}}^{N-1} x(n)W_N^{nk} \qquad (6.21)$$

Letting $n = 2r$ in the first sum and $n = 2r + 1$ in the second, Eq. (6.21) can be rewritten as

$$X(k) = \sum_{r=0}^{N/2-1} x(2r)W_N^{2rk} + \sum_{r=0}^{N/2-1} x(2r+1)W_N^{(2r+1)k} \qquad (6.22)$$

Rearranging each part of $X(k)$ into $(N/2)$-point transforms using

$$W_N^{2rk} = (W_N^2)^{rk} = \exp\left(-j\frac{2\pi}{N}\cdot 2rk\right) = \exp\left(-j\frac{2\pi}{N/2}rk\right) = W_{N/2}^{rk} \quad (6.23)$$

gives the following:

$$X(k) = \underbrace{\sum_{r=0}^{N/2-1} x(2r)W_{N/2}^{rk}}_{\substack{(N/2)\text{-point DFT} \\ \text{of even indexed} \\ \text{sequence}}} + W_N^k \underbrace{\sum_{r=0}^{N/2-1} x(2r+1)W_{N/2}^{rk}}_{\substack{(N/2)\text{-point DFT} \\ \text{of odd indexed} \\ \text{sequence}}} \quad (6.24)$$

If $G(k)$ and $H(k)$ represent the $(N/2)$-point DFTs of the even and odd indexed sequences, then $X(k)$ can be written for $k = 0$ to $N/2 - 1$ as

$$X(k) = G(k) + W_N^k H(k), \qquad k = 0, 1, \ldots, N/2 - 1 \quad (6.25a)$$

This first step in the decomposition breaks the transform into the two $(N/2)$-point transforms and the W_N^k provides the N-point combining algebra as shown in Fig. 6.5.† In evaluating Eq. (6.25a) for $k = N/2$ to N, $G(k)$ and $H(k)$ are considered periodic with period $N/2$; therefore, for $k \geq N/2$, $X(k)$ is given by

$$X(k) = G(k - N/2) + W_N^k H(k - N/2) \qquad \text{for } N/2 \leq k \leq N - 1 \quad (6.25b)$$

The total number of complex multiplications, η_1, required to evaluate the N-point transform with this first decimation becomes $N + N^2/2$, determined as follows:

$$\eta_1 = (N/2)^2 + (N/2)^2 + N = N + N^2/2$$

The first term in the sum is the number of complex multiplications for the direct calculation of the $(N/2)$-point DFT of the even-indexed sequence; the second

†Standard flow graph notation is used for convenience. Each circle with entering arrows corresponds to a value equal to the weighted sum of the quantities at the tails of the arrows where the weights are the values next to the arrows. This notation is shown below. If no value appears by the arrow the weight is by convention equal to 1.

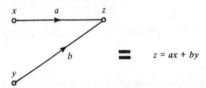

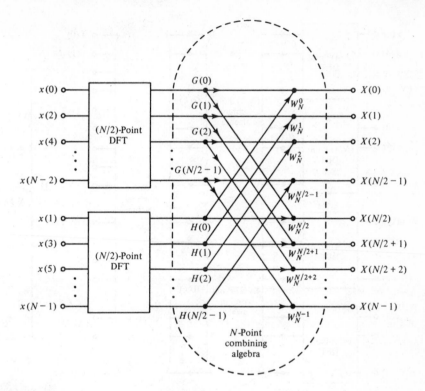

Figure 6.5 First stage of the decomposition for the decimation-in-time fast Fourier transform.

term is the number of complex multiplications for the direct calculation of the $(N/2)$-point DFT of the odd-indexed sequence; and the third term is the number of complex multiplications required for the combining algebra.

Each of the $(N/2)$-point sequences can be decimated further into two sequences of lengths $N/4$ as shown in Fig. 6.6. Each of the $(N/2)$-point combining algebras are governed by the following equation, assuming the necessary periodicity of $G(k)$ and $H(k)$:

$$X(k) = G'(k) + W_{N/2}^k H'(k), \qquad k = 0, 1, \ldots, N/2 - 1$$
$$= G'(k) + W_N^{2k} H'(k) \tag{6.26}$$

The number of complex multiplications after the second decimation, η_2, is easily seen to be

$$\eta_2 = 4(N/4)^2 + 2(N/2) + N = N^2/4 + 2N$$

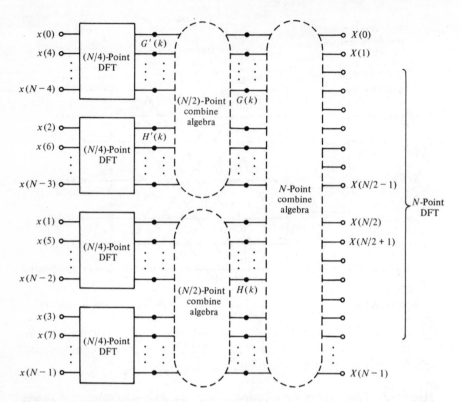

Figure 6.6 Second stage of the decomposition for the decimation-in-time fast Fourier transform.

The first term represents the number of complex multiplications necessary for a direct evaluation of the four $(N/4)$-point DFTs (Fig. 6.6), while the second and third terms are the number of complex multiplications required for the combining algebras for the first and second decimations.

Continuing this process, each $(N/4)$-point transform is broken into two $(N/8)$-point DFTs, etc. Since N is a power of 2, this process can be continued until there are v or $\log_2 N$ stages as shown in Fig. 6.7.

From Eq. (6.13), the two-point transforms can be written as follows:

$$X(0) = x(0) + W_2^0 x(1) = x(0) + x(1)$$
$$X(1) = x(0) + W_2^1 x(1) = x(0) - x(1)$$

It has been conventional to count the W_2^0 and W_2^1 as complex multiplications, even though as seen above there are no multiplications, only addition and sub-

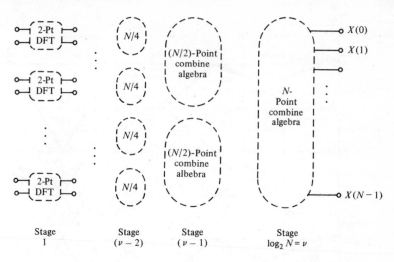

Figure 6.7 Final conceptual decomposition for the decimation-in-time fast Fourier transform.

traction. Since the combining algebra at each stage takes N complex multiplications and there are $\log_2 N$ stages, the approximate number of complex multiplications for the total decomposition becomes

$$\eta = \text{Number of complex multiplications} = N \log_2 N \qquad (6.27)$$

Therefore we have reduced the number of complex multiplications from N^2 to $N \log_2 N$ complex multiplications by the decimation-in-time approach. The flow graph for an eight-point FFT calculation is given in Fig. 6.8. In this graph, W_2^0, W_4^0, and W_8^0 and W_2^1, W_4^2, and W_8^4 are shown to emphasize symmetry even though they are $+1$ and -1, respectively.

From the flow diagram, several important observations can be made. (**1**) The input data have been shuffled. The input data appear in what is called "bit reversed" order, illustrated below for $N = 8$:

Position	Binary equivalent	Bit reversed	Sequence index
6 $\longrightarrow$	110 $\longrightarrow$	011 $\longrightarrow$	3
2 $\longrightarrow$	010 $\longrightarrow$	010 $\longrightarrow$	2

It is seen that $x(3)$ is in the sixth position and $x(2)$ is in the second position.

(**2**) The basic computational block in the diagram is called a "butterfly."

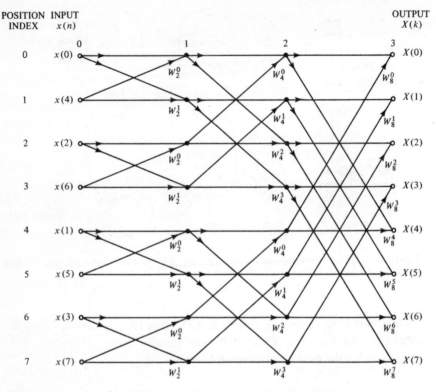

Figure 6.8 The flow graph for an eight-point decimation-in-time fast Fourier transform algorithm.

If we use m to represent the stage, and p and q to represent the position numbers in the stage, each butterfly in Fig. 6.8 can be represented as shown in Fig. 6.9, where

$$X_{m+1}(p) = X_m(p) + W_N^r X_m(q)$$
$$X_{m+1}(q) = X_m(p) + W_N^{r+N/2} X_m(q) \tag{6.28}$$

The power, r, of W_N is variable and depends upon the position of the butterfly. Also notice that $X_{m+1}(p)$ and $X_{m+1}(q)$, the outputs of the butterfly at stage $m + 1$, are calculated in terms of just $X_m(p)$ and $X_m(q)$, the corresponding values from the mth stage, and no other inputs. Therefore, if a scratch memory is available, $x_{m+1}(p)$ and $X_{m+1}(q)$ can be calculated and placed back in the storage registers for $X_m(p)$ and $X_m(q)$. This kind of computation is commonly referred to as an ''in-place'' calculation.

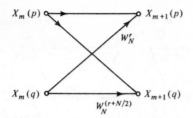

Figure 6.9 General butterfly flow graph for the DIT FFT.

(**3**) The frequency domain values, $X(k)$, are in normal order.

In the eight-point DIT FFT flow graph shown in Fig. 6.8, the W_2^0, W_2^4, and W_2^8 are not really complex numbers since they equal 1; therefore, these scale factors do not actually represent complex multiplications. Also, since W_2^1, W_4^2, and W_8^4 equal -1, they do not represent a complex multiplication—just a change in sign. As a matter of fact, the W_4^1, W_4^3, W_8^2, and W_8^6 are j or $-j$, and although they represent complex multiplications, they require only sign changes and interchanges of real and imaginary parts. If we call the multiplications corresponding to these scale factors given above trivial multiplications, there are actually only four nontrivial complex multiplications in the entire eight-point FFT shown in Fig. 6.8. As the size of the transform is increased, the proportion of nontrivial complex multiplications is reduced and the $N \log_2 N$ approximation becomes a little closer.

An example showing the calculations involved at each stage follows.

EXAMPLE 6.3

Find the DFT of the following sequence x using the DIT FFT algorithm shown in Fig. 6.8:

$$x = (1, -1, -1, -1, 1, 1, 1, -1)$$

Solution. The scale factors W_8^k are easily calculated as follows:

$$W_8^0 = 1, \qquad\qquad\qquad W_8^4 = e^{-j\pi} = -1$$
$$W_8^1 = e^{-j2\pi/8} = \sqrt{2}/2 - j\sqrt{2}/2, \qquad W_8^5 = -\sqrt{2}/2 + j\sqrt{2}/2$$
$$W_8^2 = e^{-j4\pi/8} = -j, \qquad\qquad W_8^6 = +j$$
$$W_8^3 = -\sqrt{2}/2 - j\sqrt{2}/2, \qquad\qquad W_8^7 = \sqrt{2}/2 + j\sqrt{2}/2$$

These scale factors are shown on the flow diagram below along with the transform $X(k)$, the magnitude of $X(k)$, and the arg $X(k)$.

$x(n)$ (bit reversed)					$X(k)$ (normal order)	$\lvert X(k) \rvert$	arg $X(k)$ (radians)
$x(0)$	1	2	2		0	0	—
$x(4)$	1	0	$2j$		$-\sqrt{2} + (2+\sqrt{2})j$	3.6955	+1.9635
					$\dfrac{\sqrt{2}}{2} - j\dfrac{\sqrt{2}}{2}$		
$x(2)$	−1	0	2		$2 - 2j$	2.8284	−0.7854
$x(6)$	1	−2	$-2j$		$\sqrt{2} + (\sqrt{2}-2)j$	1.5307	−0.3927
					$\dfrac{-\sqrt{2}}{2} - j\dfrac{\sqrt{2}}{2}$		
$x(1)$	−1	0	−2		4	4	0
$x(5)$	1	−2	−2		$\sqrt{2} - (\sqrt{2}-2)j$	1.5307	+0.3927
					$\dfrac{-\sqrt{2}}{2} + j\dfrac{\sqrt{2}}{2}$		
$x(3)$	−1	−2	2		$2 + 2j$	2.8284	+0.7854
$x(7)$	−1	0	−2		$-\sqrt{2} - (2+\sqrt{2})j$	3.6955	−1.9635
					$\dfrac{\sqrt{2}}{2} + j\dfrac{\sqrt{2}}{2}$		

Further Reduction The basic butterfly configuration can be further simplified to reduce the number of complex multiplications per butterfly by one. This can be shown by examining the complex multipliers in the basic butterfly configuration shown in Fig. 6.9 and noting that the horizontal scale factor can be rewritten as the negative of the diagonal scale factor as follows:

$$W_N^{r+N/2} = W_N^r \cdot W_N^{N/2} = W_N^r \cdot e^{-j(2\pi/N)(N/2)} = W_N^r \cdot e^{-j\pi} = -W_N^r \quad (6.29)$$

From Eqs. (6.28) and (6.29) the butterfly of Fig. 6.9 can be redrawn as shown in Fig. 6.10, thus illustrating a reduction to only one complex multiplication per butterfly.

By reducing one complex multiplication per butterfly, the number of complex multiplications required for the calculation of the DFT is half of that given in Eq. (6.27):

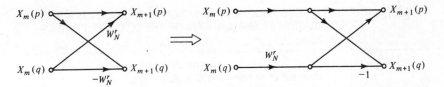

Figure 6.10 Equivalent stages for a DIT FFT butterfly producing a reduction in the number of complex multiplications.

$$\begin{array}{c} \text{Number of complex multiplications} \\ \text{required for calculating} \\ \text{the DIT FFT} \end{array} = \frac{N}{2}\log_2 N \qquad (6.30\text{a})$$

The algorithm just discussed is called the Cooley–Tukey FFT and is a very efficient way to calculate the DFT. The reduced complete eight-point decimation-in-time FFT algorithm is shown in Fig. 6.11. Since there are two

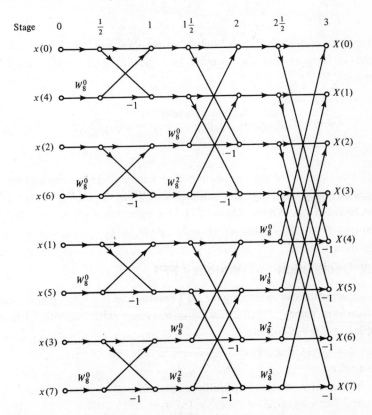

Figure 6.11 Reduced flow graph for an eight-point decimation-in-time FFT.

TABLE 6.2 NUMBER OF COMPLEX MULTIPLICATIONS FOR THE DIRECT AND COOLEY–TUKEY FFT EVALUATIONS OF THE DISCRETE FOURIER TRANSFORM

Number of stages, ν	Number of points, N	Number of complex multiplications using direct calculation, N^2	Number of complex multiplications using Cooley–Tukey FFT algorithm, $(N/2) \log_2 N$	Times faster than direct evaluation, $R = N^2/((N/2) \log_2 N)$
2	4	16	4	4
3	8	64	12	5.333
4	16	256	32	8
5	32	1,024	80	12.8
6	64	4,096	192	21.33
7	128	16,384	448	36.57
8	256	65,536	1024	64
9	512	262,144	2304	113.77
10	1024	1,048,576	5120	204.8

additions for each butterfly, $N/2$ butterflies per stage and $\log_2 N$ stages, the total number of complex additions required for evaluating a DFT using the DIT FFT shown is

$$\begin{matrix} \text{Number of complex additions} \\ \text{required for calculating} \\ \text{the DIT FFT} \end{matrix} = N \log_2 N \qquad (6.30b)$$

A comparison of the number of complex multiplications required for direct evaluation of the DFT and the number needed for the Cooley–Tukey FFT is given in Table 6.2. We see that the DIT FFT algorithm is greater than 100 times faster than the direct calculation for a 512-point DFT.

6.5.2 Decimation-in-Frequency FFT

Another algorithm for evaluating the DFT results from breaking up or decimating the transform values. The decimation-in-frequency (DIF) algorithm begins by breaking up $X(k)$ as follows:

$$X(k) = \sum_{n=0}^{N/2-1} x(n) W_N^{kn} + \sum_{n=N/2}^{N-1} x(n) W_N^{kn}$$

Letting $r = n - N/2$ in the second summation yields

$$X(k) = \sum_{n=0}^{N/2-1} x(n)W_N^{kn} + \sum_{r=0}^{N/2-1} x(r + N/2)W_N^{k(r+N/2)} \tag{6.31}$$

Since

$$W_N^{(N/2)k} = \exp\left[\left(-j\frac{2\pi}{N}\right)\frac{N}{2}k\right] = e^{-j\pi k} = (-1)^k \tag{6.32}$$

Eq. (6.31) can be written as follows:

$$X(k) = \sum_{n=0}^{N/2-1} [x(n) + (-1)^k x(n + N/2)]W_N^{nk} \tag{6.33}$$

The decimation is now obtained by taking the odd and even terms of $X(k)$. For even values of k, say $k = 2r$, and $r = 0, 1, \ldots, N/2 - 1$, Eq. (6.33) is seen to be

$$X(2r) = \sum_{n=0}^{N/2-1} [x(n) + (-1)^{2r} x(n + N/2)]W_N^{2nr}$$

$$= \sum_{n=0}^{N/2-1} [x(n) + x(n + N/2)]W_{N/2}^{nr} \tag{6.34}$$

For odd values of k, say $k = 2r + 1$ and $r = 0, 1, \ldots, N/2 - 1$, Eq. (6.33) is seen to be

$$X(2r + 1) = \sum_{n=0}^{N/2-1} [x(n) + (-1)^{2r+1} x(n + N/2)]W_N^{n(2r+1)}$$

$$= \sum_{n=0}^{N/2-1} [x(n) - x(n + N/2)]W_N^n \cdot W_{N/2}^{nr} \tag{6.35}$$

It is convenient at this point to define two functions, $g(n)$ and $h(n)$, as follows:

$$g(n) = x(n) + x(n + N/2) \tag{6.36}$$
$$h(n) = x(n) - x(n + N/2) \tag{6.37}$$

From Eqs. (6.34) and (6.35) we see that once $g(n)$ and $h(n)$ are calculated the odd and even indexed $X(k)$'s are determined from the $(N/2)$-point transforms of $g(n)$ and $h(n)W_N^n$ as shown in Fig. 6.12.

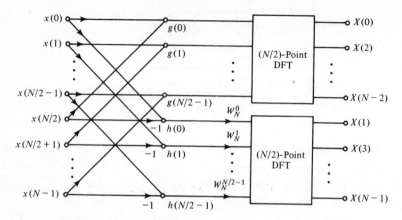

Figure 6.12 First stage of decimation-in-frequency FFT.

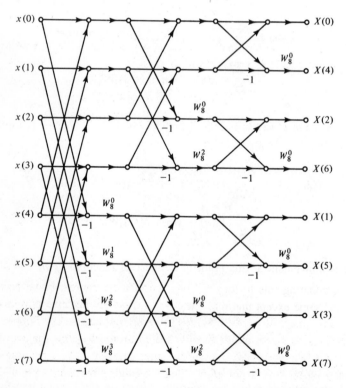

Figure 6.13 Reduced flow graph for an eight-point decimation-in-frequency FFT.

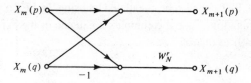

Figure 6.14   Basic butterfly for decimation-in-frequency FFT.

In a similar way, each of the $(N/2)$-point transforms can be broken into two $(N/4)$-point transforms again decimating the frequency outputs. Continuing this process down to two-point transforms, it is easily shown that $(N/2) \log_2 N$ complex multiplications and $N \log_2 N$ additions are required.

A reduced flow graph for an eight-point decimation-in-frequency FFT is shown in Fig. 6.13. For the decimation-in-frequency (DIF) algorithm it is seen that

(1) the input is in normal order,
(2) the output is in bit reversed order, and
(3) the algorithm has in-place calculations described below with the butterfly structure shown in Fig. 6.14.

$$X_{m+1}(p) = X_m(p) + X_m(q) \tag{6.38}$$

$$X_{m+1}(q) = [X_m(p) - X_m(q)]W_N^r \tag{6.39}$$

6.5.3 Computation of the Inverse DFT

The inverse DFT is given by

$$x(n) = (1/N) \sum_{k=0}^{N-1} X(k)W_N^{-kn}, \qquad n = 0, 1, \ldots, N-1 \tag{6.40}$$

In comparing this to the DFT we see that the computational procedure remains the same except that the "twiddle" factors are negative powers of W_N and the output must be scaled by $1/N$. Therefore, an inverse fast Fourier transform (IFFT) flow diagram can be obtained from an FFT flow diagram by replacing the $x(k)$'s by $X(k)$'s, scaling the input data, by $1/N$, a per stage scale factor of $\frac{1}{2}$ when N is a power of 2, and changing the exponents of W_N to negative values.

The inverse FFT can also be calculated by using the FFT algorithm directly by complex conjugation. Take the complex conjugate of the expression for $x(n)$ given in Eq. (6.40) to obtain

$$x^*(n) = \left[(1/N) \sum_{k=0}^{N-1} X(k) W_N^{-kn} \right]^* \tag{6.41}$$

But the complex conjugate of a product or sum is the product or sum of the complex conjugates. Therefore,

$$x^*(n) = (1/N) \sum_{k=0}^{N-1} X^*(k) W_N^{kn} \tag{6.42}$$

The right-hand side without the $1/N$ is recognized as the DFT of $X^*(k)$, so we can rewrite (6.42) as follows:

$$x^*(n) = (1/N) \, \text{DFT} \, [X^*(k)] \tag{6.43}$$

Taking the complex conjugate of both sides and using an FFT algorithm for the calculation of the DFT gives us the following fast computational procedure for finding the inverse DFT:

$$x(n) = (\text{FFT}[X^*(k)])^*/N \tag{6.44}$$

6.6 INTERPRETATION OF DFT RESULTS

The discrete Fourier transform takes a sequence of N complex numbers $\{x(n)\}$ and produces another sequence of complex numbers $\{X(k)\}$. The discrete Fourier transform pair is given by

$$X(k) = \sum_{n=0}^{N-1} x(n) \exp\left(-j\frac{2\pi nk}{N} \right), \qquad k = 0, 1, 2, \ldots, N-1 \tag{6.45}$$

$$X(n) = (1/N) \sum_{k=0}^{N-1} X(k) \exp\left(j\frac{2\pi nk}{N} \right), \qquad n = 0, 1, 2, \ldots, N-1 \tag{6.46}$$

When $\{x(n)\}$ is a sequence of *real* numbers, and N is even, Eq. (6.46) can be written using property 7 or 8 of Table 6.1 as a sum of sine and cosine terms as

$$x(n) = \frac{X(0)}{N} + \sum_{k=1}^{N/2-1} \frac{2}{N}\left[X_R(k)\cos\frac{2\pi kn}{N} - X_I(k)\sin\frac{2\pi kn}{N} \right] + \frac{X(N/2)}{N}\cos\pi n$$

$$(6.47)$$

In Fig. 6.15, the DFT of a real $x(n)$ is given where $x(n)$ is thought of as a sampled version of an analog signal $x_a(t)$. The axis in the transform domain has three interpretations. First, the digital frequency index k, second, the corresponding digital frequency ω_k in radians, and third, the corresponding analog frequency Ω_k in rad/sec. Their relationship is given by

$$k \longrightarrow \omega_k = k \cdot 2\pi/N \longrightarrow \Omega_k = k \cdot 2\pi/NT \qquad (6.48)$$

$$\text{Frequency index} \qquad \text{Digital frequency} \qquad \text{Analog frequency}$$
$$\text{(unitless)} \qquad\qquad \text{(radians)} \qquad\qquad \text{(radian/sec.)}$$

Since frequency indices greater than $N/2$ are redundant for real signals, the highest unaliased frequency that can be observed is related to the $N/2$ frequency index. This digital frequency $\omega_{N/2}$ equals π rad and the corresponding analog radian frequency $\Omega_{N/2}$ equals π/T rad/sec. The ω_1 and Ω_1 are called the digital and analog resolution frequencies for the N-point DFT, respectively, and are given by

$$\omega_1 = 2\pi/N\,\text{rad}, \qquad \Omega_1 = 2\pi/NT\,\text{rad/sec} \qquad (6.49)$$

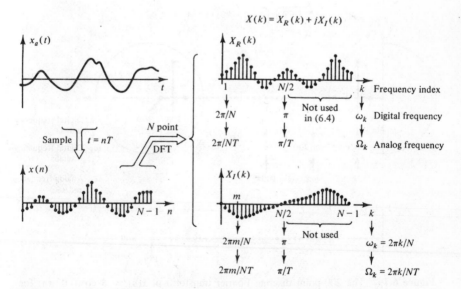

Figure 6.15 The N-point DFT of a sequence $x(n)$ which is obtained by sampling an analog signal $x_a(t)$ at a rate of $1/T$ samples per second.

The DFT of sampled sine and cosine functions that are exactly periodic throughout the sampling window are presented in the following examples. Also presented is an example of the interpretation of the inverse DFT given by Eq. (6.12) or its alternative Eq. (6.47).

EXAMPLE 6.4

Find the discrete Fourier transform of a cosine wave that has an exact number of periods within the window of samples. Let $x(n)$, as shown in Fig. 6.16, represent a sampled version of $x(t) = 3 \cos 2\pi t$ at $t = nT$ for $n = 0$–199, with $T = 0.01$. The discrete time sequence becomes

$$x(n) = 3 \cos(0.02\pi n), \qquad n = 0, 1, \ldots, 199$$

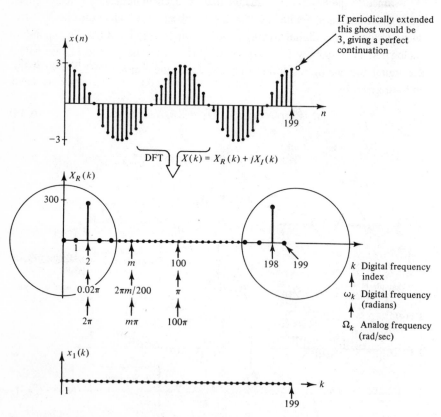

Figure 6.16 The 200-point discrete Fourier transform of $x(n) = 3 \cos(0.02\pi n)$ for $n = 0, 1, \ldots, 199$. This $x(n)$ could have been generated by sampling the analog signal $x_a(t) = 3 \cos 2\pi t$ at a rate of 100 samples/sec (Example 6.4).

Notice that $x(n)$ goes through exactly two periods of the cosine wave.

Solution. The real part $X_R(k)$ and the imaginary part $X_I(k)$ of the DFT of $x(n)$ were calculated from Eq. (6.11) and are illustrated in Fig. 6.16. It is seen that there are only two nonzero values in the frequency domain at frequency indexes 2 and $N - 2 = 198$ each with a value of 300. This value represents $AN/2$, where $A = 3$, the amplitude of the cosine wave, and N equals 200, the number of samples used. Also note that the value of $X(2)$ is entirely real as $X_I(2)$ is zero. Because of the sampling structure, the frequency index 2 corresponds to exactly two full periods of the cosine wave. This point is emphasized in the next example.

EXAMPLE 6.5

Find the DFT of a sampled version of $x(t)$ given in Example 6.4 for 1,000 samples. The $x(n)$ then becomes

$$x(n) = 3\cos(0.02\pi n), \qquad n = 0, 1, \ldots, 999$$

Solution. The $x(n)$ and its corresponding DFT are given in Fig. 6.17. The nonzero amplitude has increased by a multiple of 5 since we have

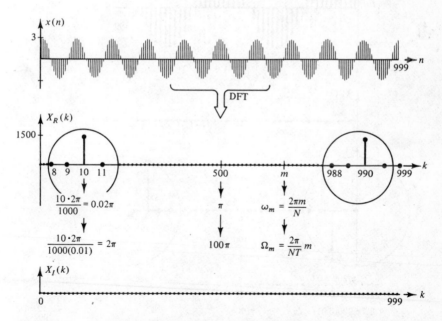

Figure 6.17 The 1000-point discrete Fourier transform of $x(n) = 3\cos(0.02\pi n)$ for $n = 0, 1, \ldots, 999$. This $x(n)$ could have been generated by sampling the analog signal $x_a(t) = 3\cos 2\pi t$ at a rate of 100 samples/sec (Example 6.5).

five times the number of samples, and the frequency index of the nonzero value has moved to $k = 10$, the number of exact periods of the cosine wave. Although the index k has increased, the corresponding digital frequency of 0.02π and analog frequency of 2π remain the same as expected.

EXAMPLE 6.6

Find the DFT of a sampled version of $x_a(t)$ given by $x_a(t) = 5 \sin 2\pi t$. Assume that 200 samples have been taken $T = 0.01$ sec apart to give $x(n) = 5 \sin 0.02\pi n$, $n = 0, 1, \ldots, 199$. These samples represent exactly two full periods of the sampled sinusoid, as shown in Fig. 6.18.

Solution. It can be shown that the real and imaginary parts of the DFT of $x(n)$ described above are as illustrated in Fig. 6.18. It should be noted that, for a sinusoid with amplitude A and exactly P periods within the

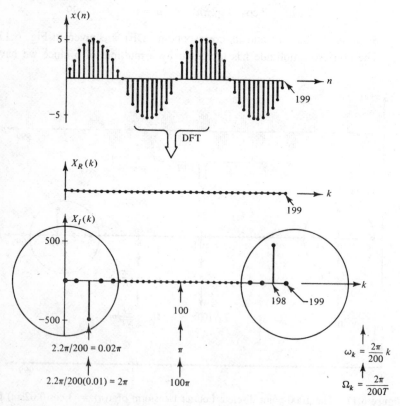

Figure 6.18 The 200-point discrete Fourier transform of $x(n) = 5 \sin(0.02\pi n)$ for $n = 0, 1, \ldots, 199$. This $x(n)$ could have been generated by sampling the analog signal $x_a(t) = 5 \sin 2\pi t$ at a rate of 100 samples/sec (Example 6.6).

window, the real part $X_R(k)$ of the DFT is zero and the imaginary part $X_I(k)$ is zero except at $k = P$ and $N - P$. For a real sequence $x(n)$, it was shown earlier that $X_I(N - k)$ equals $- X_I(k)$, which is seen for $k = 2$ and the corresponding $N - k = 198$ in Fig. 6.18. The value of $X_I(2)$ is $-AN/2$ or -500 from $A = 5$ and $N = 200$.

EXAMPLE 6.7

(a) Find the inverse DFT of $X(k)$ given in Fig. 6.19. Also (b) if the time between signal samples is $T = 0.1$ sec, give a possible corresponding analog signal $x_a(t)$.

Solution. (a) For a real $x(n)$, the alternative form [Eq. (6.47)] for $x(n)$ is given by

$$x(n) = \frac{X(0)}{N} + \sum_{k=1}^{N/2-1} \frac{2}{N}\left[X_R(k)\cos\frac{2\pi k \cdot n}{N} - X_I(k)\sin\frac{2\pi k \cdot n}{N}\right]$$
$$+ \frac{X(N/2)}{N}\cos\pi n$$

Letting $N = 8$ and using the values of $X_R(k)$ and $X_I(k)$ given, we have

$$x(n) = \frac{1.5}{8} - \frac{2}{8}(1)\sin\frac{2\pi 1n}{8} + \frac{2\cdot 2}{8}\cos\frac{2\pi 2n}{8}$$
$$- \frac{2}{8}(-1.75)\sin\frac{2\pi 2n}{8} + \frac{0.5}{8}\cos\pi n$$
$$= 0.1875 - 0.25\sin\frac{\pi n}{4} + 0.5\cos\frac{\pi n}{2}$$
$$+ 0.4375\sin\frac{\pi n}{2} + 0.0625\cos\pi n$$

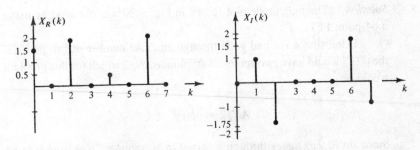

Figure 6.19 Real and imaginary parts of $X(k)$ for Example 6.7

$$\{X_R(k)\} = \{1.5, 0, 2, 0, 0.5, 0, 2, 0\},\ X_I(k) = \{0, 1, -1.75, 0, 0, 0, 1.75, -1\}$$

(b) One possibility for the analog signal that, when sampled, gives this $x(n)$, can be obtained by replacing the digital frequencies ω_k by the analog frequencies $\Omega_k = \omega_k/T$. Using this concept, $x_a(t)$ could have been

$$x_a(t) = \frac{X(0)}{N} + (2/N) \sum_{k=1}^{N/2-1} \left[X_r(k) \cos \frac{2\pi kt}{NT} - X_i(k) \sin \frac{2\pi kt}{NT} \right]$$
$$+ \frac{X(N/2)}{N} \cos \frac{\pi t}{T}$$

For the example above we have for $N = 200$ and $T = 0.1$, that $x_a(t)$ is given by

$$x_a(t) = 0.1875 - 0.25 \sin 2.5\pi t + 0.5 \cos 5\pi t$$
$$+ 0.4375 \sin 5\pi t + 0.0625 \cos 10\pi t$$

Many other analog signals could give the same sequence $x(n)$. For example, the analog signal could have been composed of higher frequencies that have been aliased. Therefore, if aliasing had taken place, the $x_a(t)$ given above would not be the actual signal sampled. There is really no way to tell if aliasing has occurred if all you have are the sampled values.

If a sinusoidal signal does not go through an exact number of periods within the sampling window, it can be shown that the discrete Fourier transform of the sequence has nonzero values for almost all values of the frequency index k. An illustration of this phenomenon, called "leakage," is given in Example 6.8.

EXAMPLE 6.8

Plot the magnitude of the 64-point DFT of the 64 points of

$$x(n) = (\tfrac{1}{32}) \sin (0.2\pi n) \qquad n = 0, 1, \ldots, 63.$$

Solution. The magnitude plot shown in Fig. 6.20 was obtained by using a 64-point FFT.

If the sine wave had gone through an exact number of full periods, the $|X(k)|$ would have two spikes of amplitude $AN/2$ which for this problem would be

$$AN/2 = (\tfrac{1}{32})\tfrac{64}{2} = 1$$

Since $\sin (0.2\pi n)$ goes through a period in 10 samples, $x(n)$ from 0 to 63 goes through 6.4 periods (a noninteger number). Note that the highest

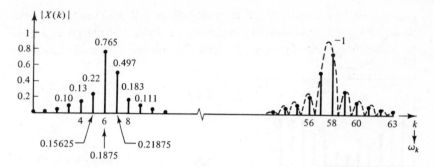

Figure 6.20 Illustration of leakage. A magnitude plot of the 64-point DFT of $x(n) = (1/32) \sin 0.2\pi n$.

value of $|X(k)|$ occurs at $k = 6$ with a value of 0.765, which is less than the 1 determined above for the case where the signal goes through an integer number of periods. To explain the above results, the relationship between the DFT and the Fourier transform of a sequence must be further explored.

6.7 DFT–FOURIER TRANSFORM RELATIONSHIPS

The Fourier transform $X(e^{j\omega})$ of an $x(n)$ which is zero outside of $n = 0, 1, 2, \ldots, N - 1$, is given for all ω by

$$X(e^{j\omega}) = \sum_{n=-\infty}^{\infty} x(n)e^{-j\omega n} = \sum_{n=0}^{N-1} x(n)e^{-j\omega n} \qquad (6.50a)$$

It is worth noting that the resulting $X(e^{j\omega})$ is a continuous function of ω.

The discrete Fourier transform (N point) of $x(n)$ is given by

$$X(k) = \sum_{n=0}^{N-1} x(n)e^{-j(2\pi kn/N)}, \qquad k = 0, 1, \ldots, N - 1 \qquad (6.50b)$$

Comparing (6.50a) and (6.50b), the DFT of $x(n)$ is seen to be the sampled version of the Fourier transform of the sequence as illustrated in Fig. 6.21 and given by

$$X(k) = X(e^{j\omega})\big|_{\omega = (2\pi k/N)}, \qquad k = 0, 1, \ldots, N - 1 \qquad (6.51)$$

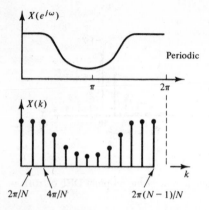

Figure 6.21 DFT–Fourier transform relationship for sequence $x(n)$ of nonzero length N. $X(k) = X(e^{j\omega})$ with $\omega = k2\pi/N$.

Thinking about this a little we see that $x(n)$ can be determined from $X(k)$ by the IDFT and $X(e^{j\omega})$ from $x(n)$ by the Fourier transform. Therefore, a way of determining $X(e^{j\omega})$ from $X(k)$ must exist. It can be shown that

$$X(e^{j\omega}) = \sum_{k=0}^{N-1} X(k)\phi\left(\omega - \frac{2\pi}{N}k\right) \tag{6.52}$$

where
$$\phi(\omega) = \frac{\exp\left[-j\omega(N-1)/2\right]}{N}\frac{\sin(\omega N/2)}{\sin(\omega/2)} \tag{6.53}$$

The formula given in Eq. (6.52) is called the DFT interpolation formula, and the $\phi(\omega)$ given in Eq. (6.53) is the interpolation function. It should be emphasized that $x(n)$ must be of finite duration for this result to hold.

Zero Padding

In this section the effect of adding or padding zeros to the end of a sequence of finite length is discussed. This zero padding might be used to fill out the sequence length so a power of 2 FFT algorithm could be used, or so a linear convolution could be performed as discussed earlier, or to provide a better-looking display of the frequency content of a finite length sequence. If N_z zeros are added to an original sequence which has a nonzero length of N_f to give a sequence of N values, and an N-point FFT taken, the sampled values of the Fourier transform are spaced $2\pi/(N_f + N_z)$ apart. In Fig. 6.22 a sequence $x(n)$ that is 1 for n equals 0–3 and zero otherwise is shown along with the magnitude of its Fourier transform. Also shown are the magnitudes of the 4-, 8-, and 16-point DFTs of the zero padded original sequence. As more zeros are added, the DFT provides closer spaced samples of the Fourier transform of the original sequence thus

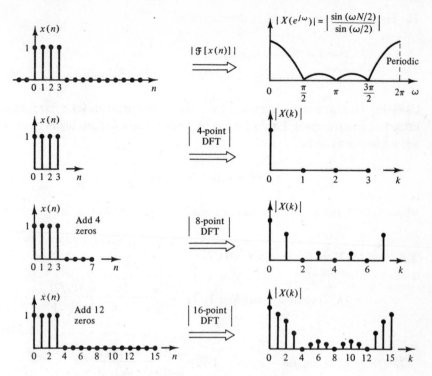

Figure 6.22   Illustration of zero padding.

giving a better displayed version of the Fourier transform. However, we really do not have any more information than could be obtained by sampling the interpolation formula [Eq. (6.52)]. There is no added resolution, just a better display of available information without going through the calculations necessary to evaluate the values of the DFT through the interpolation formula.

6.8 DISCRETE FOURIER TRANSFORMS OF SINUSOIDAL SEQUENCES

Using the fact that the DFT is the sampled version of the Fourier transform for finite length sequences, the leakage problem as described earlier can be easily explained. Suppose that a cosine sequence has been windowed to give a sequence $x(n)$ defined as follows:

$$x(n) = \begin{cases} A \cos \omega_0 n, & 0 \le n \le N - 1 \\ 0, & \text{otherwise} \end{cases}$$

The Fourier transform of $x(n)$ is determined as

$$X(e^{j\omega}) = \mathcal{F}[A\cos\omega_0 n\, w_R(n)]$$
$$= (A/2)\mathcal{F}[e^{j\omega_0 n}\, w_R(n)] + (A/2)\mathcal{F}[e^{-j\omega n}\, w_R(n)] \tag{6.54}$$

Using Eq. (1.23a) and the fact that the complex exponential causes a translation in the frequency domain, Eq. (6.54) can be written with a fair amount of algebra and trigonometry to be

$$X(e^{j\omega}) = X_R(e^{j\omega}) + jX_I(e^{j\omega})$$

where $X_R(e^{j\omega})$ and $X_I(e^{j\omega})$, the real and imaginary parts of $X(e^{j\omega})$, are given by

$$X_R(e^{j\omega}) = (A/2)\cos\left[(\omega - \omega_0)(N-1)/2\right]\frac{\sin[(\omega - \omega_0)N/2]}{\sin[(\omega - \omega_0)/2]}$$

$$+ (A/2)\cos\left[(\omega + \omega_0)(N-1)/2\right]\frac{\sin\{[\omega - (2\pi - \omega_0)]N/2\}}{\sin\{[\omega - (2\pi - \omega_0)]/2\}}$$

$$\tag{6.55a}$$

$$X_I(e^{j\omega}) = -(A/2)\sin\left[(\omega - \omega_0)(N-1)/2\right]\frac{\sin[(\omega - \omega_0)N/2]}{\sin[(\omega - \omega_0)/2]}$$

$$- (A/2)\sin\left[(\omega + \omega_0)(N-1)/2\right]\frac{\sin\{[\omega - (2\pi - \omega_0)]N/2\}}{\sin\{[\omega - (2\pi - \omega_0)]/2\}}$$

$$\tag{6.55b}$$

The real and imaginary parts, $X_R(e^{j\omega})$, and $X_I(e^{j\omega})$ of $X(e^{j\omega})$ are plotted in Figs. 6.23a and 6.23b for the discrete time cosine sequence with $\omega_0 = \pi/4$ and $N = 32$. These samples constitute four full periods of $\cos(\pi n/4)$. The corresponding real and imaginary parts, $X_R(k)$ and $X_I(k)$, of the DFT are obtained by sampling the $X_R(e^{j\omega})$ and $X_I(e^{j\omega})$ at $k2\pi/32$, as shown in Figs. 6.23c and 6.23d. Similar results for a sine sequence with $\omega_0 = \pi/4$ are shown in Figs. 6.24a–6.24d. The DFTs shown in Fig. 6.23 and 6.24 are as expected; both real and imaginary parts have zero values for all indices other than k equals 4 and 28. Yet the real and imaginary parts of $X(e^{j\omega})$ are zero except for ω equals k times $2\pi/32$.

In Figs. 6.25 and 6.26 the real and imaginary parts of $X(e^{j\omega})$ and $X(k)$ are shown for sine and cosine wave sequences with $\omega_0 = 8.8\pi/32$. It is easy to see that the samples from 0 to 31 go through 4.4 cycles. It is noticed that the sampled values of $X(e^{j\omega})$ no longer land on zeros but on nonzero values. This

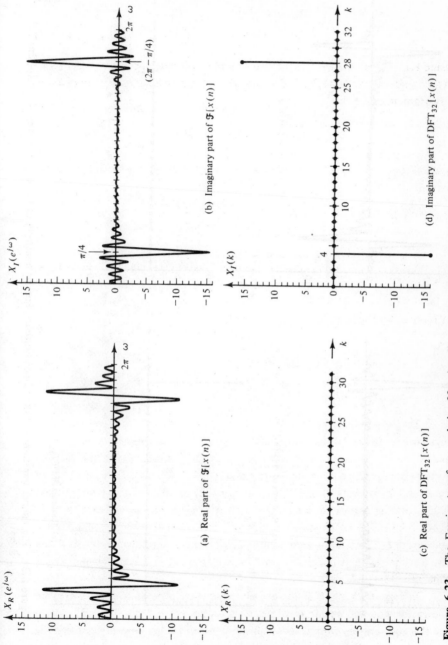

Figure 6.23 The Fourier transform and the 32-point discrete Fourier transform of the sequence $x(n) = \cos[(4)2\pi n/32]$ $[u(n) - u(n - 32)]$.

(a) Real part of $\mathcal{F}[x(n)]$

(b) Imaginary part of $\mathcal{F}[x(n)]$

(c) Real part of $\text{DFT}_{32}[x(n)]$

(d) Imaginary part of $\text{DFT}_{32}[x(n)]$

297

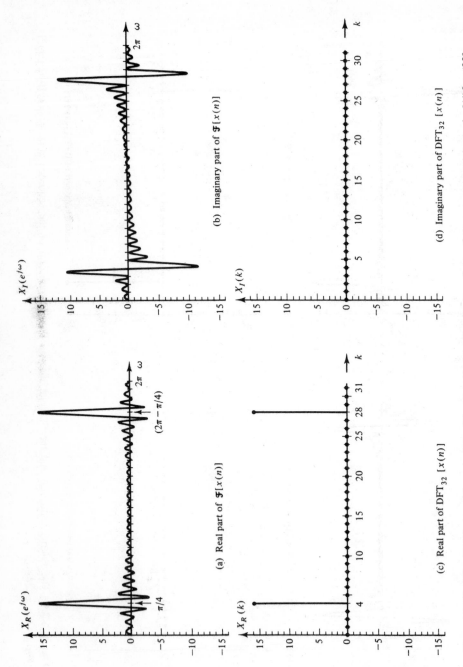

Figure 6.24 The Fourier transform and the 32-point discrete Fourier transform of the sequence $x(n) = \sin[(4)2\pi n/32][u(n) - u(n - 32)]$.

$X_R(e^{j\omega})$

$\pi/4$

2π

ω

(a) Real part of $\mathcal{F}[x(n)]$

$X_I(e^{j\omega})$

$(2\pi - \pi/4)$

2π

ω

(b) Imaginary part of $\mathcal{F}[x(n)]$

$X_R(k)$

k

(c) Real part of DFT$_{32}[x(n)]$

$X_I(k)$

k

(d) Imaginary part of DFT$_{32}[x(n)]$

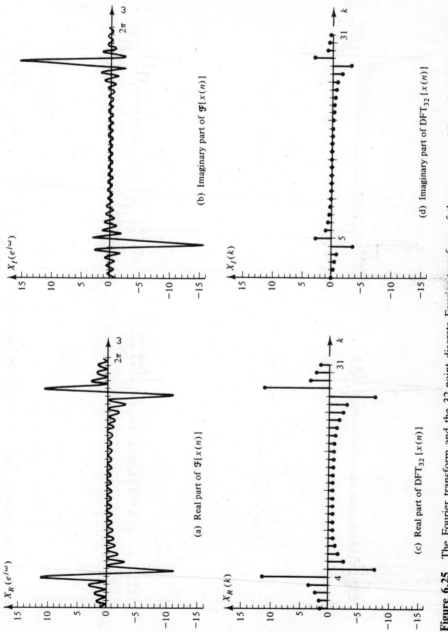

Figure 6.25 The Fourier transform and the 32-point discrete Fourier transform of the sequence $x(n) = \sin[(4.4)2\pi n/32]$ $[u(n) - u(n - 32)]$.

(a) Real part of $\mathcal{F}[x(n)]$

(b) Imaginary part of $\mathcal{F}[x(n)]$

(c) Real part of $\text{DFT}_{32}[x(n)]$

(d) Imaginary part of $\text{DFT}_{32}[x(n)]$

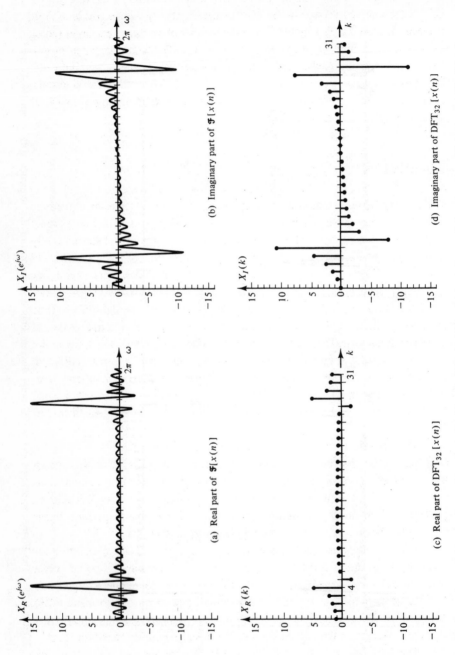

Figure 6.26 The Fourier transform and the 32-point discrete Fourier transform of the sequence $x(n) = \cos{[(4.4)2\pi n/32]}[u(n) - u(n - 32)]}$.

(a) Real part of $\mathcal{F}[x(n)]$

(b) Imaginary part of $\mathcal{F}[x(n)]$

(c) Real part of $DFT_{32}[x(n)]$

(d) Imaginary part of $DFT_{32}[x(n)]$

property is called leakage and is thus seen as a natural result of the fact that the DFT is the sampled version of the $X(e^{j\omega})$ whenever the sequence $x(n)$ is of finite duration. It should also be noted that as the number of samples gets larger (going through a larger number of cycles), spikes in $X_R(e^{j\omega})$ become larger in magnitude. Correspondingly, the smaller side peaks also get larger. The spike width gets narrower but then there are more samples for $X(k)$, and if the sinusoidal sequence does not go through an exact number of periods the minor lobes are always sampled and leakage occurs.

6.9 SUMMARY

The Fourier series expansion for continuous-time and discrete-time periodic signals has been presented in this chapter. A continuous-time periodic signal could be written as a weighted sum of a countable number of harmonically related complex exponential signals while the discrete-time periodic signal could be written as a weighted sum of a finite number of harmonically related complex exponential sequences. These expansions are special cases of the general orthonormal expansion of signals in terms of basis functions. In the discrete-time case, for example, signals could be chosen to be expanded in terms of the set of Walsh functions. The response of linear systems to Walsh functions is not easily determined, therefore their use in analyzing linear systems is limited; however, Walsh functions could be used as a basis for classification and identification in a general spectral sense. Walsh functions and other orthonormal basis functions, along with their "fast" algorithms, are nicely treated in Andrews [1].

The discrete Fourier transform was defined as the discrete Fourier series over one period. Its properties were explored, and in particular it was shown that a multiplication in the frequency domain corresponded to a circular convolution in the time domain. With proper care, circular convolution could be used to perform linear convolution, thus making it possible to perform linear filtering operations on discrete-time sequences using the DFT.

The flow diagrams for the standard decimation-in-time and decimation-in-frequency fast Fourier transform algorithms, when the number of points was a power of 2, were presented (radix 2 algorithms). When the number of points is a power of 4, radix 4 algorithms can be found, and so on (see Problem 6.24) [9]. Also, if N is a composite number or a prime, various decompositions exist that allow savings in the computation of the discrete Fourier transform [8, 9].

The DFT has been successfully used in spectral analysis [6], that is, the determination of the frequency content of a given discrete-time signal. It has been shown that since only a finite number of samples are used, like viewing a signal through a window, resolution is limited and the spectrum can be misleading due to the "leakage" effect.

REFERENCES

1. Andrews, H. C. *Computer Techniques in Image Processing.* Academic Press, New York, 1970.
2. Bloomfield, Peter. *Fourier Analysis of Time Series: An Introduction.* Wiley, New York, 1976.
3. Brigham, E. O. *The Fast Fourier Transform*, Prentice-Hall, Englewood CLiffs, NJ, 1974.
4. Enochson, L. D., and R. K. Otnes. *Digital Time Series Analysis.* Wiley, New York, 1972.
5. Enochson, L. D., and R. K. Otnes. *Programming and Analysis for Digital Time Series Data.* The Shock and Vibration Information Center, Navy Publication and Printing Service Office, 1968.
6. Harris, F. J. "On the Use of Windows For Harmonic Analysis With the Discrete Fourier Transform," *Proceedings of the IEEE*, 66, No. 1 (January 1978), 51–83.
7. Kay, S. M., and S. L. Marple, Jr. "Spectrum Analysis—A Modern Perspective," *Proceedings of the IEEE*, vol. 69 (November 1981), 1380–1419.
8. Oppenheim, A. V., and R. W. Schafer. *Digital Signal Processing.* Prentice-Hall, Englewood Cliffs, NJ, 1975.
9. Rabiner, L., and B. Gold. *Theory and Application of Digital Signal Processing.* Prentice-Hall, Englewood Cliffs, NJ, 1975.
10. Robinson, Enders A., and Manuel T. Silva. *Digital Signal Processing and Time Series Analysis*, Holden-Day, San Francisco, 1978.
11. Rabiner, L. R., and C. M. Radar, Eds. *Digital Signal Processing.* IEEE Press, New York, 1972.
12. IEEE Acoustics, Speech and Signal Processing Society, Eds. *Digital Signal Processing II.* IEEE Press, New York, 1975.
13. Aggarwal, J. K., Ed. *Digital Signal Processing.* Western Periodicals Co., North Hollywood, CA, 1979.
14. Digital Signal Processing Committee IEEE Acoustics, Speech and Signal Processing Society, Eds. *Programs for Digital Signal Processing.* IEEE Press, New York, 1979.

PROBLEMS

6.1 Compute the discrete Fourier series for the periodic $x(n)$ shown. Give both (a) the complex exponential and (b) the trigonometric expressions.

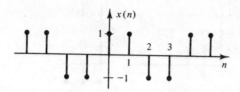

6.2 Find the discrete Fourier series representations of the periodic signals shown and plot magnitude and angle versus k for each one. How did changing the period alter the results?

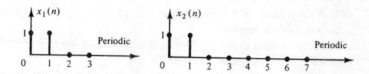

6.3 Construct the matrix for obtaining the coefficients of the Fourier series for a signal with eight-point period and put in powers of W_8^k, where k is less than eight.

6.4 Find the relationship between the $A(k)$, $B(k)$, and $X(k)$, where $X(k)$ are the exponential expansion coefficients and $A(k)$ and $B(k)$ are the trigonometric expansion coefficients for the discrete Fourier series of a real periodic sequence.

6.5 (a) Compute W_{64}^{16}.

   (b) Given $W_{24}^{16} = W_{128}^x$, solve for x.

   (c) Given $(-2) \bmod 7 = x$, where $0 \leq x \leq 6$, find x.

6.6 Suppose $x(n)$ is a sequence defined on 0–7 only as $(0, 1, 2, 3, 4, 5, 6, 7)$.
   (a) Illustrate $x[(n-2) \bmod 8]$.
   (b) If the $\mathrm{DFT}_8[x(n)] = X(k)$, what is $\mathrm{DFT}_8[x((n-2) \bmod 8)]$?

6.7 For the $x_1(n)$, $x_2(n)$, and N given, compute $x_1(n)\circledast_N x_2(n)$.
   (a) $x_1(n) = \delta(n) + \delta(n-1) + \delta(n-2)$, $N = 3$,
   $x_2(n) = 2\delta(n) - \delta(n-1) + 2\delta(n-2)$.
   (b) $x_1(n) = \delta(n) + \delta(n-1) - \delta(n-2) - \delta(n-3)$, $N = 5$,
   $x_2(n) = \delta(n) - \delta(n-2) + \delta(n-4)$
   (c) $x_1(n) = x_2(n) = \delta(n) + \delta(n-1) - \delta(n-2) - \delta(n-3)$, $N = 4$.

6.8 Given the sequences $x_1(n)$ and $x_2(n)$ below
   (a) Compute the circular convolution $x_1(n)\circledast_N x_1(n)$ for $N = 4, 5, 6, 7, 8$.
   (b) Compute the linear convolution $x_1(n) * x_2(n)$.
   (c) What value of N is necessary so that linear and circular convolution are the same on the N-point interval?
   (d) Determine the nonzero lengths of $x_1(n)$ and $x_2(n)$ and tell how you could have obtained the results of (c) without performing the calculations.

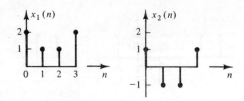

6.9 Given $x_1(n)$ and $x_2(n)$ shown below,

  (a) Find $x_1(n) \circledast_4 x_2(n)$.

  (b) What are the nonzero lengths of $x_1(n)$ and $x_2(n)$?

  (c) We wish to perform a linear convolution of $x_1(n)$ and $x_2(n)$ by taking the inverse DFT of the products of N-point DFTs of $x_1(n)$ and $x_2(n)$. What is the smallest N for which this is possible?

  (d) Describe the method to perform linear convolution using the FFT to compute the DFT. Can we use the N determined in (c)?

  (e) If an eight-point DFT is used to perform linear convolution, roughly how many complex multiplications are required using the fastest FFT algorithm discussed?

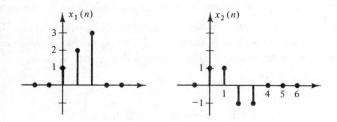

6.10 How many complex multiplications are needed to implement "fast convolution" for two sequences, each of length 32?

6.11 The response $y(n)$ of a linear system characterized by impulse response $h(n)$ can be performed by convolution. For a finite length input one way of obtaining this convolution using the discrete Fourier transform is to wait until we have the entire input signal, then compute the inverse of the product of the transforms. If the input is long, then not only must we wait until all the input is obtained, but the size of the required discrete Fourier transform can become prohibitive. Another way to perform the convolution in pseudo real time (i.e., real time with a finite delay) is by sectionalizing the input.

　　The overlap-and-add method for performing convolution is now described. Consider for illustration purposes that $h(n)$ is of finite length M defined on 0 to $M - 1$, and that $x(n)$ is of finite length L defined on 0 to $L - 1$ as shown on the facing page. Assume $x(n)$ is broken up into K sections of length M, where $K = L/M$ and each section is defined by $x_k(n)$ as follows

$$x_k(n) = \begin{cases} x(n) & kM \le n \le (k + 1)M - 1 \\ 0 & \text{elsewhere} \end{cases}$$

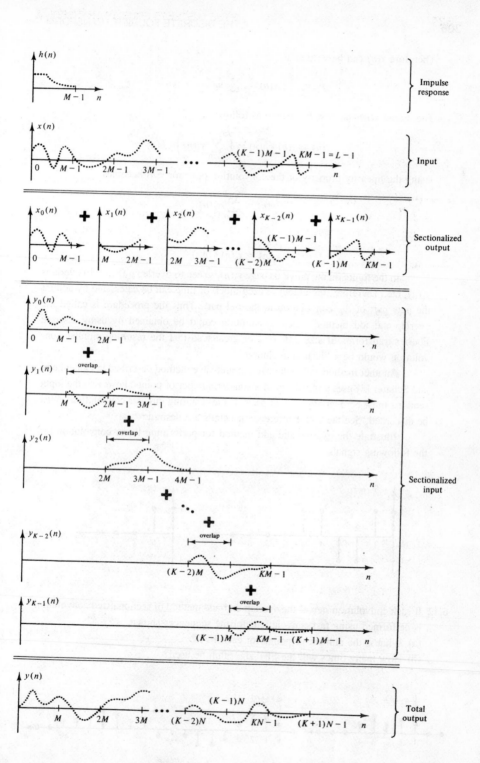

Therefore $x(n)$ can be written as

$$x(n) = \sum_{k=0}^{K-1} x_k(n)$$

The output $y(n)$ can now be written as follows

$$y(n) = x(n) * h(n) = \left(\sum_{k=0}^{K-1} x_k(n) \right) * h(n)$$

Using the linearity property of the convolution operator $y(n)$ becomes

$$y(n) = \sum_{k=0}^{K-1} y_k(n)$$

where

$$y_k(n) = x_k(n) * h(n)$$

In the figure on the previous page $y_1(n)$ is seen to overlap $y_0(n)$, $y_2(n)$ overlaps $y_1(n)$, etc. Therefore, the output at each block of time can be calculated by adding the new part of the convolution to the old part. Thus the procedure is called the overlap-and-add method. Each convolution could be obtained by using discrete Fourier transforms of size $2M - 1$ or greater so that the resulting circular convolution would be a linear convolution.

Another method called the overlap-and-save method described in Oppenheim and Schafer [8] uses a transform of a smaller number of points; however, the input sections must be overlapped, and parts of the resulting circular convolutions must be discarded. See the above reference if details are desired.

Illustrate the overlap-and-add method for performing linear convolution for the following signals:

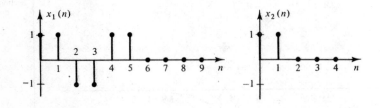

6.12 If each convolution step of the overlap-and-add method of sectionalized convolution is performed using fast convolution, for the sequences shown,

(a) What is the number of points for each FFT?

(b) How many times will the FFT algorithm be used?

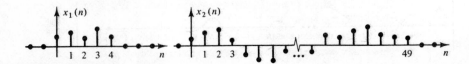

6.13 Use the eight-point DIT radix 2 FFT shown in Fig. 6.11 to find the DFT of the sequence $\{x(n)\} = \{0.707, 1, 0.707, 0, -0.707, -1, -0.707, 0\}$. Show the results in a tabular form where each column corresponds to intermediate results at each half stage.

6.14 In the basic decimation-in-time (DIT) FFT, is the input, the output, or both in bit reversed order? For a 64-point DFT obtained by the DIT FFT algorithm comparable to that of Fig. 6.8, what are the time samples corresponding to the 0, 1, 2, and 3 positions?

6.15 Suppose I want to take the DFT of a data stream of length 8192 but my analyzer has only a fixed hardware implementation for a 2048 point DFT. Assuming other storage is available along with ways of adding and multiplying, how could I get the desired 8192-point transform? (Be precise.)

6.16 What is the difference between the discrete Fourier series and the discrete Fourier transform?

6.17 If the complex multiplication time of a particular computer is 0.5 μsec and the addition time is much, much smaller, roughly how long will it take to compute a 1024-point DFT using the best FFT discussed? Assume multiplications are computed sequentially.

6.18 How many (real) storage registers are required to evaluate the DFT by an in-place FFT algorithm?

6.19 Let $X(k)$, $k = 0, 1, \ldots, N - 1$ be the discrete Fourier transform of $x(n)$. Give two methods for using an FFT to evaluate the inverse discrete Fourier transform, i.e., obtain $x(n)$ from $X(k)$.

6.20 For the eight-point FFT (DIF algorithm in Fig. 6.13),
  (a) *Exactly* how many nontrivial complex multiplications are required?
  (b) Using a formula comparable to Eq. (6.30a), roughly how many complex additions and complex multiplications are required?
  (c) Is this an "in-place" algorithm?
  (d) Is the time sequence or the frequency sequence in bit reversed order?

6.21 For an $N = 256$ DIT FFT, the input must be put in bit reversed order.
  (a) If the input index numbers are $k = 0, 1, 2, \ldots, 255$, what is the index of the sample that goes in the $k = 34$ position?
  (b) In what index k does the $n = 96$ time sample go?
  (c) Give a point other than zero that stays in its normal location, that is, the k and n indexes are equal.

6.22 Using a standard 16-point decimation in time FFT algorithm similar to Fig. 6.11 in structure,
  (a) Are the time samples or frequency samples in bit reversed order?
  (b) How many complex multiplications would be required (roughly, i.e., by formula) to compute the DFT.
  (c) How many complex additions?
  (d) How many complex multiplications for direct evaluation of a 16-point DFT?

(e) What time index samples will be put in the third, seventh, and twelfth position indices for implementing the DIT FFT (16 point)?

(f) Part of the flow diagram for the last stage is shown below. What are the twiddle factors a, b, c, d, e, f?

(g) Repeat part (f) if the structure given in Fig. 6.8 is used.

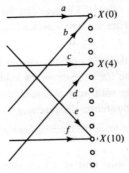

6.23 Using a standard 32-point decimation-in-time FFT algorithm,

(a) Are the time samples in normal order? Frequency samples?

(b) What time index sample will be put in the seventh position index for implementing the above transform?

(c) How many complex additions are required to compute the best DIT FFT algorithm discussed?

(d) How many complex multiplications?

(e) How many real multiplications?

(f) How many real additions?

(g) How many times faster is the FFT than the direct evaluation if complex addition times are 0.1 μsec and complex multiplication times are 1 μsec? (Use only one two-input hardware complex multiplier and one two-input complex adder.)

(h) Using the best DIT FFT, a butterfly from the last stage is isolated as shown. Determine a, b, c, d, and e.

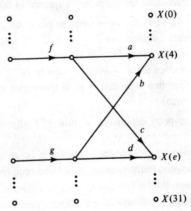

6.24 We desire an algorithm for obtaining an N-point DFT that uses four-point transforms rather than two-point transforms for its basis. The general approach will be decimation in time and the resulting algorithm is called a radix 4 fast Fourier transform. The number of points N must be assumed to be a power of 4, say 4^{α}. The first step in the approach is to break the original sequence into four $(N/4)$-point sequences as follows:

$$\begin{array}{ll}
\text{Sequence 0} & x(0), x(4), \ldots, x(N-4) \\
\text{Sequence 1} & x(1), x(5), \ldots, x(N-3) \\
\text{Sequence 2} & x(2), x(6), \ldots, x(N-2) \\
\text{Sequence 3} & x(3), x(7), \ldots, x(N-1)
\end{array}$$

The $(N/4)$-point DFT of the lth sequence is defined by $G_l(k)$ using the notation of Oppenheim and Schafer [8] and the N-point transform is obtained by combining those transforms in the following way assuming the $G_l(k)$ are periodic for $k \geq N/4$:

$$X(k) = \sum_{l=0}^{p-1} W_N^{lk} G_l(k), \qquad k = 0, 1, \ldots, N-1$$

where $p = 4^{\alpha-1}$ and $W_N = e^{-j2\pi/N}$.

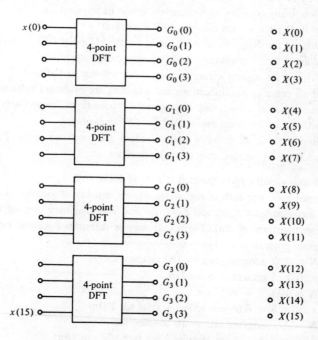

Each $(N/4)$-point DFT is then decimated into four $(N/8)$-point DFTs using the formula above with $p = 4^{\alpha-2}$ and $N = N/4$ to obtain the combination. This process is continued until $p = 4$. Using the above general approach, obtain a 16-

point algorithm by completing the following steps and answering the questions posed:

(a) Find the basic four-point algorithm and give the flow diagram.

(b) Derive the proper scale factors for the combination of the four-point transforms.

(c) Illustrate the signal flow graph for the 16-point transform giving all the "twiddle" factors in reduced form.

(d) Find the total number of complex multiplications required (do not include $+j$, $-j$, -1 as they represent change of sign and or a swap of the real and imaginary parts).

(e) Compare this number with the 16-point DIT radix 2 FFT.

(f) Are the samples for the algorithm in bit reversed order? Tell how you could find the index of the sample that would go in the jth position index; in the tenth position index.

6.25 A real analog signal $x(t)$ is sampled at a rate of 64 samples/sec for an 8-sec interval. A 512-point DFT is taken of these samples.

(a) If the best DIT FFT algorithm discussed in the text is used in the evaluation, determine roughly the number of complex multiplications and number of complex additions required.

(b) How many complex multiplications would be required for a direct evaluation of the DFT (i.e., not using an FFT algorithm)?

(c) What is the digital frequency spacing for the 512-point DFT described above?

(d) What is the corresponding analog frequency spacing?

(e) What is the highest analog frequency that will not be "aliased"?

(f) If all complex multiplications and additions are evaluated sequentially using two inputs at a time, with 2 μsec and 0.1 μsec times, respectively, how long will it take for direct evaluation? For FFT evaluation?

(g) Repeat (f) if all complex multiplications are computed in parallel, then all additions in parallel, then all multiplications, etc.?

6.26 An analog signal $x_a(t)$ is sampled at $t = nT$ for $n = 0, 1, \ldots, N - 1$ to give a sequence $x(n)$. For each of the analog signals, number of samples, and sampling rate indicated, plot $X_R(k)$ and $X_I(k)$, the real and imaginary parts of the discrete Fourier transform of $x(n)$. The results may be determined without evaluating by computer program.

(a) $N = 200$, sampling rate $= 100$ samples/sec

$x_a(t) = 2 \sin 4\pi t + 5 \cos 8\pi t$

(b) $N = 100$, $T = 10^{-3}$ sec

$x_a(t) = 3 + 0.06 \cos 100\pi t + 0.04 \sin 200\pi t$

(c) $N = 32$, sampling rate $= 16$ samples/sec

$x_a(t) = -5 + 3 \sin 10\pi t + 2 \cos 6\pi t - 2 \cos 16\pi t$

(d) $N = 128$, $T = \frac{1}{64}$ sec

$x_a(t) = \frac{1}{64} + \frac{1}{32} \cos 8\pi t + \frac{1}{16} \sin 32\pi t + \frac{1}{16} \cos 64\pi t$

6.27 **(a)** Use an FFT to evaluate and plot the real, imaginary, and magnitude of the 64-point DFT of $x(n)$ given as follows:

$$x(n) = \cos(0.5\pi n) \qquad \text{for } n = 0, 1, \ldots, 63$$

(b) Repeat (a) for

$$x(n) = \cos(0.515625\pi n) \qquad \text{for } n = 0, 1, \ldots, 63$$

(c) Discuss the difference between the results of parts (a) and (b).

(d) In general, when does the magnitude of the DFT of a real single sinusoidal sequence have just two nonzero values?

(e) When does it have less than two?

(f) When does it have more than two?

6.28 You are given the 16-pont DFT shown.

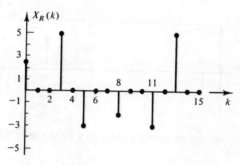

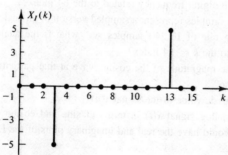

(a) Find the corresponding $x(n)$ in terms of discrete-time dc, sine, and cosine sequences.

(b) What are the digital frequencies associated with the $X_R(k)$ and $X_I(k)$ given and the digital frequency spacing?

(c) If the time samples were 0.03 sec apart, what are the corresponding analog radian frequencies and the associated analog frequency spacing?

(d) For the same sampling rate given in (c), what is the highest analog frequency in hertz that would show up "unaliased" in the DFT?

6.29 A signal $x(t)$ has been sampled at a rate of 50 samples/sec to give a total of 1000 samples. A 1000-point DFT is taken resulting in the $X_R(k)$ and $X_I(k)$ shown.
  (a) What is the so-called digital frequency resolution?
  (b) What is the corresponding analog frequency resolution?
  (c) What is the highest unaliased analog frequency that we will be able to see?
  (d) Give your educated guess for the analog signal that was sampled to give the $X(k)$ shown.

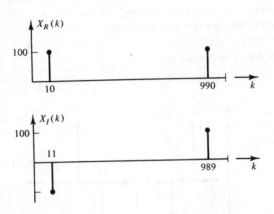

6.30 Shown are the real and imaginary parts from a 1024-point DFT.
  (a) What is k_1?
  (b) What is the digital frequency related to the 64 index?
  (c) If the time samples represent a sampled version of an analog signal, $x(t)$, using a sampling rate of 10,240 samples/sec, what is the analog frequency corresponding to the $k = 64$ index?
  (d) What is the magnitude of the cosine wave at that particular frequency? What is the phase?
  (e) What is the dc value of the signal $x(n)$?
  (f) Give an analog signal $x(t)$ in terms of sine and cosine functions that when sampled would have the real and imaginary plots of the DFT shown.

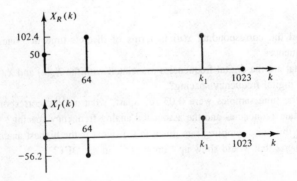

6.31 Suppose the figure below represents the magnitude of a 128-point DFT of a sampled analog signal (sampling rate of 1000 samples/sec).
(a) What do you think the analog signal that was sampled is? (Be specific.)
(b) Give two reasons why you are not certain.
(c) Repeat (a) and (b) if sampling rate is 500 samples/sec.

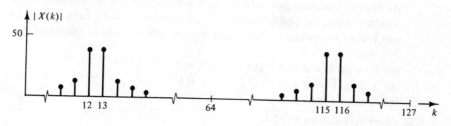

6.32 A real signal $x(t)$ has been sampled at a rate of 128 samples/sec to give a total of 512 samples. A 512-point DFT is taken to yield the real and imaginary plots shown.
(a) Find α, β, δ, and γ.
(b) What is the highest analog frequency (unaliased) that can be seen using the FFT above?
(c) Give an analog signal $x(t)$ that when sampled would have the $X_R(k)$ and $X_I(k)$ given above.

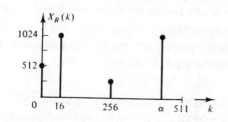

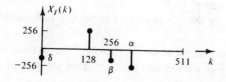

6.33 An analog signal $x_a(t) = \sin 2\pi t$ is sampled at $t = nT$ for $n = 0, 1, \ldots, 99$ and $T = 0.1$ sec to give exactly 10 full periods of a sinusoidal sequence. Use the general formula for the discrete Fourier transform [Eq. (6.11)] and orthogonality formulas from Appendix B to prove analytically the following for the 100-point DFT of those 100 samples:
(a) $X(10) = -j50$
(b) $X(90) = +j50$

(c) $X(k) = 0$ for all $k = 0, 1, \ldots, 99$ except 10 and 90.

(d) Give comparable general results for the N-point transform of a sinusoidal sequence that goes through exactly l full periods.

6.34 Define a sinusoidal sequence $x(n) = \cos(2\pi\alpha n/N)$ for $n = 0, 1, \ldots, N - 1$, where α is from the set $\{.5, 1.5, \ldots, (N - 1)/2\}$. Let $X_R(k)$ and $X_I(k)$ represent the real and imaginary parts of $X(k)$, the N-point discrete Fourier transform of the sequence. Thus, the digital frequency of the sequence is exactly halfway between two Fourier frequencies of $2\pi k/N$ and $2\pi(k + 1)N$ for some k. Determine analytically

(a) $X_R(k)$ for $k = \alpha + 0.5$ and $k = \alpha - 0.5$

(b) $X_I(k)$ for $k = \alpha + 0.5$ and $k = \alpha - 0.5$

(c) $X_I(k)$ for k other than those specified in (b)

6.35 Given the following $x(n)$:

$$x(n) = \delta(n) + \delta(n - 1) + \delta(n - 2)$$

(a) Find the Fourier transform of $x(n)$, that is. $\mathcal{F}[x(n)] = X(e^{j\omega})$, and plot the $|X(e^{j\omega})|$. Make sure your zeros are correct.

(b) Give the magnitude of the four-point DFT of the first four samples of $x(n)$.

(c) Give the magnitude of the eight-point DFT of the first eight samples of $x(n)$ above. Be sure and label axes and scales.

(d) Is the DFT the sampled version of $X(e^{j\omega})$ for (b)? for (c)?

(e) Is it possible to reconstruct the $X(e^{j\omega})$ from the eight-point DFT of part (c)?

(f) If possible, give a formula for the reconstruction. If not possible, state why you cannot reconstruct.

6.36 We have available a portion, only the first 32 samples ($n = 0, 1, \ldots, 31$), of a sinusoidal signal $x(n)$ given as follows:

$$x(n) = \tfrac{1}{64}\cos(\pi n/4)$$

(a) Find the 32-point DFT of those first 32 samples.

(b) Find the 64-point DFT of a 64-point sequence formed using those 32 samples followed (padded) by 32 zero samples.

(c) Repeat (b) for the 128-point DFT if 96 zero samples follow the original 32 samples.

(d) What can we find out by the zero padding? Can this information be determined in another way without zero padding? If so, tell how (be specific). If not, tell why not.

6.37 (a) Prove Eq. (6.47) using property 7 of Table 6.1.

(b) Prove Eq. (6.55a) and (6.55b).

6.38 How many "twiddles" will make a butterfuly "fast?"

APPENDIX A: A Short Review of Linear System Theory

Many physical systems can be modeled with linear constant coefficient differential equations. The excitation, $x(t)$, of the system can be thought of as the input, while the output, $y(t)$, is the solution of a differential equation subject to a given set of initial conditions as follows:

$$c_N \frac{d^N y(t)}{dt^N} + c_{N-1} \frac{d^{N-1} y(t)}{dt^{N-1}} + \cdots + c_0 y(t) =$$

$$d_M \frac{d^M x(t)}{dt} + d_{M-1} \frac{d^{M-1} x(t)}{dt} + \cdots + d_0 x(t)$$

initial conditions:

$$\left. \frac{d^k y(t)}{dt^k} \right|_{t=0} \qquad \text{for } k = N-1, N-2, \ldots, 0$$

Setting the initial conditions equal to zero and taking the Laplace transform of each side of the differential equation gives

$$\sum_{i=0}^{N} c_i s^i Y(s) = \sum_{i=0}^{M} d_i s^i X(s)$$

where $X(s) = \mathcal{L}[x(t)]$ and $Y(s) = \mathcal{L}[y(t)]$.

Solving the above equation in the s-domain for $Y(s)$ over $X(s)$ results in the following:

$$\frac{Y(s)}{X(s)} = \frac{\displaystyle\sum_{i=0}^{N} c_i s^i}{\displaystyle\sum_{i=0}^{M} d_i s^i}$$

This ratio of output transform to input transform is defined as the system function $H(s)$, that is

$$H(s) = \frac{Y(s)}{X(s)}$$

For a given input transform $X(s)$, the system function $H(s)$ allows the calculation of the output transform $Y(s)$ via multiplication as follows:

$$Y(s) = H(s) \cdot X(s)$$

If the input signal $x(t)$ is a unit-valued delta function, $\delta(t)$, its transform is 1, and $Y(s)$ is just $H(s)$. The inverse transform of $H(s)$ thus gives the response of the system to an impulse function and is denoted by $h(t)$:

$$h(t) = \mathcal{L}[H(s)]$$

Furthermore, if $x(t)$ is any input, the output $y(t)$ is easily seen to be the convolution with the impulse response of the system, that is,

$$y(t) = \mathcal{L}^{-1}[H(s)X(s)] = h(t) * x(t)$$

where convolution can be written in either of the following two forms:

$$y(t) = \int_{-\infty}^{\infty} h(\tau)x(t - \tau)\,d\tau = \int_{-\infty}^{\infty} x(\tau)h(t - \tau)\,d\tau$$

The fact that the output $y(t)$ to any input $x(t)$ can be obtained by convolving that input with the impulse response $h(t)$ allows the system to be characterized by either its impulse response or system function.

A system for which $h(t)$ is zero for all t less than zero is defined to be causal (realizable). Otherwise the system is noncausal.

A system is defined to be bounded input–bounded output (BIBO) stable if every input $x(t)$ such that $|x(t)|$ is less than some finite number P for all t produces an output $y(t)$ such that $y(t)$ is less than some finite Q for all t. A linear time invariant causal system is BIBO stable if all the poles of $H(s)$ are in the left half plane. The following theorem gives a condition on the impulse response $h(t)$ that is sufficient for BIBO stability.

Theorem.   A linear time invariant system is BIBO stable if and only if

$$\int_{-\infty}^{\infty} |h(t)|\,dt < \infty$$

It is also not very difficult to show for a BIBO stable system that the steady state response, i.e., after transients have died down, to a sinusoid of frequency Ω is the same frequency sinusoid whose magnitude has been multiplied by the magnitude of the transfer function $H(s)$ evaluated at $s = j\Omega$, and whose phase has been changed by the angle of $H(s)$ evaluated at $j\Omega$. Therefore $H(s)$ evaluated at $s = j\Omega$ as Ω varies form 0 to infinity, tells how the system responds to sinusoids of all frequencies and is called the frequency response of the system. That is, if

$$x(t) = A \sin \Omega t \, u(t)$$

is the input to a system with system function $H(s)$, the steady state output is given by

$$y_{ss}(t) = A|H(j\Omega)| \sin [\Omega t + \arg H(j\Omega)]$$

where the frequency response $H(j\Omega)$ is obtained from $H(s)$ as

$$H(j\Omega) = H(s)|_{s=j\Omega}$$

APPENDIX B: Useful Formulas

Finite Summation Formulas

$$\sum_{k=0}^{n} a_k = \frac{1 - a^{n+1}}{1 - a}, \qquad a \neq 1$$

$$\sum_{k=0}^{n} ka^k = a[1 - (n + 1)a^n + na^{n+1}]/(1 - a)^2$$

$$\sum_{k=0}^{n} k^2 a^k = a[(1 + a) - (n + 1)^2 a^n + (2n^2 + 2n - 1)a^{n+1} - n^2 a^{n+2}]/(1 - a)^3$$

$$\sum_{k=0}^{n} k = n(n + 1)/2$$

$$\sum_{k=0}^{n} k^2 = n(n + 1)(2n + 1)/6$$

$$\sum_{k=0}^{n} k^3 = n^2(n + 1)^2/4$$

Infinite Summation Formulas

$$\sum_{k=0}^{\infty} a^k = \frac{1}{(1-a)}, \qquad |a| < 1$$

$$\sum_{k=0}^{\infty} ka^k = \frac{a}{(1-a)^2}, \qquad |a| < 1$$

$$\sum_{k=0}^{\infty} k^2 a^k = \frac{a^2 + a}{(1-a)^3}, \qquad |a| < 1$$

Complex Variables

Rectangular Form

$$A = \alpha + j\beta, \qquad \alpha = \text{real part}$$
$$j = \sqrt{-1}, \qquad \beta = \text{imaginary part}$$

Complex Conjugate (replace j's by −j's)

$$A^* = \alpha - j\beta$$

Magnitude

$$|A| = (\alpha^2 + \beta^2)^{1/2}$$
$$|A|^2 = \alpha^2 + \beta^2 = A \cdot A^*$$

Exponential Form

$$A = Re^{j\theta}, \text{ where}$$
$$R = (\alpha^2 + \beta^2)^{1/2} \text{ (magnitude)}$$
$$\theta = \tan^{-1}(\beta/\alpha) \text{ (argument)}$$

Exponential to Rectangular

$$\alpha = R\cos\theta$$
$$\beta = R\sin\theta$$

Trigonometric

$$\cos \alpha = \frac{e^{j\alpha} + e^{-j\alpha}}{2}$$

$$\sin \alpha = \frac{e^{j\alpha} - e^{-j\alpha}}{2j}$$

$$e^{-j\alpha} = \cos \alpha - j \sin \alpha$$

$$e^{+j\alpha} = \cos \alpha + j \sin \alpha$$

Magnitude of Product and Quotient

$$|A \cdot B| = |A| \cdot |B|$$
$$|A/B| = |A| / |B|$$

Angle of Product and Quotient

$$\arg (A \cdot B) = \arg A + \arg B$$
$$\arg (A/B) = \arg A - \arg B$$

Triangular Inequality

$$|A + B| \le |A| + |B|$$

Summations Involving Trigonometric Forms

(*k*, *l*, *n* are integers)

$$\sum_{n=0}^{N-1} \cos \frac{2\pi k}{N} n = \begin{cases} 0 & \text{for } 1 \le k \le N - 1 \\ N & \text{for } k = 0, N \end{cases}$$

$$\sum_{n=0}^{N-1} \sin \frac{2\pi k}{N} n = \begin{cases} 0 & \text{for } 1 \le k \le N - 1 \\ N & \text{for } k = 0, N \end{cases}$$

Orthogonality

$$\sum_{n=0}^{N-1} \cos\frac{2\pi k}{N}n \cos\frac{2\pi l}{N}n = 0 \qquad \text{for } 1 \le k, l \le N-1, \text{ and } k \ne l$$

$$\sum_{n=0}^{N-1} \sin\frac{2\pi k}{N}n \sin\frac{2\pi l}{N}n = 0 \qquad \text{for } 1 \le k, l \le N-1, \text{ and } k \ne l$$

$$\sum_{n=0}^{N-1} \sin\frac{2\pi k}{N}n \cos\frac{2\pi l}{N}n = 0 \qquad \text{for } 1 \le k, l \le N-1, \text{ and } k \ne l$$

$$\sum_{n=0}^{N-1} \cos^2\frac{2\pi k}{N}n = \begin{cases} N/2 & \text{for } 1 \le k \le N-1 \text{ and } k \ne N/2 \\ N & \text{for } k = 0, N/2 \end{cases}$$

$$\sum_{n=0}^{N-1} \sin^2\frac{2\pi k}{N}n = \begin{cases} N/2 & \text{for } 1 \le k \le N-1 \text{ and } k \ne N/2 \\ 0 & \text{for } k = 0, N/2 \end{cases}$$

The above formulas are correct if all k, l, and n are replaced by $k \bmod N$ and $l \bmod N$.

Trigonometric Identities

$$\sin(A+B) = \sin A \cos B + \cos A \sin B$$
$$\cos(A+B) = \cos A \cos B - \sin A \sin B$$

$$\cos A \cos B = \tfrac{1}{2}[\cos(A+B) + \cos(A-B)]$$
$$\cos A \sin B = \tfrac{1}{2}[\sin(A+B) - \sin(A-B)]$$
$$\sin A \cos B = \tfrac{1}{2}[\sin(A+B) + \sin(A-B)]$$
$$\sin A \sin B = \tfrac{1}{2}[\cos(A-B) - \cos(A+B)]$$

$$\sin A + \sin B = 2\sin\tfrac{1}{2}(A+B)\cos\tfrac{1}{2}(A-B)$$
$$\cos A + \cos B = 2\cos\tfrac{1}{2}(A+B)\cos\tfrac{1}{2}(A-B)$$

$$\sin^2 A = \tfrac{1}{2}(1 - \cos 2A) \qquad\qquad \cos^2 A = \tfrac{1}{2}(1 + \cos 2A)$$
$$\sin^3 A = \tfrac{1}{4}(3\sin A - \sin 3A) \qquad \cos^3 A = \tfrac{1}{4}(3\cos A + \cos 3A)$$
$$\sin^4 A = \tfrac{1}{8}(3 - 4\cos 2A + \cos 4A) \qquad \cos^4 A = \tfrac{1}{8}(3 + 4\cos 2A + \cos 4A)$$

$$\sin 2A = 2\sin A \cos A \qquad\qquad \cos 2A = \cos^2 A - \sin^2 A$$
$$\sin 3A = 3\sin A - 4\sin^3 A \qquad\qquad \cos 3A = 4\cos^3 A - 3\cos A$$
$$\sin 4A = 4\sin A \cos A - 8\sin^3 A \cos A \qquad \cos 4A = 8\cos^4 A - 8\cos^2 A + 1$$

$$A\sin C + B\cos C = (A^2 + B^2)^{1/2}\cos(C - \tan^{-1}(A/B))$$

APPENDIX C: Table of $\mathscr{Z}$ Transform Pairs

Discrete-time sequence	$\mathscr{Z}$ transform	Region of convergence
1. $\delta(n)$	1	all z
2. $\delta(n - m)$	z^{-m}	all z except $\lvert z \rvert = 0,\ m > 0$ all z except $\lvert z \rvert = \infty,\ m < 0$
3. $u(n)$	$z/(z - 1)$	$\lvert z \rvert > 1$
4. $nu(n)$	$z/(z - 1)^2$	$\lvert z \rvert > 1$
5. $n^2 u(n)$	$z(z + 1)/(z - 1)^3$	$\lvert z \rvert > 1$
6. $n^3 u(n)$	$z(z^2 + 4z + 1)/(z - 1)^4$	$\lvert z \rvert > 1$
7. $a^n u(n)$	$z/(z - a)$	$\lvert z \rvert > \lvert a \rvert$
8. $na^n u(n)$	$az/(z - a)^2$	$\lvert z \rvert > \lvert a \rvert$
9. $n^2 a^n u(n)$	$az(z + a)/(z - a)^3$	$\lvert a \rvert > \lvert a \rvert$
10. $n^3 a^n u(n)$	$az(z^2 + 4\,az + a^2)/(z - a)^4$	$\lvert z \rvert > \lvert a \rvert$
11. $na^{n-1}u(n)$	$z/(z - a)^2$	$\lvert z \rvert > \lvert a \rvert$
12. $\dfrac{n(n - 1)}{2!}\, a^{n-2} u(n)$	$z/(z - a)^3$	$\lvert z \rvert > \lvert a \rvert$
13. $\dfrac{n(n - 1)(n - 2)}{3!}\, a^{n-3} u(n)$	$z/(z - a)^4$	$\lvert z \rvert > \lvert a \rvert$
14. $\dfrac{n(n - 1)\cdots(n - (k - 2))}{k!}\, a^{n-k+1} u(n)$	$z/(z - a)^k$	$\lvert z \rvert > \lvert a \rvert$
15. $\sin \omega_0 n\, u(n)$	$z \sin \omega_0/(z^2 - 2z \cos \omega_0 + 1)$	$\lvert z \rvert > 1$
16. $\cos \omega_0 n\, u(n)$	$z(z - \cos \omega_0)/(z^2 - 2z \cos \omega_0 + 1)$	$\lvert z \rvert > 1$
17. $\sin(\omega_0 n + \theta)\, u(n)$	$z[z \sin \theta + \sin(\omega_0 - \theta)]/(z^2 - 2z \cos \omega_0 + 1)$	$\lvert z \rvert > 1$
18. $\cos(\omega_0 n + \theta)\, u(n)$	$z[z \cos \theta - \cos(\omega_0 - \theta)]/(z^2 - 2z \cos \omega_0 + 1)$	$\lvert z \rvert > 1$
19. $\sinh(\omega_0 n)\, u(n)$	$(z \sinh \omega_0)/(z^2 - 2z \cosh \omega_0 + 1)$	$\lvert z \rvert > e^{\omega_0}$
20. $\cosh(\omega_0 n)\, u(n)$	$z(z - \cosh \omega_0)/(z^2 - 2z \cosh \omega_0 + 1)$	$\lvert z \rvert > e^{\omega_0}$

Index

A/D converter, 3, 4, 168
A/D-digital filter-D/A structure, 3, 5–6, 61, 66, 168, 179
  design of digital filter for, 168–179
  equivalent analog response, 62, 63
  implementation, 16, 252–253
Aliasing, 3, 63, 64, 65
Aliasing point, 66
All-pass filter, 117
Analog design using digital filters, 179–181
Analog filters
  Butterworth, 122–127, 129–134
  Chebyshev, 134–146
  elliptic, 146–163
  equivalent, 62, 63, 168
  See also specific type
Analog filter specifications, 121, 129, 132, 177, 180
Analog signal: *see* Continuous-time signals
Analog signal processing, 2
Analog systems: *see* Continuous-time systems
Analog-to-analog transformations, 127, 128
Analog-to-digital transformations
  backward difference approach, 168–170
  bilinear transformation approach, 172–175
  forward difference approach, 207
Anti-aliasing filter, 3

Backward differences, 169, 170
Band-limited signal, 58
Band-pass filter, 120, 121
Bandstop filter, 121
Bartlettt window, 197, 199
BIBO: *see* Bounded input–bounded output stability
Bilinear transformation, 172–175
  definition, 173
  use in filter design, 175–179
  prewarping, 174–175
  properties of, 173
  warping, 174, 175
Biquadratic forms, 14, 15, 220, 222, 225, 229
Bit reversed order, 277, 285
Blackman window, 197, 198
Block diagram notation, 218
Bounded input–bounded output stability, 35–38
  conditions for, 36, 87, 111
Butterfly, 279, 281, 285
Butterworth filter (analog)
  design of low-pass, 129–131
  design of band-pass, 132–134
  determination of order n, 130
  magnitude squared frequency response, 122, 124

Butterworth filter (analog) (*Continued*)
  maximally flat, 122
  normalized low-pass, 125, 127
  properties, 122, 127
  tables standard and factored form, 127
Butterworth filter (digital)
  design of low-pass, 181–187
  determination of order n, 186
  normalized low-pass tables, 184
Butterworth polynomials, 125, 127

Cascade combination, 35
Cascade realizations
  FIR filter, 248–249
  IIR filter, 224–228
Canonic biquadratic section, 14
Cauchy integral theorem, 87
Cauchy's formula, 88
Causal sequence, 38, 111
Causal system, 38, 111
Characterizations of systems: *see* System
      function, relationship to
Chebyshev filer, 134–146
  design of low-pass, 139
  determination of order n, 139
  equiripple property, 136
  magnitude squared frequency response,
      136
  normalized low-pass tables, 140–143
  properties of, 136
Chebyshev polynomial definition, 135
Chebyshev polynomial table, 135
Circular convolution, 265, 267–268
Circular translation, 265–266
Coefficient ROM, 16, 252–253
Complex convolution property, 81–82, 196
Complex exponential Fourier series
  continuous-time, 258
  discrete-time, 259
Complex sequence, 20
Complex variables (useful formulas), 318
Computation of
  discrete Fourier series, 259–262
  discrete Fourier transform, 270–272
  frequency response, 104
  inverse DFT, 285–286
Continuous-time Fourier series
  complex exponential representation, 258
  trigonometric form, 258
Continuous-time signals, 19
  Nyquist rate for, 59

review, 315
  sampling of, 57
Continuous-time systems, 19
  Comparison with discrete-time, 111
  short review for time invariant systems,
      315–317
  summary for time-invariant systems, 111
Convergence, region of, 76, 84
Convolution, 28–34
  circular, 265, 267–268
  complex, 81
  continuous-time, 111, 316
  discrete-time, 28, 111
  duration of, 34
  equality: circular and linear, 270–271
  Fourier transform of, 56
  frequency domain, 81–82
  linear, 28
  operator, 29
  overlap-and-add method, 304, 305–306
  properties of, 29
  Z transform of, 81
Controller, 252–253
Critical frequencies, 175, 180
Cut-off frequency, 121

D/A converter (digital-to-analog), 3, 168
DATA RAM, 252–253
DC gain, 46
Decimation-in-frequency FFT, 282–285
  butterfly, 285
  complex additions required, 282
  complex multiplications required, 281
  flow graph for, 284
  ordering of samples, 285
  in place computation, 285
Decimation-in-time FFT, 273–282
  bit reversed order, 277
  butterfly, general, 279
  butterfly, reduced, 281
  complex additions required, 282
  complex multiplications required, 281
  flow graph for, 278
  in place computation, 278
  radix 2 algorithm, 301
  radix 4 algorithm, 301, 309
  reduced form flow graph, 281
  relative speed, 282
Design of analog Butterworth filters
  band-pass design, 132–134
  low-pass design, 129–131

normalized design tables, 127
transformation table, 128
See also Butterworth filter (analog)
Design of analog Chebyshev filters
normalized design tables, 140–143
low-pass design, 139
transformation table, 128
See also Chebyshev filter
Design of analog elliptic filters
low-pass design, 161
normalized design tables, 149–160
transformation table, 128
See also Elliptic filters
Design of digital Butterworth filters, 186–187
bilinear transformation approach, 175–179
digital-to-digital transformation, 181–183
normalized design tables, 184
See also Butterworth filter (digital)
Design of digital filters
using the bilinear transformation, 175–179
using digital-to-digital transformations,
181–187
impulse invariant, 187–190
Design of finite impulse response (FIR) filters, 191–206
computer techniques, 205–206
design steps, 201–202
frequency response, 195
properties of, 194, 201
table for design, 201
windowing technique, 196–205
Design of infinite impulse response (IIR) filters, 168–191
bilinear transformation approach, 175–179
digital-to-digital transformation approach,
181–187
impulse invariant approach, 187–190
minimization of mean squared error,
190–191
DFT: *see* Discrete Fourier transform
DIF FFT: *see* Decimation-in-frequency FFT
Difference equations: *see* Linear constant
coefficient difference equations
Differential equation, numerical solution of,
168–175
Digital filters
in A/D–$H(z)$–D/A structure, 61, 179–181,
189–190
advantages, 17
Butterworth, 181–187; *see also* design of
digital Butterworth filters
definition, 3

design of: *see* Design of digital filters
equivalent analog frequency response, 62,
63
finite impulse response (FIR), 43
frequency response for, 45
implementation of, 16, 252–253
infinite impulse response (IIR), 43
linear phase, 192
realization: *see* Realizations of digital filters
specifications for, 175, 180
Digital frequency, 62, 287
Digital processor, 2–17
general structure, 3, 5, 7, 16, 253
use as equivalent analog filter, 3
Digital signal, 21
Digital signal processing, 1–16
definition, 1
example, 2
special structures, 7–16
Digital specifications, 175, 177, 180, 183,
191
Digital system, 21
Digital-to-analog converter, 3, 4, 168
Digital-to-digital transformations, 181–183
Direct form I realization
FIR filters, 246–247
IIR filters, 217–218
Direct form II realization, IIR filters,
219–222
Direct programming realization, 236–238
Discrete Fourier series pair, 259
Discrete Fourier transform (DFT), 262–272
decimation-in-frequency FFT, 282–285
decimation-in-time FFT, 273–282
definition, 263
direct computation of, 272
fast computation of: *see* Fast Fourier transform
finite duration sequences, 34–35
interpolation formula, 294
inverse, 263
linear convolution by, 271
pair, 263
properties
circular convolution, 265, 267–268
circular translation, 265, 266
conjugation, 265
linearity, 265
multiplication by exponential, 265, 266
product time domain, 265
for real signals, 265

Discrete Fourier transform (DFT) (*Continued*)
  table, 265
  relation to discrete Fourier series, 264
  relation to Fourier transform, 264,
    293–294
  samples of the Fourier transform, 293–295
  symmetry for real signals, 265
Discrete Fourier transform of sinusoidal se-
    quences, 288–293, 295–301
Discrete-time systems, 20, 26–55
  causal, 38, 111
  bounded input–bounded output stability of,
    35–38, 87, 111
  deterministic, 20
  frequency domain representation, 44
  linear, 26
  linear shift-invariant, 26
  realizable, 38
  representations
    difference equation, 40, 111
    frequency response, 44–45, 103, 111
    impulse response, 29, 44, 103, 111
    state variable, 231–234, 237–246
    system function, 103, 111
  response to a complex exponential se-
    quence, 44
  response to a sinusoidal signal, 45
  shift-invariant, 26
  stability, 35–38, 87, 111
  state variables, 236–241
  time-invariant, 26
Discrete-time filters, 167, 168
  See also Digital filters
Discrete-time Fourier series, 259–262
  complex exponential form, 259
  fundamental frequency, 259
  matrix formulation, 262
  relationship to discrete-Fourier transform,
    262–263
  relationships between forms, 260
  trigonometric forms, 260
Discrete-time Fourier transform, 55
Discrete-time signals, 21–26
  causal, 38
  exponential, 22
  left-hand, 85
  right-hand, 84
  sinusoidal, 22, 23
  unit sample, 22
  unit step, 22
Distortion (phase), 196, 197
DIT FFT: *see* Decimation-in-time FFT

Elliptic filters
  definition, 146
  design of, 161
  determination of order, 148
  normalized, 147
  ripple property, 146
  tables for, 149–160
Energy of sequences, 24
Energy spectrum density, 56
Equivalent analog frequency response, 3,
    5–6, 62, 63, 66, 180
Exponential sequence, 22
  multiplication by, 265, 266
  z transform of, 77

Fast Fourier transform (FFT): *see* Decima-
    tion-in-frequency FFT; Decimation-
    in-time FFT
FFT: *see* Fast Fourier transform
Filters
  all-pass, 117
  analog Butterworth, 122–127, 129–134
  analog Chebyshev, 134–146
  analog elliptic, 146–163
  band-pass, 120, 121
  causal, 38, 111
  digital Butterworth, 182–187
  finite impulse response (FIR) system, 43
  high-pass (HP), 120
  ideal, 121
  infinite impulse response (IIR) system, 43
  linear phase, 191, 192, 194, 199–205
  maximally flat, 122
  pass-band, 120, 121, 147
  stop-band, 120, 121, 147
  transition band, 120, 121, 147
  See also Design of analog Butterworth fil-
    ters; Design of analog Chebyshev fil-
    ters; Design of analog elliptic filters;
    Design of digital Butterworth filters;
    Design of digital filters
Final value theorem, 82
Finite duration sequence, 34
Finite impulse response (FIR) system
  compared to IIR system, 206
  definition, 43
  design: *see* Design of finite impulse re-
    sponse (FIR) filters
  difference equation representation, 205
  multiple input–single filter, 11, 13
  multiple input–multiple filter, 11, 13

realizations of, 14, 15, 246–252
single input–single filter, 8
single input–multiple filter, 9, 10
Finite summation formulas, 317
FIR: *see* Finite impulse response system
Flowgraph notation, 274
Forward differences, 207
Fourier representations, pictorial summary, 264
Fourier series
  continuous-time, 258
  discrete-time, 259–262
Fourier transform
  continuous-time, 61
  discrete-time, 61
  inverse for continuous-time, 61
  inverse for discrete-time, 61
  inverse for sequences, 55
  relationships, 61
  of sequences, 55
Fourier Transform pairs
  continuous-time signals, 57, 61, 264
  discrete-time signals, 61, 264
  sequences, 55, 264
Frequency domain representation, 44–55
Frequency response
  computation of, 104
  continuous-time systems, 111
  convergence of, 53
  discrete-time systems, 45
  equivalent analog systems, 3, 5–6, 62, 63, 66, 180
  inverse relationship, 53
  properties of, 51
  of rectangular window, 51
Frequency transformations
  analog-to-analog, 127–129
  analog-to-digital, 170–173
  digital-to-digital, 181–183
Fundamental frequency, 258, 259

Guard band, 66
General filter forms, 163, 164
  magnitude squared frequency response, 163

Hamming window, 197, 198, 200
Hanning window, 197, 298
Hardware implementation of digital filters, 16, 252–253

Ideal filters, 121
Ideal low-pass filter, digital, 54, 55
IIR: *see* Infinite impulse response filters
Impulse, discrete-time, 22
  convolution with, 29
Implementation of digital filters, 16, 252–253
Impulse-invariant design, 187–190
Impulse response
  continuous-time, 111, 316
  convolution, 111, 316
  discrete-time, 101–103
  relationship to frequency response, 101, 111
Infinite impulse response (IIR) filters
  comparison to FIR filters, 210
  definition, 43
  design by
    bilinear transformation method, 175–179
    digital-to-digital transformation, 181–183
    minimization of mean squared error, 190–191
  implementation of, 16, 252–253
  realizations
    cascade, 224–228
    direct, 217–224
    parallel, 228–231
    state variable, 231–246
  specifications for, 175, 177, 180, 183, 191
Infinite summation formulas, 318–320
Initial conditions
  continuous-time, 315
  discrete-time, 40, 41, 106–107
Initial value theorem, 82
In-place computation, 278, 285
INPUT/OUTPUT interface, 252, 253
Interpolation formula, 59, 61, 294
  sampling of continuous signals, 59
  sampling of Fourier transform, 294
Interpretation of DFT results, 287
Inverse discrete Fourier transform, 263, 285
Inverse Fourier transform (continuous), 61
Inverse Fourier transform (sequences), 55, 61
Inverse Z transform, 87–100
  complex inversion, 87–89
  division, 98
  partial fraction expansions, 91–98
  table (general), 321

Inverse $\mathcal{Z}$ transform *(Continued)*
   table for repeated roots, 95
   table (useful sequences), 83

Kaiser window, 197

Leakage
   definition, 292
   example of, 292, 293
   explanation, 295–301
Left-hand sequences, 85
   region of convergence for, 86
Lengths of sequences, 34–35
Linear constant coefficient difference equations
   definition, 40, 111
   relationship to
     frequency response, 103, 111
     impulse response, 44, 103, 111
     system function, 103, 111
   solution by
     classical methods, 105
     repeated substitutions, 40
     $\mathcal{Z}$ transforms, 105
   state variable representations, 231–241
Linear convolution: *see* Convolution
Linear phase filter
   definition, 191
   design of, 199–205
   sufficient conditions for, 192, 194
Linear shift-invariant discrete-time systems, 111
   causality, 111
   frequency domain representation, 111
   impulse response of, 111
   linear constant coefficient difference equation, 111
   stability, 111
Linear system, 26

Magnitude response, 45, 46
Main lobe, 196, 197, 201
Maximally flat filter, 122
Minimization of mean squared error IIR filter design, 190–191
Multiple input–multiple filter, 11, 13
Multiple input–single filter, 11, 12
Multiplier/accumulator (MAC), 252–253

Nested programming realization, 238–243
Nonanticipatory, 38
Numerical solution of differential equations, 168–175, 207, 209
Nyquist sampling theorem, 59
Nyquist rate, 59

Operations on sequences, 24
Orthogonality summations, 320
Overlap–add method of convolution, 305, 306

Parallel combination, 35
Parallel realizations, 14, 228–230
Parseval's theorem, 71
Partial fraction expansions, 91
Passband, 120, 121, 147
Passband critical frequency, 120, 121
Periodic sequence, 20, 259
Phase response, 45–46
Physical realizable: *see* Causal system
Poles, 78
Poles, 78
Prefilter, 66
Prewarping, 175
Processing, definition, 1
   analog signal example, 2
   digital signal example, 2–7
Prototype filter, 134

Quantizer, 3

Realizable, 38
Realization block diagram, 218
Realizations of digital filters
   cascade form (FIR), 14, 15, 248–249
   cascade form (IIR), 224–228
   direct forms (IIR), 217–224
   general FIR filters, 11–15, 246–248
   linear phase forms, 248–252
   parallel forms, 14, 15
   state variable realizations, 231–246
     direct programming, 236–238
     nested programming, 238–241
     transformed forms, 243–246
Rectangular window
   definition, 51–53

frequency response, 51–53, 197, 198
use in digital filter design, 196–205
Reconstruction filter, 3
Reconstruction of band-limited signal, 59
Region of convergence (ROC) for Z transforms
definition, 76
left-hand sequence, 84
relationship to stability for system functions, 86
right-hand sequence, 84
table for a few useful sequences, 83
table for Z transform pairs, 321
Relationships between system representations, 100–102, 103
Residue for poles, 88
Resolution
analog frequency, 287
digital frequency, 287
Right-hand sequence, 84
ROC: *see* Region of convergence

Sampling of continuous-time signals, 57, 277–291
Sampling theorem, 59
Sequences
bounded, 36
causal, 38
complex, 20
energy of, 24
exponential, 22
finite duration, 34
Fourier transform of, 55
left-hand, 85
operations, 24, 25
periodic, 20, 23, 262, 263
random, 20
real-valued, 21
right-hand, 84
sinusoidal, 22, 23
unit sample, 22
unit step, 22
Shifting property, time domain, 81
Shift-invariant system, 26
Side lobe, 197
Sifting property, 23
Signals, 1
analog, 19
band-limited, 58
continuous-time, 19
definition, 1

digital, 21
discrete-time, 20
random, 20
sampled, 21
See also Sequences
Signal flow graph, 274, 278, 281, 284
Single input–single filter, 8
Single input–multiple filters, 9, 10
Sinusoidal signals
continuous-time, 64, 258, 288–290
discrete-time, 22, 23, 47, 259–260, 265, 287–293, 295–300
discrete-Fourier transforms of, 288–293, 295–301
response to, 45, 46; *see also* Steady state response
Z transform of, 83
Specifications
analog, 121, 129, 132, 177, 180
digital, 175, 177, 180, 183, 191
Stability: *see* Bounded input–bounded output stability
State of system, 232
State transition matrix, 233
State vector, 233
State variable realizations, 231–246
direct programming realizations, 236–238
nested programming realizations, 238–241
transformed state vector realizations, 243–246
Steady state response
continuous-time, 44, 317
for digital filters, 46, 48, 110
discrete-time, 46, 48, 110
equivalent analog, 62, 63
for first order system, 47, 48, 108
Stop band, 120, 147
Stop-band critical frequency, 120
Structure of special digital signal processors, 7–11
multiple input–multiple filter, 11, 13
multiple input–single filter, 11, 12
single input–multiple filter, 9, 10
single input–single filter, 8
Summation formulas, 317–320
finite, 317–318
orthogonality, 320
trigonometric form, 319–320
Systems: *see* Continuous-time systems; Discrete-time systems
System function
analog (continuous time), 111

System function (*Continued*)
  discrete-time, 86, 111
  relationship to
    difference equation, 101, 103
    frequency response, 101–102, 103
    impulse response, 86, 101, 103

Table for repeated roots, 95
Time invariant, 26
Transformed state vector realizations,
  243–246
Transient response of first order system, 47,
  48, 108, 110
Transition band, 121, 147
Trigonometric identities, 320
Trigonometric form of Fourier series
  continuous-time, 258
  discrete-time, 259, 260
Twiddle factors, 285

Unit sample response, 29, 44, 103, 111
Unit sample sequence, 22, 23, 29
Unit step sequence, 22

Warping, 175
Windows
  Bartlett, 197, 198
  Blackman, 197, 198
  Hamming, 197, 198, 200
  Hanning, 197, 198
  Kaiser, 197

rectangular, 51–53, 197, 198
See also Design of finite impulse response
  (FIR) filters, windowing techniques

$\mathcal{Z}$ transform, 76–111
  calculation
    by definition, 76–77
    use of properties, 80–84
    by tables, 83, 321
  definition, 76
  of exponential sequences, 77
  inverse by
    complex inversion integral, 87–91
    division, 98–100
    partial fraction expansion, 91–98
    table for repeated roots, 95
  properties of, 81–82
    convolution (time domain), 81
    convolution ($\mathcal{Z}$ domain), 81
    final value theorem, 82
    initial value theorem, 82
    linearity, 81
    multiplication by an exponential, 81
    multiplication by a ramp, 81
    translation, 81
  region of convergence, 76, 84
  solution of linear constant coefficient dif-
    ference equations, 40–42, 105–107
  table for a few useful sequences, 83
  table of $\mathcal{Z}$ transform pairs, 321
  of truncated delayed signals, 104–105
Zero, 78
Zero padding, 294

Printed and Bound by KIN KEONG PRINTING CO. PTE. LTD. - Republic of Singapore.